Sake

National Sake that attracts the world

Tokyo University of Agriculture,
Guide to Sake Breweries and Famous Sake

Tokyo University of Agriculture

Sake
National Sake that attracts the world
Tokyo University of Agriculture, Guide to Sake Breweries and Famous Sake

CONTENTS

4 On Publication
    Kanju Osawa, President, Tokyo University of Agriculture

6 List of Sake Breweries

11 Sake in the World
    Ken Hosaka, Professor,
    Department of Brewing Science, Faculty of Applied Biological Sciences,
    Tokyo University of Agriculture

14 The World's Favorite Nihonshu from Tokyo University of Agriculture (SAKE)

18 Sake Brewery's Recommendations "Dishes to go with Sake"

25 Basic knowledge of Japanese Sake
1) Main production areas of sake rice nationwide
2) Geographical Indication (GI) Protection System
3) Sake brewing is based on "first koji, second yeast, third brewing".
4) Sake Classifications
5) Sake terminology

40 How to Enjoy Japanese Sake with Washoku
    Sayuri Akuzawa, Professor,
    Department of Food Safety and Health, Faculty of Applied Biological Sciences,
    Tokyo University of Agriculture

43 Guide to Sake Breweries and Famous Sake Brands

Hokkaido	44	Tokai	209
Tohoku	48	Kinki	244
Kanto	98	Chugoku	279
Koshinetsu	143	Shikoku	307
Hokuriku	187	Kyushu	320

発刊にあたって

学校法人東京農業大学 理事長　**大澤 貫寿**

　日本酒の歴史は古く、今から千年以上も前に日本で造られ、日本の風土と食生活に相まって独自に発展してきたものです。西暦千二百年頃、室町時代には個人々が自前の酒蔵を持ち、様々な地酒を造り販売し始めたとされています。その後、国内各地の神事や祭事の際に用いられるようになり、多くの造り酒屋が誕生し広く庶民にまで普及していきました。海外で初めて日本酒が出品されたのはオーストリア万国博覧会（1873年）です。それ以前は、オランダとの交易を通してアジアの各地で日本酒が飲まれていたようです。

　その後、時代と共に種々様々な日本酒が製造され、今では千五百軒の醸造元で数千の銘柄を製造、販売するに至っています。それらは、豊富な湧水と人々の好みに合わせ原料である米や酵母の品種や醸造技術の改良によって、特徴ある香りと味を各銘柄に持たせています。

　東京農業大学は、1953年に醸造学科を設立し、日本酒製造に関する科学・技術的研究を進めながら造り酒屋の後継者養成にも努めています。本学醸造科学科は、日本酒に関する教育研究の中心的役割を担う人材養成機関です。本学の卒業生は広く醸造界で活躍し、国内はもとより広く海外でも活躍しています。特に国内における造り酒屋の5割を本学卒業生が占めており、そこで生産された日本酒は、日本を代表する銘柄となっています。

　和食が世界文化遺産に指定された今、世界各地の食に合う日本酒のよさを世界の多くの食通に知ってもらうために本書を刊行しました。

On Publication

Kanju Osawa, President, Tokyo University of Agriculture and Technology

Sake has a long history. It was first brewed in Japan more than 1,000 years ago, and has developed uniquely in harmony with the Japanese climate and eating habits. It is said that around the year 1200, individuals began to establish their own sake breweries and produce and sell various kinds of sake. After that, sake began to be used for Shinto rituals and festivals in various parts of the country. This led to even more sake breweries and the spread of sake to the general public. The first time sake was exhibited overseas was at the World Exposition in Austria in 1873. Even before that, Japanese trade with the Netherlands had already taken it to many parts of Asia.

Since then, many different types of sake have been produced, and today there are over 1,500 breweries producing and selling thousands of brands. Each brand has its own distinctive aroma and taste, thanks to the abundance of spring water, the variety of rice and yeast used as ingredients, and improvements in brewing techniques to suit people's tastes.

The Tokyo University of Agriculture established the Department of Brewing Science in 1953 to promote scientific and technical research on sake production, and to train successors for the brewing industry. The Department of Brewing Science plays a central role in education and research on sake. Our graduates are active in the brewing industry, both in Japan and abroad. Graduates of the university are brewers in half of all sake breweries in Japan, and the sake produced at these breweries include the major Japanese brands.

Japanese food (washoku) has been designated as a World Heritage, so we are publishing this book to let the world's gourmets know about the delicious taste of sake that goes well with food from all over the world.

List of Breweries

Here are the sake breweries where graduates of Tokyo University of Agriculture and Technology (TUAT) are active. From north to south, there are 296 companies in total.

Hokkaido region

Hokkaido	Kobayashi Shuzo	45
	Nippon Seishu	46
	Fukutsukasa Shuzo	47

Tohoku region

Aomori	Ozaki Shuzo	49
	Kikukoma Shuzo	50
	Takenami Shuzo	51
	Narumi Jozoten	52
	Hatomasamune	53
	Momokawa	54
	Morita Shobei	55
Iwate	Akabu Shuzo	56
	Asabiraki	57
	Azumamine Shuzoten	58
	Iwate Meijo	59
	Nanbu Bijin	60
	Hirota Shuzoten	61
	Washinoo	62
Miyagi	Ichinokura	63
	Kakuboshi	64
	Saura	65
	Niizawa Jozoten	66
	Hagino Shuzo	67
	Yamawa Shuzoten	68
Akita	Akita Shurui Seizo	69
	Akita Meijo	70
	Kariho Shuzo	71
	Dewatsuru Shuzo	72
	Tenju Shuzo	73
	Fukurokuju Shuzo	74
	Maizuru Shuzo	75
Yamagata	Eau de Vie Shonai	76
	Koikawa Shuzo	77
	Kouzaka Shuzo	78
	Goto Shuzoten	79
	Koya Shuzo	80
	Sato Sajiemon	81
	Shindo Shuzoten	82
	Suzuki Shuzoten Nagaikura	83
	Takagi Shuzo	84
	Tatenokawa Shuzo	85
	Dewazakura Shuzo	86
Fukushima	Kaitou Otokoyama Shuzo	87
	Kitanohana Shuzojo	88
	Suzuki Shuzoten	89
	Takahashi Shosaku Shuzoten	90
	Tatsuizumi Shuzo	91
	Tsurunoe Shuzo	92
	Niida Honke	93
	Bandai Shuzo	94
	East Japan Shuzo Cooperative Association	95
	Minenoyuki Shuzojo	96
	Yamatogawa Shuzoten	97

Kanto region

Ibaraki	Urazato Shuzoten	99
	Okabe Gomei-gaisha	100
	Hirose Shoten	101
	Hagiwara Shuzo	102
	Buyu	103
	Yamanaka Shuzoten	104

	Yoshikubo Shuzo	105
	Raifuku Shuzo	106
Tochigi	Inoue Seikichi Shoten	107
	Utsunomiya Shuzo	108
	Kobayashi Shuzo	109
	Sagara Shuzo	110
	Shimazaki Shuzo	111
	Tentaka Shuzo	112
	Tomikawa Shuzoten	113
	Matsui Shuzoten	114
	Morito Shuzo	115
	Watanabe Shuzo	116
Gunma	Asama Shuzo	117
	Ohtone Shuzo	118
	Kondo Shuzo	119
	Shimaoka Shuzo	120
	Nagai honke	121
	Yanagisawa Shuzo	122
	Yamakawa Shuzo	123
Saitama	Igarashi Shuzo	124
	Kamaya	125
	Koyama honke Shuzo	126
	Shinkame Shuzo	127
	Bukou Shuzo	128
	Kitanishi Shuzo	129
	Yao Honten	130
Chiba	Inahana Shuzo	131
	Iinuma honke	132
	Kameda Shuzo	133
	Kidoizumi Shuzo	134
	Kubota Shuzo	135
	Nabedana	136
	Baba honten Shuzo	137

	Miyazaki Honke	138
Tokyo	Ishikawa Shuzo	139
	Ozawa Shuzo	140
Kanagawa	Kumazawa Shuzo	141
	Nakazawa Shuzo	142

Koushinetsu region

Yamanashi	Ide Jozoten	144
	Takenoi Shuzo	145
	Tanizakura Shuzo	146
	Yamanashi meijyo	147
Nagano	Kasuga Shuzo	148
	Daishinnsyu Shuzo	149
	Okazaki Shuzo	150
	Ono Shuzoten	151
	Kadoguchi Shuzoten	152
	Kurosawa Shuzo	153
	Shusen kurano	154
	Sakenunoya honkin Shuzo	155
	Shinsyu meijyo	156
	Senjyo	157
	Daisekkei Shuzo	158
	Chikumanishiki Shuzo	159
	Tsuchiya Shuzoten	160
	Toshimaya	161
	Totsuka Shuzo	162
	Hanno Shuzo	163
	Nanawarai Shuzo	164
	Nishiiida Shuzoten	165
	Miyasaka Jozo	166
	Yukawa Shuzoten	167
	Yonezawa Shuzo	168

Niigata	Asahi Shuzo	169
	Ikeura Shuzo	170
	Ikedaya Shuzo	171
	Ishimoto Shuzo	172
	Echigozakura Shuzo	173
	Koshi meijo	174
	Kaganoi Shuzo	175
	Kikusui Shuzo	176
	Kiminoi Shuzo	177
	Kusumi Shuzo	178
	Takeda Shuzoten	179
	Tochikura Shuzo	180
	Hakkai Jozo	181
	Masukagami	182
	Miyao Shuzo	183
	Murayu Shuzo	184
	Yukitsubaki Shuzo	185
	Watanabe Shuzoten	186

Hokuriku region

Toyama	Sansyouraku Shuzo	188
	Kiyoto Shuzojo	189
	Takasawa Shuzojo	190
	Tateyama Shuzo	191
	Fukutsuru Shuzo	192
Ishikawa	Kobori Shuzoten	193
	Kano Shuzo	194
	Kikuhime	195
	Kuze Shuzoten	196
	Sakurada Shuzo	197
	Hakutou Shuzoten	198
	Hashimoto Shuzo	199
	Higashi Shuzo	200
	Fukumitsuya	201
	Yoshida Shuzoten	202
Fukui	Kokuryo Shuzo	203
	Tajima Shuzo	204
	Nanbu Shuzojo	205
	Manaturu Shuzo	206
	Miyakehikozaemon Shuzo	207
	Yasumoto Shuzo	208

Tohkai region

Gifu	Adachi Shuzojo	210
	Kaba Shuzojo	211
	Kawasiri Shuzojo	212
	Kuramoto Yamada	213
	Komachi Shuzo	214
	Sugihara Shuzo	215
	Chigonoiwa Shuzo	216
	Tenryou Shuzo	217
	Hakusen Shuzo	218
	Hayashi honten	219
	Nunoya Hara Shuzojo	220
	Harada Shuzojo	221
	Hirase Shuzoten	222
	Michisakari	223
	Watanabe Shuzoten	224
Shizuoka	Isojiman Shuzo	225
	Takashima Shuzo	226
	Sugii Shuzo	227
	Doi Shuzojo	228
	Fujinishiki Shuzo	229
	Yamanaka Shuzo	230

Aichi	Sawada Shuzo	231
	Sekiya Jozo	232
	Naito Jozo	233
	Maruishi Jozo	234
	Banjo Jozo	235
	Yamazaki	236
	Watanabe Shuzo	237
Mie	Ota Shuzo	238
	Ogawa honke	239
	Takijiman Shuzo	240
	Hayakawa Shuzo	241
	Gensaka Shuzo	242
	Moriki Shuzojo	243

Kinki region

Shiga	Emisiki Shuzo	245
	Kitajima Shuzo	246
	Furukawa Shuzo	247
Kyoto	Kizakura	248
	Kinosita Shuzo	249
	Gekkeikan	250
	Takeno Shuzo	251
	Saito Shuzo	252
	Fujioka Shuzo	253
	Masuda tokubee Shoten	254
	Mukai Shuzo	255
Hyogo	Ibaraki Shuzo	256
	Okada honke	257
	Okamura Shuzojo	258
	Okuto Shoji	259
	Kikumasamune Shuzo	260
	Konishi Shuzo	261
	Konotomo Shuzo	262
	Sakuramasamune	263
	Sawanotsuru	264
	Hakutsuru Shuzo	265
	Tajime	266
	Tatsuuma honke Shuzo	267
	Nadagiku Shuzo	268
	Nihonsakari	269
	Hakutaka	270
	Honda Shoten	271
	Yaegaki Shuzo	272
Nara	Umenoyado Shuzo	273
	Kita Shuzo	274
	Kuramoto Shuzo	275
Wakayama	Ozaki Shuzo	276
	Takagaki Shuzo	277
	Yoshimura Hideo Shoten	278

Chugoku region

Tottori	Inata honten	280
	Yamane Shuzojo	281
	Akana Shuzo	282
	Oki Shuzo	283
	Okuizumo Shuzo	284
	Nihonkai Shuzo	285
	Fuji Shuzo	286
	Rihaku Shuzo	287
Okayama	Isochidori Shuzo	288
	Kamikokoro Shuzo	289
	Sanko Masamune	290
	Miyashita Shuzo	291
	Ochi Shuzojo	292

	Toshimori Shuzo	293
	Watanabe Shuzo Honten	294
Hiroshima	Imada Shuzo Honten	295
	Kanemitsu Shuzo	296
	Koizumi Honten	297
	Saijotsuru Jozo	298
	Nakao Jozo	299
	Nakano Koujiro Honten	300
	Fujii Shuzo	301
	Houken Shuzo	302
Yamaguchi	Sumikawa Shuzojo	303
	Nakashimaya Shuzojo	304
	Nagayama Shuzo	305
	Horie Sakaba	306

Shikoku region

Tokushima	Tsukasagiku Shuzo	308
	Honke Matsuura Shuzojo	309
	Miyoshikiku Shuzo	310
Kagawa	Kawatsuru Shuzo	311
Ehime	Ishizuchi Shuzo	312
	Kondo Shuzo	313
	Sakurauzumaki Shuzo	314
	Suto Shuzo	315
	Seiryo Shuzo	316
Kouchi	Kameizumi Shuzo	317
	Takagi Shuzo	318
	Tsukasabotan Shuzo	319

Kyushu region

Fukuoka	Asahikiku Shuzo	321
	Ishikura Shuzo	322
	Isonosawa	323
	Takahashi Shoten	324
	Hiyokutsuru Shuzo	325
	Morinokura	326
Saga	Amabuki Shuzo	327
	Sachihime Shuzo	328
	Koyanagi Shuzo	329
	Tenzan Shuzo	330
	Madonoume Shuzo	331
	Kiyama Shoten	332
	Matsuuraichi Shuzo	333
	Yamato Shuzo	334
Nagasaki	Aimusume Shuzo	335
	Fukuda Shuzo	336
	Mori Shuzojo	337
	Yoshidaya	338
Kumamoto	Kameman Shuzo	339
	Kawazu Shuzo	340
	Zuiyo	341
	Chiyonosono Shuzo	342
	Tuzyun Shuzo	343
Ooita	Kuncho Shuzo	344
	Komatsu Shuzojo	345
	Sanwa Shurui	346
	Nakano Shuzo	347
	Bungo meijyo	348
	Yatsushika Shuzo	349

Sake in the world

In recent years, there has been a growing movement in Japan to make sake "the world's sake" as a crystallization of the Japanese character, symbolizing the Japanese climate and the patience, careful work, and sensitivities of the Japanese people.

This can be seen in the steadily growing overseas market for sake. In 1989, Japan exported 6,700 kiloliters of sake. In 2008, it was 12,000 kiloliters the export amount of sake was about 6,700 kl, but due to the steady increase in exports, the amount increased to about 12,000 kl in 2008. In the past twenty years, the volume has almost doubled, but even this will most likely be exceeded. In 2010, 12,800 kiloliters were exported, growing to 16,200 kiloliters in 2013 after the Japanese cuisine washoku was registered as a UNESCO World Intangible Cultural Heritage. Looking at the Ministry of Finance's statistical data, this period seems to have been a major turning point: in the 25 years between 1989 to 2013, the amount of exports increased by 9,500 kl (an average annual growth rate of about 4%). However, the increase since 2013 has been even greater: about 16,300 kiloliters in 2014, 18,200 in 2015, and 21,800 kl in 2020. This is actually an increase of about 6,000 kiloliters in the seven years following 2014. In addition, the annual average rate of increase for the seven years from 2014 is about 10%, a constant increase. This is more than twice as great as the increase in the twenty-five years previous. Sake was already being drunk around the world. The difference came after washoku was registered as an intangible cultural heritage and became known to the world. Export sales are trending in the same way as the quantity, from 3.92 billion yen in 2003 to 8.96 billion yen in 2012, an increase of about 5 billion yen over the past ten years. This grew to 10.52 billion yen in 2013, 11.51 billion yen in 2014, reaching 24.14 billion yen in 2020. This is an increase of 12.5 billion yen per year since 2013, partly due to an increase in sales volume, but even more to an increase in the amount of high-grade sake (Ginjoshu and Daiginjoshu), which has a higher unit price per bottle. The countries that support this consumption have not changed much, with the U.S. in first place, China (including Hong Kong) in second, and South Korea in third.

In addition, judging panels in the U.S., U.K., and France are playing a major role in raising the profile of sake. In the United States, the largest exporter of sake, the number of entries in the National Sake Competition (headquartered in Hawaii, USA) has been increasing every year since 2001. Sake is attracting a great deal of attention overseas, including in the United States. Joy of Sake (a tasting event for general consumers) is held in cities in Japan and the United States, and the appeal of sake is disseminated through this event. In Europe, we are working with the IWC (International Sake Competition). In Europe, the SAKE category of the International Wine Challenge (IWC, headquartered in the UK) was established in 2007, and since then many Japanese sakes have been entered. Needless to say, since the sake category was added to the wine competition, it has attracted a lot of attention and is regarded as a prestigious award. The impact of the competition has been so great that orders for winning sake brands have been received both domestically and internationally. In addition, the Kula Master (a judging panel of sommeliers and chefs, headquartered in France) is attracting attention as a new judging panel for sake that goes well with French cuisine. This competition is unique in that the judges are people who handle sake. This is why the number of new sake breweries participating in this competition is increasing.

These are just a few of the factors that are driving the overseas expansion of sake, but we are also seeing new orders from overseas. In Europe in particular, demand is expected to grow, and the concept of "terroir," which can be seen in agricultural products and the products made from them, especially in France. In the case of sake, the term "terroir" has never been a concern, but it has become an important concept in France. For example, in the case of rice, " environment, variety and cultivation are important. In France and other wine cultures, viticulture and winemaking are one and the same.

The same is true for sake. The same concept applies to sake. It depends on how far a sake brewing company can go, but this way of thinking will become necessary when considering the rest of the world.

Since sake was registered under the Geographical Indication (GI) protection system, more and more sake breweries are working on domestic rice, water, and production. I believe that if sake breweries rooted in their communities can actively develop strategies that demonstrate their regional characteristics, they will be able to proudly present themselves as world-class sake.

Sake is becoming a part of the export strategies of Cool Japan, but there are some issues, such as the need for strategy tailored to the different market structure of the partner country, and the fact that the selling price in the export destination is three to five times higher than in Japan due to tariffs, etc., putting it at a disadvantageous product when competing with foreign products. Consumption, however, is increasing despite the high price, especially among the wealthy segments of export markets, as the awareness of the aforementioned screening committee increases, indicating that people understand the merits of the products. Nevertheless, there is still a need for effective promotion strategies and the development of export systems for sake that take advantage of regional and local characteristics.

Some say there is a limit to the growth of the domestic market, but considering the increasing recognition of sake in other countries, it is inevitable that it will be marketed with overseas markets in mind. On the other hand, the increase in overseas recognition is turning the attention of the domestic market back to the merits of Japanese sake. The opportunities for people around the world to drink sake, which is the very essence of traditional Japanese culture, are expanding greatly, and I hope that sake will spread its wings to the world.

Ken Hosaka

(Professor, Department of Brewing Science, Faculty of Applied Biological Sciences, Tokyo University of Agriculture)

Loved around the World
Sake from Tokyo University of Agriculture

Sake brewed by graduates of Tokyo University of Agriculture is drunk around the world as **SAKE**. Here is a map of the breweries and the countries to which they export. **SAKE** will continue to amaze the world's gourmets.

1. Kobayashi Shuzo (Hokkaido)
2. Nihon Seishu
3. Ozaki Shuzo
4. Narumi Jozoten
5. Hatomasamune
6. Morita Shobei
7. Akabu Shuzo
8. Iwate Meijo
9. Nanbu Bijin
10. Hirota Shuzoten
11. Washinoo
12. Ichinokura
13. Kakuboshi
14. Saura
15. Niizawa Jozoten
16. Hagino Shuzo
17. Akita Shurui Seizo
18. Akita Meijo
19. Kariho Shuzo
20. Tenju Shuzo
21. Maizuru Shuzo
22. Koikawa Shuzo
23. Kouzaka Shuzo
24. Goto Shuzoten
25. Koya Shuzo
26. Sato Sajiemon
27. Shindo Shuzoten
28. Suzuki Shuzoten Nagaikura
29. Tatenokawa Shuzo
30. Dewazakura Shuzo
31. Tsurunoe Shuzo
32. Niida Honke
33. Bandai Shuzo
34. Yamatogawa Shuzoten
35. Hirose Shoten
36. Yamanaka Shuzoten
37. Yoshikubo Shuzo
38. Raifuku Shuzo
39. Inoue Seikichi Shoten
40. Utsunomiya Shuzo
41. Kobayashi Shuzo (Tochigi)
42. Shimazaki Shuzo
43. Tentaka Shuzo
44. Matsui Shuzoten
45. Watanabe Shuzo
46. Asama Shuzo
47. Kondo Shuzo
48. Kamaya
49. Koyama honke Shuzo
50. Kitanishi Shuzo
51. Yao Honten
52. Iinuma honke
53. Kameda Shuzo
54. Kidoizumi Shuzo
55. Nabedana
56. Ishikawa Shuzo
57. Ozawa Shuzo
58. Kumazawa Shuzo
59. Nakazawa Shuzo
60. Takenoi Shuzo
61. Tanizakura Shuzo
62. Daishinnsyu Shuzo
63. Okazaki Shuzo
64. Kurosawa Shuzo
65. Shusen kurano
66. Shinsyu meijyo
67. Senjyo
68. Chikumanishiki Shuzo
69. Tsuchiya Shuzoten
70. Hanno Shuzo
71. Nanawarai Shuzo
72. Nishiiida Shuzoten
73. Miyasaka Jozo
74. Yukawa Shuzoten
75. Yonezawa Shuzo
76. Asahi Shuzo
77. Ishimoto Shuzo
78. Echigozakura Shuzo
79. Kaganoi Shuzo
80. Kikusui Shuzo
81. Kiminoi Shuzo
82. Kusumi Shuzo
83. Hakkai Jozo
84. Masukagami
85. Miyao Shuzo
86. Watanabe Shuzoten
87. Sansyouraku Shuzo
88. Takasawa Shuzojo
89. Tateyama Shuzo
90. Kobori Shuzoten
91. Kano Shuzo
92. Kikuhime
93. Sakurada Shuzo
94. Hakutou Shuzoten
95. Hashimoto Shuzo
96. Higashi Shuzo
97. Fukumitsuya
98. Yoshida Shuzoten
99. Kokuryu Shuzo
100. Tajima Shuzo
101. Nanbu Shuzojo
102. Manaturu Shuzo
103. Miyakehikozaemon Shuzo
104. Kaba Shuzojo
105. Kuramoto Yamada
106. Komachi Shuzo
107. Sugihara Shuzo
108. Chigonoiwa Shuzo
109. enryou Shuzo
110. Hakusen Shuzo
111. Hayashi honten
112. Harada Shuzojo
113. Hirase Shuzoten
114. Michisakari
115. Doi Shuzojo
116. Yamanaka Shuzo
117. Sawada Shuzo
118. Sekiya Jozo
119. Banjo Jozo
120. Yamazaki
121. Ota Shuzo
122. Takijiman Shuzo
123. Hayakawa Shuzo
124. Gensaka Shuzo
125. Emisiki Shuzo
126. Kitajima Shuzo
127. Kinosita Shuzo
128. Gekkeikan
129. Saito Shuzo
130. Masuda tokubee Shoten
131. Mukai Shuzo
132. Ibaraki Shuzo
133. Konotomo Shuzo
134. Sakuramasamune
135. Sawanotsuru
136. Tajime
137. Tatsuuma honke Shuzo
138. Nadagiku Shuzo
139. Hakutaka
140. Honda Shoten
141. Yaegaki Shuzo
142. Umenoyado Shuzo
143. Kita Shuzo
144. Inata honten
145. Yamane Shuzojo
146. Akana Shuzo
147. Oki Shuzo
148. Okuizumo Shuzo
149. Nihonkai Shuzo
150. Fuji Shuzo
151. Rihaku Shuzo
152. Kamikokoro Shuzo
153. Sanko Masamune
154. Miyashita Shuzo
155. Toshimori Shuzo
156. Imada Shuzo Honten
157. Kanemitsu Shuzo
158. Saijotsuru Jozo
159. Nakao Jozo
160. Fujii Shuzo
161. Sumikawa Shuzojo
162. Nakashimaya Shuzojo
163. Horie Sakaba
164. Tsukasagiku Shuzo
165. Honke Matsuura Shuzojo
166. Miyoshikiku Shuzo
167. Kawatsuru Shuzo
168. Ishizuchi Shuzo
169. Sakurauzumaki Shuzo
170. Seiryo Shuzo
171. Kameizumi Shuzo
172. Takagi Shuzo
173. Tsukasabotan Shuzo
174. Asahikiku Shuzo
175. Ishikura Shuzo
176. Takahashi Shoten
177. Hiyokutsuru Shuzo
178. Morinokura
179. Amabuki Shuzo
180. Sachihime Shuzo
181. Tenzan Shuzo
182. Madonoume Shuzo
183. Yamato Shuzo
184. Aimusume Shuzo
185. Fukuda Shuzo
186. Mori Shuzojo
187. Yoshidaya
188. Kawazu Shuzo
189. Zuiyo
190. Chiyonosono Shuzo
191. Tuzyun Shuzo
192. Kuncho Shuzo
193. Komatsu Shuzojo
194. Sanwa Shurui
195. Nakano Shuzo
196. Bungo meijyo
197. Yatsushika Shuzo

Loved around the World Sake from Tokyo University of Agriculture

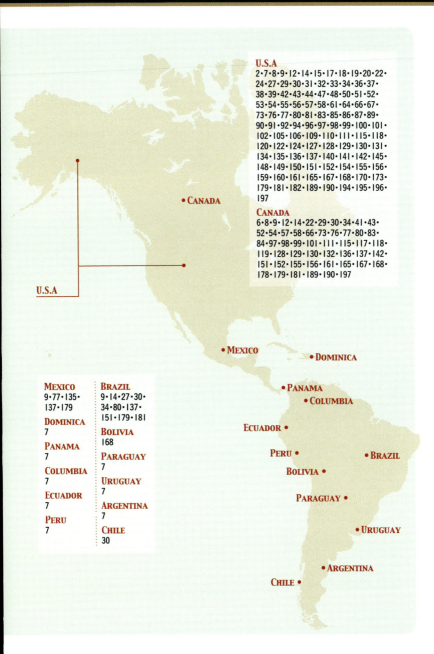

U.S.A
2·7·8·9·12·14·15·17·18·19·20·22·
24·27·29·30·31·32·33·34·36·37·
38·39·42·43·44·47·48·50·51·52·
53·54·55·56·57·58·61·64·66·67·
73·76·77·80·81·83·85·86·87·89·
90·91·92·94·96·97·98·99·100·101·
102·105·106·109·110·111·115·118·
120·122·124·127·128·129·130·131·
134·135·136·137·140·141·142·145·
148·149·150·151·152·154·155·156·
159·160·161·165·167·168·170·173·
179·181·182·189·190·194·195·196·
197

CANADA
6·8·9·12·14·22·29·30·34·41·43·
52·54·57·58·66·73·76·77·80·83·
84·97·98·99·101·111·115·117·118·
119·128·129·130·132·136·137·142·
151·152·155·156·161·165·167·168·
178·179·181·189·190·197

MEXICO
9·77·135·137·179

DOMINICA
7

PANAMA
7

COLUMBIA
7

ECUADOR
7

PERU
7

BRAZIL
9·14·27·30·34·80·137·151·179·181

BOLIVIA
168

PARAGUAY
7

URUGUAY
7

ARGENTINA
7

CHILE
30

EUROPE

RUSSIA
9·19·30·109·115·137·181

FINLAND
7

SWEDEN
19·21·30·33·43·55·73·127·130·178·179

NORWAY
19

DENMARK
33·73·159·178

LITHUANIA
9·105

POLAND
30

CZECH
7

HUNGARY
14

GERMANY
9·14·22·30·33·35·39·42·43·50·54·64·69·73·75·76·80·94·96·101·107·115·119·127·130·

135·137·141·143·144·145·151·161·167·173·178·179·181·192·195

NETHERLANDS
9·14·30·33·43·50·54·80·87·89·115·118·124·

127·131·137·155·160·167·178·179·186·197

BELGIUM
9·33·73·80·101·110·137·178·197

LUXEMBOURG
50

AUSTRIA
33·43·80·130·178·179

SWITZERLAND
9·11·19·30·43·50·58·73·80·106·110·119·129·144·149·159·160·167·168·173·178·179·197

BULGARIA
7

GREECE
7·137

ITALY
9·14·19·22·30·33·42·43·58·73·80·97·114·130·148·160·178·179·181

SPAIN
9·14·19·30·33·43·58·73·80·98·141·151·165·173·178·197

PORTUGAL
7·165

FRANCE
4·7·9·11·12·22·30·33·35·36·37·38·42·43·50·73·76·79·80·96·97·107·110·111·114·115·118·119·129·136·137·141·144·145·151·152·160·165·167·171·173·174·178·179·180·181·193·195·197

U.K.
9·12·14·15·17·19·27·29·30·34·40·42·43·50·54·57·62·66·67·68·73·75·76·77·80·83·86·97·98·99·107·110·111·114·115·118·127·128·130·

132·133·135·136·137·142·153·157·163·165·167·168·171·173·176·179·181·192·197

PALESTINE
7

ISRAEL
9·14·19·80·133

LEBANON
9·19·30

QATAR
7

U.A.E (DUBAI)
9·14·30·43·78·80·89·115·151·163·179·181

INDIA
9·30·50

NIGERIA
7

UGANDA
9·30

SOUTH AFRICA
137

Loved around the World Sake from Tokyo University of Agriculture

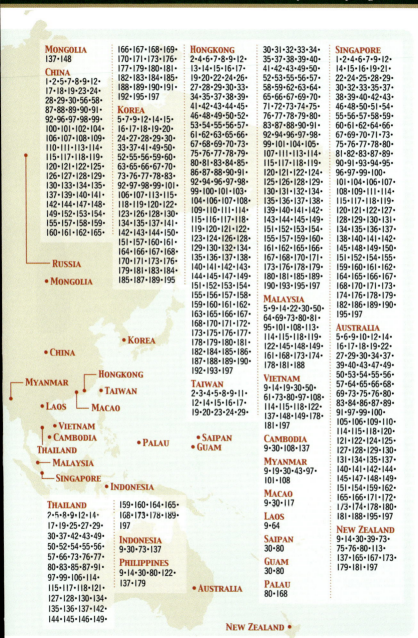

MONGOLIA
137·148

CHINA
1·2·5·7·8·9·12·
17·18·19·23·24·
28·29·30·56·58·
87·88·89·90·91·
92·96·97·98·99·
100·101·102·104·
106·107·108·109·
110·111·113·114·
115·117·118·119·
120·121·122·125·
126·127·128·129·
130·133·134·135·
137·139·140·141·
142·144·147·148·
149·152·153·154·
155·157·158·159·
160·161·162·165·

RUSSIA

MONGOLIA

166·167·168·169·
170·171·173·176·
177·179·180·181·
182·183·184·185·
188·189·190·191·
192·195·197

KOREA
5·7·9·12·14·15·
16·17·18·19·20·
24·27·28·29·30·
33·37·41·49·50·
52·55·56·59·60·
63·65·66·67·70·
73·76·77·78·83·
92·97·98·99·101·
106·107·113·115·
118·119·120·122·
123·126·128·130·
134·135·137·141·
142·143·144·150·
151·157·160·161·
164·166·167·168·
170·171·173·176·
179·181·183·184·
185·187·189·195

KOREA

CHINA

MYANMAR

HONGKONG

TAIWAN

LAOS **MACAO**

VIETNAM
CAMBODIA

THAILAND

MALAYSIA

SINGAPORE

INDONESIA

THAILAND
2·5·8·9·12·14·
17·19·25·27·29·
30·37·42·43·49·
50·52·54·55·56·
57·66·73·76·77·
80·83·85·87·91·
97·99·106·114·
115·117·118·121·
127·128·130·134·
135·136·137·142·
144·145·146·149·

HONGKONG
2·4·6·7·8·9·12·
13·14·15·16·17·
19·20·22·24·26·
27·28·29·30·33·
34·35·37·38·39·
41·42·43·44·45·
46·48·49·50·52·
53·54·55·56·57·
61·62·63·65·66·
67·68·69·70·73·
75·76·77·78·79·
80·81·83·84·85·
86·87·88·90·91·
92·94·96·97·98·
99·100·101·103·
104·106·107·108·
109·110·111·114·
115·116·117·118·
119·120·121·122·
123·124·126·128·
129·130·132·134·
135·136·137·138·
140·141·142·143·
144·145·147·149·
151·152·153·154·
155·156·157·158·
159·160·161·162·
163·165·166·167·
168·170·171·172·
173·175·176·177·
178·179·180·181·
182·184·185·186·
187·188·189·190·
192·193·197

TAIWAN
2·3·4·5·8·9·11·
12·14·15·16·17·
19·20·23·24·29·

PALAU

SAIPAN
GUAM

159·160·164·165·
168·173·178·189·
197

INDONESIA
9·30·73·137

PHILIPPINES
9·14·30·80·122·
137·179

AUSTRALIA

30·31·32·33·34·
35·37·38·39·40·
41·42·43·49·50·
52·53·55·56·57·
58·59·62·63·64·
65·66·67·69·70·
71·72·73·74·75·
76·77·78·79·80·
83·87·88·90·91·
92·94·96·97·98·
99·101·104·105·
107·111·113·114·
115·117·118·119·
120·121·122·124·
125·126·128·129·
130·131·132·134·
135·136·137·138·
139·140·141·142·
143·144·145·146·
151·152·153·154·
155·157·159·160·
161·162·165·166·
167·168·170·171·
173·176·178·179·
180·181·185·189·
190·193·195·197

MALAYSIA
5·9·14·22·30·50·
64·69·73·80·81·
95·101·108·113·
114·115·118·119·
122·145·148·149·
161·168·173·174·
178·181·188

VIETNAM
9·14·19·30·50·
61·73·80·97·108·
114·115·118·122·
137·148·149·178·
181·197

CAMBODIA
9·30·108·137

MYANMAR
9·19·30·43·97·
101·108

MACAO
9·30·117

LAOS
9·64

SAIPAN
30·80

GUAM
30·80

PALAU
80·168

SINGAPORE
1·2·4·6·7·9·12·
14·15·16·19·21·
22·24·25·28·29·
30·32·33·35·37·
38·39·40·42·43·
46·48·50·51·54·
55·56·57·58·59·
60·61·62·64·66·
67·69·70·71·73·
75·76·77·78·80·
81·82·83·87·89·
90·91·93·94·95·
96·97·99·100·
101·104·106·107·
108·109·111·114·
115·117·118·119·
120·121·122·127·
128·129·130·131·
134·135·136·137·
138·140·141·142·
145·148·149·150·
151·152·154·155·
159·160·161·162·
164·165·166·167·
168·170·171·173·
174·176·178·179·
182·186·189·190·
195·197

AUSTRALIA
5·6·9·10·12·14·
16·17·18·19·22·
27·29·30·34·37·
39·40·43·47·49·
50·53·54·55·56·
57·64·65·66·68·
69·73·75·76·80·
83·84·86·87·89·
91·97·99·100·
105·106·109·110·
114·115·118·120·
121·122·124·125·
127·128·129·130·
131·134·135·137·
140·141·142·144·
145·147·148·149·
151·154·159·162·
165·166·171·172·
173·174·178·180·
181·188·195·197

NEW ZEALAND
9·14·30·39·73·
75·76·80·113·
137·165·167·173·
179·181·197

NEW ZEALAND

17

Sake Brewery's Recommendations
"Dishes to go with Sake"

An introduction to sake and cuisine recommended by the sake breweries featured in this book.
Dishes that make sake more appealing. Sake that makes the food taste better.
When they are perfectly matched, the food will be many times more enjoyable.

Tuna and Okayama Leek Farce
Farce means "to stuff" in French. Tuna is stuffed into chives that look like a serving dish. It is a local dish in Okayama.

Kamikokoro Shuzo
Kamikokoro Junmai Ginjo Muroka Namazake Fuyu no Tsuki
(P289)

Fresh seafood and potherbs escabeche
This is a western style Nanbanzuke, a dish of marinated seafood with potherbs in white wine vinegar and olive oil.

Miyashita Shuzo
Kiwamihijiri Junmai Ichiban Shizuku
(P291)

Sashimi
A crisp, dry sake that enhances the flavors of sashimi, side dishes, and simply seasoned dishes.

Takijiman
Junmai Karakuchi Hayase
(P240)

Sake Brewery's Recommendations "Dishes to go with Sake"

Imoni
Imoni is a Yamagata dish often eaten during the taro harvest season from autumn to winter, and goes perfectly well with sake.

Dewazakura Shuzo
Dewazakura Ichiro
(P86)

Fruit Tomato Caprese with Salted-koji Mozzarella Cheese
Caprese is a type of salad originating in Italy. It can be served at a party as an appetizer or snack with sparkling sake.

Yamanashi Meijo
Shichiken Sparkling Yama no Kasumi
(P147)

Roast beef
Meat goes very well with Junmai Daiginjo, a clear and refreshing sake. Tuck into the finest Kobe roast beef.

Tatsuma Honke Shuzo
Kuromatsu Hakushika Junmai Daiginjo
(P267)

Teriyaki Yellowtail
A sharp aftertaste unique to Honjozoshu goes well with thickly seasoned Japanese food and Western food with butter and olive oil.

Umenoyado Shuzo
Umenoyado Honjozo
(P273)

Shrimp Ajillo
Garlic-flavored olive oil is delicious on bread. Ajillo is a perfect match for Daiginjoshu.

Nanbu Shuzojo
Kyukyoku no Hanagaki
(P205)

Kaiseki Cuisine
Dishes made with fish from Hakata go well with sake made in Hakata breweries. Junmai Daiginjoshu with its refreshing aftertaste is the perfect sake for a meal.

Ishikura Shuzo
Junmai Daiginjo Hyakunengura
(P322)

Milt Tempura
Tempura of cod milt is crispy outside and creamy inside. This rich, creamy dish is the perfect accompaniment for sake.

Sawanotsuru
Junmai Daiginjo Zuicho
(P264)

Gyoza (pot stickers)
Junmai Daiginjoshu is best served lukewarm, around 40°C. Utsunomiya is famous for its gyoza, characterized by its vegetable-based filling.

Utsunomiya Shuzo
Shikisakura
Junmai Daiginjo Imai Shohei
(P108)

Sake Brewery's Recommendations "Dishes to go with Sake"

Oden
This is a local sake from Kaga City in Ishikawa Prefecture, where the winters are harsh and cold. A glass of hot sake with oden will warm your body and soul.

Hashimoto Shuzo
Judaime Umajuku Honjozo
(P199)

Wagyu Roast Beef
Sake made from Omachi rice with rhododendron flower yeast using the kimoto-zukuri method, is a perfect match for food. Roast beef is best served with salt.

Amabuki Shuzo
Amabuki Kimoto
Junmai Daiginjo Omachi
(P327)

Cheese
Sake and cheese are both fermented foods made by fermenting rice and milk, respectively. When combined, they have a synergistic effect and increase in flavor.

Yamazaki Limited Partnership
Yumesansui Juwari Oku VINTAGE
Junmai Ginjo Genshu
(P236)

Scallops and Sake Lees with Cream Sauce
Make by pulling apart the sake lees and putting them in a pot. Add milk, salt, and miso paste, and stir until thickened.

Ishimoto Shuzo
Koshino Kanbai
Kinmuku Junmai Daiginjo
(P172)

Hinai Jidori Yakitori

Hinai chickens have been raised in the northern part of Akita Prefecture since ancient times. Along with kiritanpo, this is an Akita dish and goes well with Akita sake.

Kariho Shuzo
Kariho Ichihozumi Junmai Ginjo
(P71)

Traditional Cuisine
Keichan

Chicken is seasoned with garlic, ginger and soy sauce or miso, and then stir-fried with vegetables such as cabbage and bean sprouts.

Nunoya Hara Shuzojo
Genbun Hana Kobo Sakura
(P220)

Grilled Squid

Junmaishu with a strong, dry flavor goes well with seafood. Grilled squid and simmered mackerel with miso are delicious.

Ozaki Shuzo
Ando Suigun Tokubetsu Junmai
(P49)

Seasonal Vegetables

Junmai Ginjoshu with an elegant Ginjo aroma that mellows out, is served with vegetables. The dish shown in the lower right of the photo is grilled eggplant, tomato, and boiled edamame.

Ono Shuzoten
Yoake Mae "Nama Ippon"
(P151)

Sake Brewery's Recommendations "Dishes to go with Sake"

Beef Griller
Iga sake made with Ukonnishiki sake rice has a delicious rice flavor and a sharp aftertaste. It is also goes well with fatty sashimi.

Ota Shuzo
Hanzo Tokubetsu Junmaishu Igasan Ukonnishiki
(P238)

Seafood
Junmai Ginjoshu has a delicate and elegant taste. It goes well not only with seafood, but also with light dishes with a gentle flavor such as quiche.

Saura
Junmai Ginjo Urakasumi Zen
(P65)

Saba Heshiko Ginjo Jikomi
Heshiko is a dish made by pickling bluefish in salt and then in nukazuke (salted ricebran paste). It is a preserved food for winter.

Kano Shuzo
Jokigen Yamahai Junmai
(P194)

Zakuzaku
A dish of Fukushima Prefecture. It is called zakuzaku because various vegetables, such as taro, carrot, and radish are cut into small pieces and boiled.

East Japan Sake Brewers Association
Junmai Daiginjo Shizuku Sake Kinnojo
(P95)

Eel
Junmai Taruzake (cask sake) with a refreshing aroma of Yoshino cedar and a sharp flavor is delicious with richly seasoned dishes such as eel.

Kikumasamune Shuzo
Kikumasamune Junmai Taruzake
(P260)

Simmered Sand Lance
A specialty of Kobe, simmered sand lance is a 2 to 4cm long sand lance cooked with sugar, soy sauce, mirin and ginger.

Hakutsuru Shuzo
Hakutsuru Josen Junmai Nigorisake Sayuri
(P265)

Grilled Fish
A Junmai Daiginjoshu that lets you enjoy the flavor of the Omachi sake rice goes well with grilled fish, sashimi, carpaccio, Spanish mackerel tataki, and sukiyaki.

Toshimori Shuzo
Akaiwa Omachi Junmai Daiginjo
(P293)

Cheese, Dried Fruits, Nuts
Junmai Daiginjoshu has a mellow aroma and full flavor with spring water from Mt. Fuji. Delicious with cheese, dried fruits and nuts.

Ide Jozoten
Kai no Kaiun Junmai Daiginjo Furoku
(P144)

Sake Basics

This section introduces the reader to some basic information about sake. Learn how it is made and what the geographical indication(GI) protection system is.
A little knowledge will probably make sake taste even better.

Sake Basics (1)

Major Sake Rice Cultivation Areas in Japan

This map shows what varieties of sake rice are grown in the different regions of Japan.

Koushinetsu region

Ipponjime, Kamenoo, Kinmonnishiki, Koshiibuki, Koshitanrei, Gohyakumangoku, Sankeinishiki, Shirakabanishiki, Takanenishiki, Tamasakae, Hattannishikinigou, Hitogokochi, Miyamanishiki, Yumesansui

Hokuriku region

Ishikawasake 30 gou, Ishikawamon, Oyamanishiki, Kamenoo, Kinmonnishiki, Koshinoshizuku, Koshihikari, Gohyakumangoku, Shinriki, Nagaikimai, Tominoka, Hanaechizen, Fukunohana, Hokuriku 12 gou, Miyamanishiki, Yamadanishiki, Yukinosei

Kinki region

Akitsuho, Inishinomai, Iwai, Kyonokagaki, Ginfubuki, Shiragiku, Shinriki, Takanenishiki, Tajimagouriki, Tamasakae, Nadanishiki, Nihonbare, Hakutsurunishiki, Hinohikari, Hyogokitanishiki, Hyogokoinishiki, Hyogoyumenishiki, Fukunohana, Mizukagami, Murasakikomachi, Yamadanishiki

Chugoku region

Akihikari, Akebono, Asahi, Omachi, Kaminomai, Gouriki, Koiomachi, Kokuryomiyako, Sakanishiki, Saitonoshizuku, Senbonishiki, Nakateshinsenbon, Hatanso, Hattannishiki

Shikoku region

Oseto, Omachi, Kazanaruko, Ginnosato, Ginnoyume, Sanukiyoimai, Shizukuhime, Tosaurara, Tosanishiki, Matsuyamamii

Kyushu region

Oitamii 120 gou, Ginnosato, Saganohana, Shinriki, Hanakagura, Hananishiki, Yumeikkon, Wakamizu

Major Sake Rice Cultivation Areas in Japan

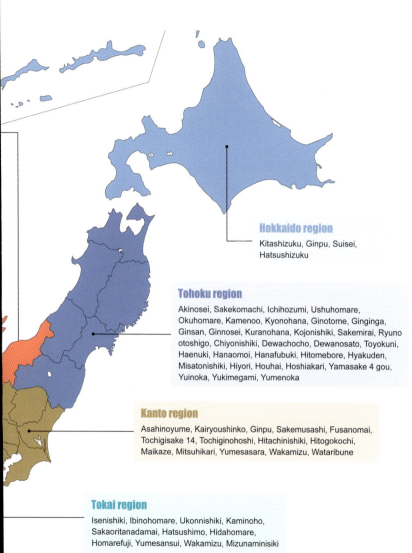

Hokkaido region
Kitashizuku, Ginpu, Suisei, Hatsushizuku

Tohoku region
Akinosei, Sakekomachi, Ichihozumi, Ushuhomare, Okuhomare, Kamenoo, Kyonohana, Ginotome, Ginginga, Ginsan, Ginnosei, Kuranohana, Kojonishiki, Sakemirai, Ryuno otoshigo, Chiyonishiki, Dewachocho, Dewanosato, Toyokuni, Haenuki, Hanaomoi, Hanafubuki, Hitomebore, Hyakuden, Misatonishiki, Hiyori, Houhai, Hoshiakari, Yamasake 4 gou, Yuinoka, Yukimegami, Yumenoka

Kanto region
Asahinoyume, Kairyoushinko, Ginpu, Sakemusashi, Fusanomai, Tochigisake 14, Tochiginohoshi, Hitachinishiki, Hitogokochi, Maikaze, Mitsuhikari, Yumesasara, Wakamizu, Wataribune

Tokai region
Isenishiki, Ibinohomare, Ukonnishiki, Kaminoho, Sakaoritanadamai, Hatsushimo, Hidahomare, Homarefuji, Yumesansui, Wakamizu, Mizunaminisiki

Sake Basics (2)

Geographical Indication (GI) Protection System

── What is the Geographical Indication (GI) protection system? ──

The Geographical Indication (GI) protection system, which is often seen in the EU and other countries, allows alcoholic beverages and agricultural products to bear the name of their place of origin (regional brand) if their established quality, social reputation, or other characteristics can be attributed primarily to the geographical origin of the product.

── Designation of Nihonshu (Japanese sake) as a Geographical Indication (GI) ──

The National Tax Agency, which has jurisdiction over the liquor industry, designated Nihonshu (Japanese sake) as a geographical indication on December 25, 2015. This was done with an eye to improving brand value and promoting exports.

Effects of designation
Only domestically produced rice is used to make Nihonshu, and only sake produced in Japan can be called Nihonshu.
1. Liquor produced overseas or made from rice produced overseas cannot be labeled as sake, even when distributed in Japan. This makes it easy for consumers to distinguish Nihonshu (sake) from other liquors.
2. When sold abroad, an appeal can be made for sake as a high-quality and reliable alcohol.
3. By differentiating between sake and similar liquors produced outside Japan, its brand value will increase.

The term Nihonshu will contribute to demand, both domestically and overseas.

Geographical Indications (GI) for Liquor Products in Japan designated by the Commissioner of the National Tax Agency

Name	Date of designation	Region	Sake category
Iki	June 30,1995	Iki City, Nagasaki Prefecture	Spirits (Shochu/Awamori)
Kuma	June 30, 1995	Kumagun and Hitoyoshishi, Kumamoto Prefecture	Spirits
Ryukyu	June 30, 1995	Okinawa Prefecture	Spirits
Satsuma	December 22, 2005	Kagoshima Prefecture (excluding Amami City and Oshima County)	Spirits
Hakusan	December 22, 2005	Hakusan City, Ishikawa Prefecture	Seishu (sake)
Yamanashi	July 16, 2013	Yamanashi Prefecture	Wine
Nihonshu (Japanese sake)	December 25, 2015	Japan	Seishu
Yamagata	December 16, 2016	Yamagata Prefecture	Seishu
Hokkaido	June 28, 2018	Hokkaido	Wine
Nadagogo	June 28, 2018	Wards of Nada and Higashinada in Kobe, cities of Ashiya and Nishinomiya in Hyogo Prefecture	Seishu
Harima	March 16, 2020	Cities of Himeji, Aioi, Kakogawa, Ako, Nishiwaki, Miki, Takasago, Ono, Kasai, Shiso, Kato, Tatsuno, and Akashi, towns of Taka, Inami, Harima, Ichikawa, Fukusaki, Kamikawa, Taishi, Kamigori and Sayo in Hyogo Prefecture	Seishu
Mie	June 19, 2020	Mie Prefecture	Seishu
Wakayama Umeshu (plum wine)	September 7, 2020	Wakayama Prefecture	Other liquor
Tone Numata	January 22, 2021	Numata City, towns of Katashina, Kawaba, and Showa in Tone County, Gunma Prefecture	Seishu

Sake Basics (3)

Sake Making
First koji, second yeast, third brewing

The general process of sake production consists of five stages: rice polishing, making koji, making yeast, brewing unrefined sake, and bottling from the top of the tank. All of these processes require a great deal of attention, but as has been said since ancient times, "First koji, second yeast, third brewing". The parts that require careful handling of microorganisms are the most important. Understanding the process, will make the sake you drink taste even better.

What rice to use and how to handle it
—the first step in sake brewing

The rice used to make sake is referred to as *sakamai*, varieties of rice more suitable for sake brewing than the regular rice people eat (see page 24).

The type of rice used (variety, place of origin, cultivation method, etc.) and how it is handled (polishing, soaking, steaming, etc.) will greatly affect the flavor of the finished sake.

 Brown rice — Brown rice is rice with the husks removed. The next step in milling requires grains that are large and soft, making them easier to polish.

 Polishing rice — The surface layer of brown rice contains fiber, fat, and protein which can have an adverse effect on sake quality, so it is all shaved off. This is called rice polishing.

 Rice — Rice is polished over two full days. If polishing is done too quickly, the rice kernel is crushed and the frictional heat causes rice to deteriorate.

 Rice washing → Soaking → Steaming — The powdered shavings (bran) remaining on the surface of polished rice is washed off. After that, rice is soaked to absorb the necessary amount of water.

 Steamed rice — After the soaked rice is drained, it is steamed. It is important to steam rice so that the outside is hard and the inside is soft.

Sake Making First koji, second yeast, third brewing

The first and most important step in the koji making process.

As we learned on the previous page, koji is the first step in making sake. It is also said that 70% of the sake making process is completed when good koji is made, so you can see that this is a very important part of the process. Time and temperature are carefully controlled to allow rice and microbes to work well together.

Koji comes out of the collaboration of rice and microorganisms.

蒸米 Steamed rice

Yukamomi (floor mashing)

Steamed rice is spread out on the floor of the koji room and sprinkled with koji (*Aspergillus oryzae*).

Hikikomi (pulling in)

Steamed rice that has been cooled to about 35°C is put into the koji room where the temperature and humidity are kept constant.

Mori (heaping)

To make rice easier to work with, the rice is divided into small portions in a flat wooden box called kojibuta, and piled up in a heap in the center of each box.

Kirikaeshi (cutting back)

After floor mashing, rice is left in a mound for 10–12 hours, after which it is broken down, and rice is mixed well and spread out to make sure bacteria grows evenly.

Shimaishigoto (final work)

About 40 hours after hikikomi, the piles of steamed rice in the kojibako are broken down and churned, and then repositioned. Temperature of the rice is kept at 42 to 43°C.

Sekigae, nakashigoto (repiling and middle work)

In order to keep temperature and bacteria growth even, the positions of the kojibako are changed, and the piles of steamed rice inside them are broken and remixed.

Dekoji, Karashijo

About 48 hours after hikikomi, rice is taken out of the koji room (dekoji), and cooled in the dry the room (karashijo).

In the process of koji growth, the starch in the white rice is saccharified by enzymes. At the dekoji stage, it has the aroma of roasted chestnuts and a mildly sweet taste. It has finally become rice koji. This koji, which is no longer growing koji bacteria, is used in the subsequent process of making the yeast and fermentation mash.

米麹 Rice koji

二酛 Second Yeast

The process of making the "origin" or "mother" of sake

The second most important process is the yeast, which in Japanese means "sake mother". Rice malt, brewing water, and steamed rice are mixed together, and yeast starter is added.

Lactic acid prevents the growth of harmful bacteria and cultivates a large amount of excellent yeast. There are several methods, such as letting lactic acid bacteria produce lactic acid (raw yeast, Yamahai yeast) or adding lactic acid in advance (rapid brewing yeast).

The power of lactic acid is important

This yeast is used to cultivate the yeast that will direct the development of fermentation mash, an indispensable element for producing high quality sake. The power of lactic acid is required to do this.

米麹 / Rice Koji: Koji rice accounts for about 30–33% of the total amount of rice used in yeast.

蒸米 / Steamed rice: Steamed rice accounts for 67-70% of the total amount of rice used in yeast.

汲水 / Water: If the amount of rice used for the sake mother is 100%, water is used at a ratio of 110%.

酒母 / Yeast mash: This "mother of sake," is used in the subsequent brewing process to make the fermentation mash.

乳酸 / Lactic acid: 7ml of yeast is used for every liter of water.

酵母 / Yeast: Pre-cultivated yeast is put into the sake mother tank as starter and cultivated purely with the power of lactic acid.

Kimoto is a method of producing yeast using lactic acid bacteria.

Water, rice malt, and steamed rice are divided into several hangirioke containers (about 200 liters in capacity), and the steamed rice and koji are piled up in mounds. The mounds are then pulled apart with oar-like paddles to produce a large amount of lactic acid. The strong acidity of the lactic acid kills most of the harmful bacteria and the lactic acid bacteria itself, allowing only the yeast to be purely.

If this process, called yamaoroshi, is not used and the rice grains are dissolved only by the power of koji enzymes to bring out the power of lactic acid bacteria to make yeast mash that is the same as raw yeast, it is called Yamahai yeast. This rapid brewing sake mash (yeast mash) does not wait for the natural formation of lactic acid bacteria, but rather cultivates yeast in a short period of time.

Sake Making First koji, second yeast, third brewing

Alcohol is produced by the power of yeast.

The third part of the process, called brewing, is the production of fermentation mash. Koji, steamed rice, and water are added to the yeast mash, and it turns sugar into alcohol.

The fermentation mash is characterized by parallel double fermentation and the breakdown of rice starch into sugar by koji enzymes. The yeast in the mash converts the sugar into alcohol.

Preparation

When adding koji, steamed rice, and water to a yeast mash are divided, it is called three-stage preparation. Dividing it into four steps is called four-stage preparation. The mash causes a slow change in environment in a way that maintains its functions.

Cultivation

Careful temperature and data control very important for cultivating fermentation mash to create a good balance between saccharification and fermentation.
Regular sake is cultivated at 15 degrees for about 15 days. Daiginjo is usually about 10 degrees for about 35 days.

Filtering the Moromi, which is the goal ingredient, is called josou or shibori. New sake that has just been pressed in this process is cloudy with undigested starch particles, yeast, and koji mold mycelium, and this cloudiness is called ori.

Ori is left to settle in a cool and dark place for about 10 days. This process is called orihiki or orisage. After this process, the sake is filtered, heated, stored, and watered added before being bottled and shipped.

Sake is made by a process of first koji, second yeast, third brewing.

In the world of sake, the word "brewing" is used to describe all these processes.

In some breweries the brewer is involved in choosing water and rice. Other breweries begin their work with rice cultivation.

Sake Basics (4)
Classifications of Sake

Sake such as Junmaishu and Ginjoshu is called Tokutei Meishoshu, and is divided into eight types. The criteria for classification are mainly determined by the rice polishing ratio and the ingredients used, including brewing alcohol.

Since there are no strict standards for these labels, each brewer decides at their own discretion, but in general, if the sake is made to have a splendid aroma, it is labeled Junmai Ginjoshu or Ginjoshu. If it has a cool taste, it is often labeled Tokubetsu Junmai or Tokubetsu Honjozo Sake.

It is important to note, therefore, that the labeling is only a guide, and does not indicate the rank of the flavor or the rarity of the sake.

Specific Name Classifications

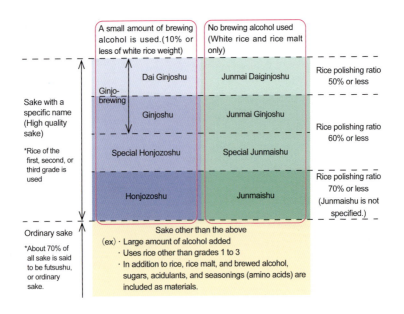

Classifications of Sake

General classification	Overview	Specified name	Indication of ingredients	Rice polishing ratio
Junmaishu (pure rice sake)	Made from white rice with a milling ratio (*1) of 70% or less, rice malt, and water; literally "rice" alone. It has a rich, full-bodied flavor with a high acidity level.	Junmaishu	Rice, rice malt	No restrictions
		Special Junmaishu		60% or less
Honjozoshu	White rice with a milling ratio of 70% or less, rice malt, and water. A small amount of brewing alcohol (*2) is added. Mostly refreshing, light and dry.	Honjozoshu	Rice, rice malt and brewing alcohol	70% or less
		Special Honjozoshu		60% or less
Ginjoshu	Ginjozukuri (*3) is made from white rice with a milling ratio of 60% or less, rice malt, water, and brewing alcohol (not used in the pure rice type). Sake made from highly polished rice with a milling ratio of 50% or less is further classified as "Daiginjo" or "Junmai Daiginjo," the highest rank.	Ginjoshu	Rice, rice malt and brewing alcohol	60% or less
		Daiginjoshu		50% or less
		Junmai Ginjoshu	Rice, rice malt	60% or less
		Junmai Daiginjoshu		50% or less

*1 Milling ratio: A value that indicates the ratio of white rice (white rice that has undergone the polishing process of removing the surface layer of the brown rice to remove miscellaneous flavors) to brown rice (rice that has only had its outer husk removed). The lower the value, the more time and money is required to mill the rice. For example, if the milling ratio is 60%, it means that 40% of the surface of the brown rice has been removed. Sake, even ordinary sake that is not classified as "sake with a specific name" is made with a milling ratio of about 75% or less. White rice eaten at home is generally milled to about 92%.

*2 Brewing alcohol: Alcohol distilled from starchy or sugar-containing materials through fermentation. It is used in the sake production process to adjust the alcohol content, quality, and flavor. In the case of Honjozo and Ginjo sake, the maximum amount added is limited to 10% or less of the weight of the white rice used as ingredients.

*3 "Ginjozukuri" means "brewed with scrutiny," and refers to a special production method using carefully selected, high-quality ingredients and refined techniques, with great care and attention to detail.

A combination chart for "aroma" and "flavor."

There are two words to describe sake: "splendid" and "mild" for the aroma, and "rich" and "light" for the flavor. The chart below shows the combination of "aroma" and "flavor" as a guide. When you taste sake, try to describe it as you wish.

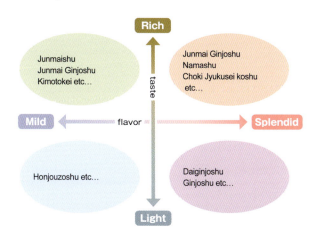

Sake Basics (5)

Sake Terminology
To deepen your understanding of sake

This section explains basic sake terminology that will be of interest to you as you read through this book and become familiar with sake.

> **Terms assigned according to the production process, time of production, storage method, etc.**
>
> On sake labels, the brand name is often followed by the classification of sake as "Junmai," "Ginjo," or some other specific name, and other words are sometimes added to indicate the differences in the sake's manufacturing process and taste. Here are some typical examples.

Kan-zukuri

Kanzukuri is the process of making sake during the cold season from November to March. The cold season is considered the best time for sake brewing because it is difficult for bacteria to grow and good sake can be produced.

Kiippon

The term kiippon is used when the sake is made in one's own brewery and is also pure rice wine.

Kimoto-zukuri

Malted rice, steamed rice, and water are placed in a shallow vat at a low temperature, and then stirred with an oar-like paddle.
In this way, natural lactic acid bacteria grows and fermentation is promoted. This allows natural lactic acid bacteria to proliferate and grow a large amount of excellent, powerful yeast, while suppressing bacteria under strong acidity. The result is a robust, dense flavor.

Yamahai, Yamahai jikomi

This is the process of kimoto without the paddle stirring. The taste is as rich and full-bodied as the kimoto process. In addition to kimoto and yamahai jikomi, there is also sokujomoto, in which lactic acid is added, but this tends to produce a lighter taste.

Genshu

A type of sake that has been pressed and bottled without adding water (except when the alcohol content is adjusted to less than 1%) and without reducing the alcohol content.

Ordinary sake has an alcohol content of 14 to 16 degrees, but undiluted sake has a higher alcohol content of 18 to 20 degrees.

Koshu (old sake)

Sake made in the previous year, or even earlier. It is classified by brewing year (BY = brewery year) and labeled as 14BY (produced in 2002) or 26BY (produced in 2014). The taste settles down and becomes mellow with aging. Sake that has been brewed for more than five years can be labeled as "treasure sake."

Shinshu (new sake)

Sake made in the current year (between July 1 and June 30 of the following year). Because it has not yet matured, it retains its characteristic young aroma. New sake is usually sold during the winter months of December through February. It contrasts with koshu (old sake).

Kijoshu

When brewing sake, half (or part) of the water used is sake. Characterized by a rich sweetness and often drunk as an aperitif.

Dakushu, Doburoku (unrefined, unfiltered sake)

A sake made without straining the unrefined sake, or after straining it lightly with a coarse cloth. The rice grains and koji remain in it. The yeast is still alive and contains carbon dioxide gas, so be careful when opening the bottle, as the contents will come out with great force.

Nigorizake (unrefined, filtered sake)

Nigorizake is the white, cloudy sake that is produced when the unrefined sake is pressed with a coarse cloth or net during the sake production process. It is often namashu (activated sake), but some are Junmaishu" or Honjozoshu, depending on the ingredients.

Orizake

The koji and yeast that remain after the fermentation mash are collected and bottled as is. It is also called "origarami".

Kasseishu (activated sake)

Sake that has not been heat-treated and has only been

filtered through a coarse cloth. The yeast is still alive, so it often contains carbon dioxide.

Taruzake (barrel sake)

Sake that has been stored in cedar barrels for a certain period of time and has the aroma of cedar wood. Some sake is shipped in barrels, while others are bottled from the barrels.

Namazake (raw sake)

Normally, sake is sterilized twice before shipment (during storage and bottling), but sake that is not sterilized twice is called Namazake. It has the fresh taste and aroma of freshly pressed sake.

Namachouzoushu (raw stored sake)

This is sake that has been stored at low temperatures without being heated, and is only heated once, just before shipping. It is the sake closest to its original state where the original flavor of the sake is still alive.

Hiyaoroshi (packed cold)

Hiyaoroshi is sake that has been heated once and matured through storage during the summer, and then bottled without heating for shipment in the fall. It has the fresh taste of new sake.
This is a term derived from the fact that in the old days sake was packed cold (raw) in barrels.

Tobindori, Tobingakoi

The part of sake which is the best, squeezed out of a suspended bag into an 18-liter bottle. It is the highest quality sake for the brewery.

Arabashiri

When making high-grade sake, the process of making the fermentation mash includes a method of pressing the unrefined sake by piling up a tankful of sake bags and letting the weight of the bags push the sake out naturally, without applying added pressure.
There are three names (ara, naka, seme) for the sake that flows out, depending on the order in which it is pressed, and the sake that comes out first is called Arabashiri. It is clouded and has a fresh taste and a splendid aroma.

Nakadori, Nakagumi

The sake that comes out after Arabashiri, described above, is called Nakadori or Nakagumi. It is often bottled using only this portion and submitted to appraisal societies. The last part, which is squeezed out by applying pressure, is called seme or oshikiri.

Terms used to describe the flavor of sake.

Sake metre value (SMV)

One measure of whether sake is sweet or dry. The higher the sugar content, the sweeter the taste, and the lower the sugar content, the drier the taste. Sake with a high sugar content is expressed as a minus value, while sake with a low sugar content is expressed as a plus value.

The higher the negative value, the sweeter the sake, and the higher the positive value, the drier the sake. However, the actual sweetness can vary depending on the alcohol content, acidity, amino acid content, aroma, and the temperature chosen when drinking.

Japanese alcohol content table

Very dry	+6.0 and above
Dry	+3.5 – +5.9
Slightly dry	+ 1.5 – + 3.4
Normal	- 1.4 – + 1.4
Slightly swee	- 1.5 – - 3.4
Sweet	- 3.5 – - 5.9
Very sweet	- 6.0 or more

Acidity

The total amount of acid contained in sake is expressed in milliliters of sodium hydroxide solution required to neutralize 10 ml of sake. The higher the acidity, the spicier the sake and the darker the taste. 1.5 or more is considered rich, while less than 1.5 is considered light and refreshing.

Amino acidity

This is the numerical value of the amount of amino acids contained in sake, which number in the dozens. This value determines how rich or light a sake is. 1.0 is the standard, and the higher the value, the richer the taste and the lower the value, the lighter the taste. Daiginjoshu and Junmai Daiginjoshu are made with a higher rice polishing ratio to remove any miscellaneous flavors, so even at around 1.0, the taste is refreshing and rich.

Yeast

Yeast is a general term for single-celled microorganisms that work to break down sugar into alcohol and carbon dioxide. The fruit-like aroma of Ginjo sake is due to the work of this yeast. There are many types, including "association yeast," which is distributed by the Japan Brewers Association, and "proprietary yeast," which has been developed independently by local governments and breweries. Some typical examples are introduced here.

(Table on the right)

Sake brewing year

This is an original classification of the sake brewing year, which runs from July 1 to June 30 of the following year.

Rice polishing ratio

The percentage of white rice remaining after rice polishing, expressed as a percentage. The smaller the number, the higher the rice polishing ratio.

About sake rice

Rice suitable for sake brewing

Sake rice is rice that has been improved for sake brewing. Among the different varieties, rice that has been designated by the Ministry of Agriculture, Forestry and Fisheries in terms of production area and breeding is called rice suitable for sake brewing.

The amount of rice produced is limited and the cost is high, so it is used for special sake such as Ginjoshu. Characterized by large grains and the large shinpaku (core) in the center of the rice, it absorbs water well and has a low protein content, which makes it easy for koji mold to attach to it, resulting in high quality sake.

The following is a list of typical sake brewing rice varieties that appear frequently in this book.

Yamadanishiki

Yamadanishiki is the most popular rice for sake brewing, and is also known as the king of sake rice. It has a large core and is best suited for Daiginjo sake with a milling rate of 50% or less. It produces sake with a deep umami flavor and a rich aroma.

Miyamanishiki

Miyamanishiki is a relatively new type of rice suitable for sake brewing, originating in Nagano Prefecture in 1978. It was named Miyamanishiki because it has a core like the snow at the top of the Northern Alps. It is characterized by a refreshing mouthfeel and elegant taste.

Gohyakumangoku

Gohyakumangoku is the most widely grown sake brewing rice in Niigata and Hokuriku regions. It has a straightforward taste, and is used in various classes of sake, such as Junmai, Ginjo and Daiginjo.

Omachi

Omachi has a large grain and core, so it is used as a hybrid in many areas because of its superiority as a rice suitable for sake brewing. It is said that more than

	Yeast	Feature
Main Kyokai Yeast	Kyokai No. 6	The oldest yeast, may be described as the father of Kyokai yeast. It was cultivated from the yeast produced at the Aramasa Shuzo in Akita Prefecture, and used to establish the brewing method of ginjo, which is commonly used today. It has a much lower acidity than other brewer's yeast.
	Kyokai No. 7	Cultivated from the yeast produced at the Miyasaka Jozo in Nagano Prefecture. It is characterized by its strong fermentation and gorgeous ginjo aroma. This strong fermentation is often used to brew ordinary sake.
	Kyokai No. 9	It is called "kouro yeast" and used by many breweries to make ginjo. It has low acidity, high aroma, and strong fermentation at low temperatures.
	Kyokai No. 10	Often used for ginjo and junmai, and characterized by its low acidity and high ginjo aroma. Its low acidity makes it suitable for junmai, and its high ginjo aroma makes it suitable for ginjo.
	Kyokai No. 11	This yeast is a mutation from Kyokai No. 7 and highly resistant to alcohol. It is characterized by low amino acid content and high malic acid content. It can be used to make dry and sharp sake with high alcohol content.
	Kyokai No. 14	It has low acidity, strong fermentation, and a high ability to produce ginjo aroma. These characteristics make it suitable for making clean and light ginjo.
	Kyokai No. 1501	Developed in 1991 as Akita-style flower yeast in Akita Prefecture. This non-foaming yeast was registered as Kyokai yeast in 1996, and produces a ginjo aroma with high ethyl caproate.
	Kyokai No. 1601	This non-foaming yeast is a high ethyl caproate-producing strain, and was distributed in 2001. It is used for daiginjo.
	Kyokai No. 1801	One of the most aromatic yeasts in the current Kyokai yeast. It produces a large amount of ethyl caproate, a component of ginjo aroma, making it suitable for daiginjo.
Major prefectural unique developments	Aomori Yeast	It was jointly developed by the Hirosaki Regional Technical Research Institute of the Aomori Industrial Research Center and the Brewing Society of Japan. It includes "Mahoroba Hana Yeast".
	Iwate Yeast	It was jointly developed by the Iwate Industrial Research Institute and the Iwate Biotechnology Research Center. It includes "Ginjo No. 2".
	Akita-style Flower Yeast	Original yeast of Akita Prefecture. It is characterized by a strong, gorgeous aroma and low acidity, which is suitable for daiginjo and junmai daiginjo.
	Yamagata Yeast	Developed at the Yamagata Research Institute of Technology. It has the ability to ferment even in Yamagata's cold climate, and is characterized by its refreshing fruit aroma.
	Utsukushima Yume Yeast	Developed at the Aizu Wakamatsu Technical Support Center of the Fukushima Technology Centre. It produces a gorgeous and mild taste with low acidity.
	Alps Yeast	Developed at the Nagano Prefecture Food Industrial Research Institute. It is characterized by its high production of ethyl caproate, an element of ginjo aroma.
	Tochigi Yeast	Many "Tochigi Yeasts" were developed for ginjo, but the new Tochigi Yeast "T-ND42" produces more acid and is more suitable for making thicker sake.
	Shizuoka Yeast	Developed at the Shizuoka Prefecture Industrial Research Institute. It produces a lot of isoamyl acetate, which is characteristic of ginjo aroma, and has a melon-like aroma.
	Hiroshima Yeast	One of the yeasts is "Hiroshima Ginjo Yeast" and produces a lot of gorgeous apple-like aroma (ethyl caproate).
	Kochi Yeast	Five types of "Kochi Yeast" have been developed at the Kochi Prefectural Industrial Technology Center and are widely used by sake brewers in the prefecture.

half of the best rice for sake brewing has Omachi in its background. The result is a sake with a rich and robust flavor. Bizen Omachi and Akaiwa Omachi in Okayama Prefecture are particularly famous.

Dewasansan

This rice was developed by Yamagata Prefecture over an 11-year period for use in ginjo, and produces sake with a light, soft, and broad flavor.

Hattannishiki

A sake rice produced mainly in Hiroshima Prefecture. It is characterized by its deep, rich, and mellow flavor. The aroma is subdued, and the combination of yeast and rice allows for a wide range of flavors.

Ginpu

This sake rice was developed from Hokkaido rice. In recent years, sake made with Gimpu has won awards at fairs. It brings out the harmony of the splendid aroma and taste of the rice.

How to drink sake (temperature)

Temperature

There are three main drinking temperatures for sake: chilled, room temperature, and heated. Some descriptions are hitohada (slightly warm), nurukan (lukewarm), atsukan (hot), and so on. The actual temperature is not clear. In fact, the serving temperature of sake is often described very vaguely, even though it can vary much more than other liquors. It can be drunk at a wide range of temperatures, from 0°C (freezing) to 55°C (hot sake).

The aroma and taste change in an extremely complex and diverse manner due to even a slight temperature difference.

0 to 5°C Ice temperature, on the rocks

For those who don't like the unique aroma of sake, cooling it to the extreme makes it easier to drink. Also, if you find it too sweet, chilling it will make it less sticky. This is not recommended for Ginjoshu, which has a pleasant aroma.

5°C to 10°C Chilled

To make the most of the coolness of plain sake, Honjozo, Genshu and Namachozo, it is recommended they be drunk chilled. For Ginjoshu, be careful not to chill it too much, as you will not be able to enjoy the aroma.

8°C to 15°C Lightly cooled

This is the best temperature range for Ginjo and Daiginjo. The harmony of the splendid aroma and taste of rice is brought out.

Around 20°C Room temperature

Drink at room temperature, not chilled or warmed. This is the best temperature range for Junmaishu, which has a distinct rice flavor.

Around 35°C Warm on the skin

Lightly warmed to a temperature close to the human body temperature. Recommended for Junmaishu, but you can also try this temperature for Ginjoshu with less aroma and more pronounced flavor.

Lukewarm around 40°C

Slightly warmer than body temperature, about the temperature of a lukewarm hot spring. This temperature brings out the aroma and flavor of the rice characteristic of Junmaishu.

About 50°C Hot

A popular way to drink regular sake and Honjozo in winter.

However, sweet sake may be too sticky, so dry sake works better. Junmaishu is also recommended for richer types, such as the Kimoto and Yamahai brewing. If it is too hot, the alcohol will evaporate and the taste will be unbalanced.

Around 60°C Piping hot

Add fugu (blowfish) fins, fish bones which have roasted, etc. to add a little extra flavor. This is also a good way to enjoy sake that has been open for a few days or that doesn't suit your palate.

How to Enjoy Japanese Sake and Washoku

Washoku, Japanese cuisine, is based on rice, and there is no reason why sake, made from rice, should not go well with it. Sake and washoku are a perfect combination that bring out the best in each other. Another advantage of sake is that it can be paired with a surprisingly wide variety of foods, including meat, beans, seafood, mushrooms, and cheese dishes. Here I would like to discuss how to enjoy Japanese sake and washoku.

The ever-changing world of washoku

Washoku is made up of several factors: "cuisine", which can easily be imagined as a part of Japanese food culture, from fine cuisine served at ryotei (traditional Japanese restaurants) to local and home-style cooking; the "manner" in which the food is eaten; and the "dining table scenery", which reflects all the dishes on the table. Washoku, which is often associated with vague images, has been changing as cultural exchange with other countries has become more active.

The basic format of washoku consists of rice (the staple food), soup and okazu (side dishes). This was the form of food that was developed in the samurai society during the Kamakura period (1185-1333). Later, the concept of ichijusansai (one soup three dishes) became the basic form of washoku, and this has been handed down to the present. The okazu in a meal are prepared to make the rice taste good, while the okazu for enjoying sake are renamed shukou (sake snacks) even though they may be the very same thing.

Making the most of the taste and shape of ingredients

At the root of Japanese food culture, there is the notion of maximizing the taste and the form of the ingredients themselves. Such values (preferences) have led to the choice of simple cooking methods, such as cooking "rice" with white grains, eating fish raw (sashimi) or boiling fish in water and soy sauce. In addition, the way ingredients are cut and the way okazu (main dish) are arranged on the plate are considered to make the dish colorful and visually pleasing. Some of the examples of okazu are, sashimi, simmered dishes, grilled dishes, tempura and so on. In a sense, okazu is like an a la carte meal, with a high degree of freedom.

Along with such a wide variety of okazu, there is also a wide repertoire of sake to choose from; Seishu or Nigorizake (there is a difference in clarity), sweet sake or dry sake, and also alcohol content varies. In order to get the most out of sake, it can be served at different temperatures, such as cold or warm. In addition to the characteristics of the ingredients themselves, such as taste and texture, the psychology of the eater—eating habits and preferences—plays a role in what constitutes deliciousness. In other words, deliciousness can be obtained through free personal choice. Nothing makes a person happier than discovering the food and sake they truly enjoy.

The presentation of tableware and sake cups enhance the taste

In addition to taste, there are other ways to enjoy washoku and sake. One of them is through the beauty of tableware and sake cups in terms of material, shape, color, and pattern. Japanese food culture places great importance on food as well as the tableware used to serve them, and they complement each other. The tableware and the sake cups are stoneware, china, lacquerware, glass, etc., and have developed to reflect the different textures of the materials when held in the hand, as well as the shapes, colors, and patterns. They are associated with the four seasons, such as chrysanthemum flowers and bamboo leaves and go by names, such as Aritayaki, Shikki, Edokiriko, and so on. Washoku and sake are enjoyable in terms of taste, but there is more to it; the beauty created by the combination of food and tableware, and the beautiful scene of the entire dining table laden with food and sake. One of the ways to enjoy washoku and sake is not only to taste it with your tongue, but also to see it with your eyes and let it fill your heart with pleasure.

Sake is a perfect match for local cuisine

Finally, I would like to mention the compatibility between sake and local cuisine. Local cuisine uses ingredients that are nurtured by the climate and natural aspects of the region, using a cooking method that has been handed down over the years, which also expresses the characteristics of the region. In 2007, the Ministry of Agriculture, Forestry and Fisheries (MAFF) held an event called 100 Best Local Dishes of Agriculture, Forestry, and Fisheries, and since then local ingredients and traditional dishes have been attracting a lot of attention using the terms, "local gourmet" and "grade B gourmet".

Needless to say, sake is made from local water, rice, and natural environment, and

reflects the culture. As the location changes, so does the sake. Which type of sake goes best with delicious food gathered from the mountains or the sea? The answer to this question can be found in local cuisine. Hokkaido's chanchanyaki, Yamagata's tongarajiru, Chiba's sesame-marinated sardines, Ishikawa's jibuni, Yamaguchi's fugu dish, Kumamoto's horse sashimi, Okinawa's ikasumijiru, and the list goes on and on. With the development and diversification of information transmission methods, the climate, history, food culture, and local cuisine of each region are introduced in a fast, broad, and attractive manner. The recipes of some dishes are available, which helps you to cook them at home. The most luxurious experience is to visit the region and savor the sake and food along with the local climate, but it is also possible to have a moment of pleasure at a distance.

Under the title of Japanese Sake: Sake that Captivates the World, our graduates introduce some of the breweries where they are working hard to make sake in a unique and sincere way, and also some of the local cuisine. If you are interested in the exquisite combination of sake and snacks while thinking about the local cuisine, you will be able to taste the local sake more deeply.

Sayuri Akuzawa
(Professor, Department of Nutritional Science and Food Safety,
Faculty of Applied Biosciences, Tokyo University of Agriculture)

Guide to Sake Breweries and Famous Sake

It is no exaggeration to say that the sake breweries in which the graduates of Tokyo University of Agriculture are active are representative of sake breweries in Japan.

Sake brewing is becoming more sophisticated and evolved as advanced research results are combined with traditional techniques .
We continue to refine our brewing skills in order to pass on and further advance our local customs as well as the culture of Japanese sake – we preserve traditions, and we love our land.

This guide introduces renowned sake produced by graduates of Tokyo University of Agriculture, along with a peak into the breweries.

This book is based on a survey conducted between November 2020 and March 2021.
The collection of sake bottles in the background of this page are displayed at the Tokyo University of Agriculture Museum of Food and Agriculture.

Guide to Sake Breweries and Famous Sake

Hokkaido Region

Food Culture of Hokkaido Region

Hokkaido's northern region: At the northern tip of Hokkaido, the Monument to the Northernmost Point of Japan at Cape Soya, Asahikawa, and Furano attract many visitors. This region is a major producer of hairy crabs. Teppo-jiru, or miso soup made from chopped crab, and Migaki-nishin, or preserved herring with its internal organs removed and dried in the sun, and pickled herring with vegetables, are popular home delicacies in this region.

East Hokkaido: Along with the fishing industry, large-scale agriculture such as field crops, dairy farming, and pig farming has developed in this region, the history of which dates back to the frontier days. Dishes using pork, such as Pork Don, have become local specialties. The region is also known for Sampei-jiru – a hot pot dish with preserved fish and vegetables with salty flavor -- which has become popular throughout Hokkaido. In the east of Hokkaido, salted salmon is used in this dish.

Central Hokkaido: Hokkaido is surrounded by the Pacific Ocean, the Sea of Japan, and the Sea of Okhotsk, and is blessed with seafood, but its vast plains also produce crops. Hokkaido is famous for its potatoes, and there is a local dish called "Imo Mochi" simply made with just potatoes. The potatoes are steamed, mashed, and shaped into a bun, and then baked. It is a simple dish with a subtle aroma and sweet taste.

The southern part of Hokkaido: This region is famous for hosting a port of call for the Kitamae Ship, as well as for a dish called Matsumae pickles. It is a preserved food made by marinating local squid and kelp in soy sauce. The umami of the squid and kelp complement each other well, and is favored as a side dish for rice or sake. Ikameshi, or squid stuffed with Uruchi rice or mochi rice cooked with seasonings, is not only a homestyle dish, but also a popular Ekiben (lunch box sold at train stations) that fills the stomach.

| Water source | Ishikari River system |
| Water quality | Soft |

Kobayashi Shuzo

3-109, Nishiki, Kuriyama-cho, Yubari-gun, Hokkaido 069-1521　TEL.0123-72-1001
http://www.kitanonishiki.com/

Famous sake from a coal mining town

The power of coal makes brewing possible in severe coldness.

Kobayashi Shuzo Co., Ltd. was founded as a sake brewer in Sapporo in 1878 (Meiji 11). In 1900, the brewer moved its headquarters to Kuriyama-cho, Yubari-gun. During the Meiji era (1868-1912), Sake brewers in Hokkaido were forced to brew sake under extremely freezing temperatures. The climate was so harsh that the brewers were worried that the fermentation process would stop. The second-generation head of the Kobayashi family, Yonezaburo, saw the potential of coal as the energy source of the future and brought in many roof tile craftsmen to build a brick warehouse with excellent thermal insulation. The 10-year project was a success, and the warehouse grew in size with an increase in demand for coal. The company's foresight in utilizing coal led to the current prosperity of Kobayashi Brewery.

Kitanonishiki
Tokubetsu Junmaishu Maruta

"A rich taste with a strong overall umami"

A good balance with acidity, can be enjoyed warm.

Special Junmaishu	1,800ml/720ml
Alc.	16%
Rice type	Ginfuu
Sake meter value	+5.0
Acidity	1.8

Kitanonishiki
Kuraikanjyuku Secret Junmai

"An old sake with a refreshing taste"

Aged for three years or more to reduce the amino acids and balance the acidity with the ripe sweetness of the rice. Can be served warm or cold.

Ginjoshu	1,800ml/720ml
Alc.	17%
Rice type	Kitashizuku
Sake meter value	-5.0
Acidity	2.1

Nippon Seishu

Water source	subsoil water from the Toyohira River
Water quality	medium-hard

2, Minami 3-jo Higashi 5-chome, Chuo-ku, Sapporo-shi, Hokkaido 060-0053 TEL.011-221-7106
E-mail: jouzoubu@fukutsukasa.jp   https://www.fukutsukasa.jp/

The brewery that kept pace with Hokkaido's pioneering history

A sake brewery that continues to flap its wings in northern Japan like the Tanchigetsuru crane dancing in the sky.

Founded in 1872, this is the only sake brewery in Sapporo that has kept pace with the history of Hokkaido's development. Hokkaido's winters are long and severe. The vast land is covered with pure white snow, turning it into a world of pure white. In the spring, the pure white snow melts and soaks into the rich soil, creating pure subsoil water over a long period of time. Rice cultivation has benefited from this natural environment and Hokkaido has become one of the leading rice-producing regions in Japan in terms of quality and quantity. Nippon Seishu aims to express the richness of Hokkaido's nature in their sake by carefully observing the growing conditions of rice harvested each year. Their main brand Chitose Zuru mainly uses Kitashizuku, a locally-grown rice suitable for sake brewing in Hokkaido, and has been highly evaluated at various sake competitions.

Chitose Zuru
Junmai Daiginjo Zuishou

"Sake that goes well with a wine glass"

This sake has a noble, delicate aroma and mellow taste that is matured during the process of long-term, low-temperature fermentation. The delicate aroma is reminiscent of fresh Japanese pears and flower honey.

Junmai Daiginjoshu	1,800ml/720ml/300ml
Alc.	16-17%
Rice type	Kitashizuku
Sake meter value	-2.0
Acidity	1.2

Chitose Zuru Shibata
Karakuchi Junmai

"The king of Junmaishu"

It has a clean and refreshing taste, but you can also enjoy the flavor of rice. It is best served lukewarm to enhance the sweetness and umami.

Junmaishu	1,800ml/720ml
Alc.	15-16%
Rice type	Ginfuu
Sake meter value	+6.0
Acidity	1.5

| Water source | Kushiro Marsh Subsoil Water |

Fukutsukasa Shuzo

2-13-23 Sumiyoshi, Kushiro-shi, Hokkaido 085-0831　TEL.0154-41-3100
E-mail: jouzoubu@fukutsukasa.jp　https://www.fukutsukasa.jp/

What we continue to aim for is sake to accompanies the local food culture

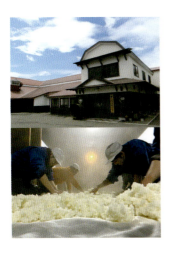

The Daiginjo brewed with the most flavorful Ginpu rice grown in Hokkaido.

In 1919, Chotaro Yanase founded a company, Shikishima Shokai, in Kushiro City for the purpose of wholesaling liquor, soft drinks, miscellaneous goods, and food products. In 1921, the company was moved to its current location. The brewery has been running the entire time. As the only sake brewery in Kushiro, it pursues the original flavor of sake and compatibility with Hokkaido's local ingredients. In recent years, the brewery has also developed products unique to the region. For example, there is Sokojikara, sake which is stored in Japan's only operating coal mine tunnel. The brewery also sells Minanikori, a yogurt liqueur made only from milk produced in Hokkaido. In response to the changing food culture in Hokkaido, the brwery continues to aim to make sake that matches the local food culture.

純米酒

Fukutsukasa Junmaishu

"Sake with meals, good for pairing with local Hokkaido dishes"

Refreshing Junmaishu brewed with Ginpu rice, one of Hokkaido's most flavorful varieties.

Junmaishu	720ml
Alc.	14%
Rice type	Ginfuu
Sake meter value	+4.0
Acidity	1.5

大吟醸酒

Fukutsukasa Daiginjo

"Good as a gift, good as a treat for yourself"

Brewed in the low-temperature, long-term fermentation process. This daiginjo is nurtured by Hokkaido's cool climate, and only a small number of bottles are produced.

Daiginjoshu	720ml
Alc.	15%
Rice type	Suisei
Sake meter value	+2.0
Acidity	1.3

Guide to Sake Breweries and Famous Sake
Tohoku Region
Aomori, Iwate, Miyagi, Akita, Yamagata, Fukushima

Food Culture of the Tohoku Region

Aomori: With its thriving fishing industry, Aomori boasts a variety of soups made from fish and shellfish. There is squid sushi, boiled squid stuffed with squid legs, shredded and salted carrots and cabbage, and pickled in vinegar. The prefecture is also known for a New Year's dish that uses a whole codfish. Oma tuna, caught using a single line off the coast of Oma, and scallops cultivated in Mutsu Bay, are also well-known.

Iwate: Off the coast of Sanriku, where the Kuroshio Current flows, is a rich fishing ground and a leading producer of sea urchins, ascidians, oysters and scallops. It is also known as Japan's leading producer of seaweed and abalone. The prefecture is also a producer of matsutake mushrooms. For weddings and funerals, it is customary to eat various kinds of rice cakes stuffed with ingredients such as bean paste or walnuts inside.

Miyagi: Rice production thrives in Miyagi prefecture. It is known for a dish called Salmon Oyakodon with rice cooked in broth made from salmon fillets with salmon roe on top. Yaki-haze Zouni and Zundamochi are eaten around New Year's and other celebrations. The prefecture is also known for oysters, Sasakamaboko (bamboo fish cake) made from Alaska pollack surimi, and beef tongue.

Akita: The prefecture's grouper fishery is famous. A hot-pot dish with fish and vegetables cooked in Shottsuru (a kind of fish sauce made from grouper) is a local specialty. It is also famous for many hot pot dishes such as Tara-nabe with chopped codfish and milt, Hinai Jidori chicken, Kiritanpo, and other preserved foods such as Iburi-Gakko. The young shoots of junsais, which can only be grown in clean water, are pinched and used for vinegared dishes and soup.

Yamagata: Cherries, pears, and other fruits are widely grown in Yamagata. It is also known for preserved foods and soups that use a lot of vegetables and fish. The custom of Imoni-kai (communities meeting to enjoy sweet potato stew) is said to have originated back when it was difficult to preserve taro in the winter, so that everyone could eat it before the weather turned cold. Dada-cha-mame beans, Yonezawa beef, and tamakonnyaku (konnyaku boiled in soy sauce) are also famous.

Fukushima: Dobu-jiru, a specialty in Fukushima Prefecture, is a soup that does not use water. It is made with fried "ankimo" (monkfish liver), fish meat and vegetables seasoned with miso. It is also famous for preserved foods such as Shimi-mochi as well as dishes using Migaki-nishin (herring) and Shio-Kujira (salted whale meat). Dried herring pickled with Japanese sansho pepper is often served at izakaya restaurants.

Water source	Shirakami Mountains
Water quality	Soft

Ozaki Shuzo

30, Ryoshimachi, Ajigasawa-machi, Nishitsugaru-gun, Aomori 038-2744 TEL.0173-72-2029
E-mail: andou-suigun@ozakishuzo.com   http://www.ozakishuzo.com

Sake brewery at the foot of Shirakami Mountains, where the gods reside.

Sake brewed with traditional techniques nurtured by the majestic Sea of Japan and Shirakami Mountain Range.

Ozaki Shuzo was founded 162 years ago, in 1860. The first generation of the family immigrated to Ajigasawa from Wakasa (present-day Fukui Prefecture), where they worked as fish brokers and pawnbrokers before starting their own sake brewing business. It has been 155 years since the family established itself in this area. The brewery, located in a wonderful place surrounded by the magnificent Japan Sea and the divine Shirakami Mountain Range, is well-established and has maintained traditional sake brewing methods. Initially, the brewery produced the sake brand Shiragiku, and it is now known for the brand called Ando Suigun, first produced in 1988, and becoming its signature sake. The brewery uses plenty of subterranean water from the Shirakami-Sanchi region, a World Natural Heritage Site, and locally grown rice. Most of the sake brewed here is dry and matches well with the seafoods and crops grown in the Ajigasawa region.

特別純米酒
Ando Suigun
Tokubetsu Junmai

"Junmaishu with clear and light taste"

This sake is named after Ando Suigun, the leader of northern Japan who came seeking a grand, new land. Dry sake with a rich flavor and a clean aftertaste.

Special Junmaishu	1,800ml/720ml
Alc.	15%
Rice type	Hanaomoi/Masshigura
Sake meter value	+2.5
Acidity	1.8

大吟醸酒
Daigin Kaminoza

"Gorgeous aroma and a rich flavor"

This Daiginjoshu has a splendid aroma and a rich, deep flavor. Unblended sake good served on the rocks.

Daiginjoshu	1,800ml/720ml
Alc.	17%
Rice type	Hanaomoi
Sake meter value	+1.6
Acidity	2.0

Kikukoma Shuzo

Water source	Hakkoda Mountains
Water quality	Soft

12 Aza-kawaramachi, Gonohe-machi, Sannohe-gun, Aomori 039-1554 TEL.0178-62-2323
E-mail: kikukoma@yellow.plala.or.jp http://www.kikukoma.com

Brewed in the northern climate
A brewery with a century-long history

Sake must be loved by the locals. The quality of color, flavor, and taste in every drop reflects the passion of the brewery.

Kikukoma Shuzo Brewery has been in business since 1910, making it a century-old brewery. The slogan of the brewery is "Excellence in sake is only possible when it is loved by the local community." The reason why tasteless and odorless rice can create different tastes of sake when brewed by various breweries is not only because of the characteristics of the climate and water, but also because of the skills of the toji (master brewer) who makes the most of the living ingredients, koji and yeast. Kikukoma Shuzo had a master brewer, Ikuo Fujita, who is now retired. Mr. Fujita was active in this field for 50 years. The brewery puts every effort and all its spirit into producing beautiful color, aroma, and taste in every single drop of sake. One of the advantages of the famous Kikukoma brand is its M2 yeast, which is characterized by low acidity. It takes a long time to brew, and needs care. It is brewed in the coldest part of the country, where the temperature drops to minus 10 degrees in winter.

純米酒
Kikukoma Junmaishu

"Dry, fragrant Junmaishu"

This Junmaishu has a mild, full-bodied flavor with a gentle ginjo aroma. It can be enjoyed in a wide range of temperatures, from cold to warm.

Junmaishu	1,800ml
Alc.	15-16%
Rice type	Hanafubuki
Sake meter value	+4.0
Acidity	1.2

大吟醸酒
Kikukoma Daiginjo

"A drop of the finest fragrance of traditional techniques"

A sake brand made by the Nanbu Toji, a master brewer who put his heart and soul into brewing it during the cold season. It has a fruity aroma and a delicate, refreshing taste.

Daiginjoshu	720ml
Alc.	15-16%
Rice type	Yamadanishiki
Sake meter value	+2.0
Acidity	1.2

Water source	Iwaki River subsoil water
Water quality	Soft

Takenami Shuzo

21-4 Ikuyozaki, Numazaki, Inagaki-cho, Tsugaru-shi, Aomori 037-0106 TEL.0173-23-5053
E-mail: takenami@takenami.biz http://www.takenami-shuzoten.com/

Brewed by the oldest brewery in Aomori
A powerful, dense, and delicious sake

Loved by the locals in Aomori, a great place to enjoy good food.

Enjoy hot junmaishu. Since its establishment in 1645, the Takenami Shuzo has been in Tsugaru, seeking to create sake that can be enjoyed during meals. It has aimed to produce good Junmaishu that can be enjoyed hot. The 17th-generation master brewer, Yoshiaki Takenami, believes that the true taste of sake can be experienced when it's served hot. In Aomori, where delicious food is available throughout the four seasons, drinking hot Junmaishu has become an old and new way of enjoying sake. The leading brand is Shichirobei. The brewery offers a wide range of sake lineups, from Junmaishu to Junmai Daiginjo. The flavors are simple, rich, and powerful. Aomori's oldest and smallest sake brewery continues to brew sake with deep, refined skills.

Shichirobei Tokubetsu Junmaishu

"A perfect match for food"
Special Junmaishu that has matured to create a harmonious blend of umami and mellowness. Serving at around 60 degrees is recommended.

Special Junmaishu	1,800ml/720ml
Alc.	15.5%
Rice type	Hanafubuki
Sake meter value	+9.0
Acidity	1.8-1.9

Shichirobei Junmai Ginjo Yamadanishiki

"Ships after a year of aging"
Junmai Ginjoshu that can be enjoyed at room temperature or hot. Serving at 50 degrees is recommended.

Junmai Ginjoshu	1,800ml/ 720ml
Alc.	15.5%
Rice type	Yamadanishiki
Sake meter value	+8.0
Acidity	1.6-1.8

Narumi Jozoten

Water source	Mt.Minamihakkoda subsoil water
Water quality	Soft

1-1 Oaza-nakamachi, Kuroishi-shi, Aomori 036-0377 TEL.0172-52-3321
E-mail: kikunoi@beach.ocn.ne.jp https://narumijozoten.com/

Rooted in the local community
A long-established company in Tsugaru, Kuroishi

A long-established sake brewery established in 1806, passing on the beauty of sake brewed in the climate of Tsugaru.

Narumi Jyouzouten, founded in 1806, is a small sake brewery located in a corner of the local Nakamachi Komise Dori, an arcade-like street that has been around since the early Edo period (1603-1868) and was designated as traditional architecture in 2005. The brewery itself was designated as a cultural asset by the city in April 1998. The representative brands of sake include "Kikunoi", "Inamuraya Bunshiro", and "Inamuraya", which are favored by the locals. The brewery was rebuilt at the beginning of the Taisho era (1912-1926) and is now over 100 years old, but it is well suited for sake brewing as it is cool in summer and protected from the cold outside in winter. The brewery uses soft well water from the southern Hakkoda Mountains and high quality rice from Aomori Prefecture to make sake, combining the traditional brewing techniques cultivated over the years with new ones.

純米大吟醸酒
Junmai Daiginjo Inamuraya Bunshiro

"The king of sake named after the master"

This Junmai Daiginjoshu is fruity and fragrant, with a full-bodied umami flavor enhanced by the sweetness of rice and a crisp finish. Chill in the refrigerator.

Junmai Daiginjoshu ... 1,800ml/720ml
Alc. 16.4%
Rice type .... Yamadanishiki
Sake meter value .... -1.0
Acidity 1.3

純米吟醸酒
Junmai Ginjo Inamuraya

"A new starting point for Inamuraya"

The gorgeous aroma of Junmai Ginjoshu is maintained, while the soft water from Mount Hakkoda creates a soft, light, umami taste.

Junmai Ginjoshu 1,800ml/720ml
Alc. 16%
Rice type .... Hanaomoi
Sake meter value .... -1.0
Acidity 1.5

Water source	Subsoil water from the Oirase River system of Mt.Hakkoda
Water quality	Soft

Hatomasamune

176-2 Aza-inayoshi, Sanbongi, Towada-shi, Aomori 034-0001   TEL.0176-23-0221
E-mail: sake@hatomasa.jp   https://www.hatomasa.jp/

Local water, local rice, and local Sake

Fusing tradition and innovation to produce sake that meets the needs of the times.

Hatomasanune was founded in 1899 (Meiji 32). Back then, the brewery was known for Inaomasamune, a brand named after the Inao River. In the early Showa period (1926-1989), the brewery changed its name to Hatomasamune in honor of a white pigeon that lived in the brewery's shrine as a guardian deity. The brewery's motto is "Local sake is the crystallization of local food culture." Traditionally, a Nambu Toji master brewer was in charge of brewing sake here, but since 2004, only brewers from the Towada region have been taught the brewing techniques. Since then, the brewery aimed for "reproducible sake brewing" by combining intuition gained from experience with quantified data. Early Hatomasamune brewers had a passionate desire to "brew authentic local sake using local water, rice, and a local master brewer." Today, Hatomasamune is dedicated to "brewing terroir (local) sake."

Hatomasamune Junmai Daiginjo Hanaomoi

"Refreshing and rich flavor"

This sake has a rich taste, well-balanced with a splendid aroma. Best to serve it cold with dishes like sashimi of Japanese flounder or scallops.

Junmai Daiginjoshu	1,800ml
Alc.	16%
Rice type	Hanaomoi
Sake meter value	±0
Acidity	1.5

Hatomasamune Tokubetsu Junmaishu Hanafubuki

"Goes with anything - good served both warm and cold"

This Special Junmaishu features a mild aroma, fine acidity, and a mellow, rich flavor that draws out the full flavor of the rice.

Special Junmaishu	1,800ml
Alc.	15-16%
Rice type	Hanafubuki
Sake meter value	±0
Acidity	1.6

Momokawa

Water source	Oirase River system
Water quality	Soft

112 Kamiakedo, Oirase-cho, Kamikita-gun, Aomori 039-2293 TEL.0178-52-2241
https://www.momokawa.co.jp

Since it was established, the motto has been "quality first"

Our philosophy is to make people happy through delicious sake.

Momokawa was first brewed in 1889. The company name at the time was Murai Shuzoten, but in 1944, a number of manufacturing companies merged to form Nikita Shuzo Co., Ltd. Later, in 1984, brands other than Momokawa were spun off, and the company was renamed Momokawa Co., Ltd. For over 30 years, Momokawa has been striving to make good sake with the catchphrase, "Morning knows good sake" and with the philosophy of making people happy through sake. The catchphrase means that the time you spend with good sake is a moment of happiness, and that such a moment gives us energy for tomorrow and allows us to have a good morning.

純米大吟醸酒
Daiginjo Junmai Hanaomoi

"A perfect match with sea urchin from Sanriku"

This sake has a splendid rising aroma, and when sipped, the flavor of the rice is fully present and the rich aroma spreads out.

Junmai Daiginjoshu	1,800ml/720ml
Alc.	15-16%
Rice type	Hanaomoi
Sake meter value	+3.0
Acidity	1.4

純米吟醸酒
Junmai Ginjoshu brewed with wine yeast

"is recommended for people not fond of sake"

Sake brewed with wine yeast. Enjoy the elegant acidity of white wine and the fruity flavor of ginjoshu.

Junmai Ginjoshu	720ml
Alc.	13-14%
Rice type	Domestic rice
Sake meter value	-23.0
Acidity	4.3

| Water source | Takase River subterranean well water from Hakkoda Mountains |
| Water quality | Soft |

Morita Shobei

230 Aza-shichinohe, Shichinohe-machi, Kamikita-gun, Aomori 039-2525 TEL.0176-62-2010
E-mail: morishou@morishou.co.jp http://www.morishou.co.jp

Single-minded sake brewing in Shichinohe, the village of the derby

In Shichinohe, the village of the derby, the brewery has continued to produce honest sake.

This local sake brewery is located in Shichinohe, at the eastern foot of Mount Hakkoda in Aomori Prefecture at the northern tip of Honshu. It uses only locally grown sake rice such as Hanaomoi and Hanafubuki. Shichinohe was home to Omi merchants beginning in the early Edo period, and commerce flourished there. This brewery also has its roots in an Omi merchant from Noda Village in Shiga Prefecture, and was founded in 1777. Shichinohe is also known as the birthplace of the Nambu horse in the Middle Ages and the Derby horse in modern times. The brewery's main brand, Komaizumi, was named after the folklore that pure water springs in the village of horses. One of the Komaizumi sake is named Magokoro (sincerity) reflecting the brewery's desire to brew sake with honesty. Komaizumi has been handed down from generation to generation by adhering to the traditional methods of the Nanbu style, while at the same time incorporating state-of-art technology.

特別純米酒

Komaizumi Tokubetsu-junmai
Special contract farming Sakuta

"Umami produced by rice grown in special contract farming"

The lightness and richness of this special Junmaishu harmonizes well with a wide range of dishes. Drinking it in a wide-mouthed glass will enhance its fullness.

Special Junmaishu	1,800ml/720ml
Alc.	14.3%
Rice type	Hanafubuki/Reimei
Sake meter value	+3.0
Acidity	1.4

純米吟醸酒

Komaizumi Junmaiginjo
Shichiriki

"Local sake made by seven breweries"

The melon-like aroma and the rich taste of the rice. The good balance of acidity and bitterness, and the sharpness that goes down well are unique qualities of sake produced by the Nanbu Toji.

Junmai Ginjoshu	1,800ml/720ml
Alc.	16-17%
Rice type	Hanaomoi

Akabu Shuzo

Water source	Morioka
Water quality	Soft

1-8-60 Kitaiioka, Morioka-shi, Iwate 020-0857　TEL.019-681-8895
https://www.akabu1.com

A new history has began at Akabu Shuzo

With its new brewery built in 2013, Akabu Shuzo aims to brew sake that fits the times.

Akabu Shuzo, a long-established sake brewery founded in 1896 in Otsuchi-cho, Iwate Prefecture, was devastated by the Great East Japan Earthquake on March 11, 2011, and lost a brewery that had been run for more than a century. Encouraged by customers and many others, Akabu vowed to revive its business, and built a new sake brewery in Morioka City, Iwate Prefecture in 2013. A group of young people, led by the sixth-generation head of the brewery, came together under the slogan, "Create a new history for Akabu Shuzo". Since then, they have worked together to learn how to make sake that meets the needs of the contemporary age, and have continued to brew sake "from Iwate with passion, love, and guts" without compromise. AKABU Junmaishu, brewed with "Ginginga" rice produced in Iwate Prefecture and sake yeast developed by the prefecture, won the Gold Award at the International Wine Challenge (IWC). It has a fresh taste with the flavor typical of Junmaishu.

純米酒
Akabu Junmaishu

"Fresh fruit flavors is the key"

Has a fresh aroma reminiscent of soft white peaches and a clear fruit aroma like grapefruit on the finish.

Junmaishu	720ml
Alc.	15%
Rice type	Ginginga
Sake meter value	private
Acidity	private

純米大吟醸酒
AKABU Gokujyounokire Junmai daiginjo

"Iwate's top-quality sake rice Yuinoka"

This Junmai Daiginjoshu is made from 35% polished Yuinoka, Iwate Prefecture's highest quality sake rice. It is fermented at a very low temperature and pressed at the perfect time.

Junmai Daiginjoshu	720ml
Alc.	15%
Rice type	Yuinokaori
Sake meter value	private
Acidity	private

Water source	Daijishimizu
Water quality	Soft

Asabiraki

10-34, Daijiji-cho, Morioka-shi, Iwate 020-0828 TEL.019-652-3111
E-mail: info@asabiraki-net.jp http://www.asabiraki-net.jp

Local sake made by modern master brewers

The pride of Southern Iwate: Iwate's local sake made with local rice and local water.

In a country with a rich culture, there is rich sake that is unique to that country. Iwate Prefecture has long been known as the home of the Nanbu Toji brewing method of sake production. Since its establishment in 1871, Asakai sake has been nurtured by the people who love it, refined by the rich nature in this region blessed with rice, water, and skills. Now, more than 150 years later, it is still highly regarded locally, nationally, and internationally as a brand representing Iwate, the home of sake. Mr. Masahiko Fujio, the Nanbu Toji master brewer at this brewery, was selected as a Contemporary Master Brewer by the Ministry of Health, Labor and Welfare in 2005. He continues to brew sake with honesty, sparing no effort with the motto of "always sticking to the basics." The brewery has won the gold prize at the National New Sake Competition 22 times (since 2001).

Nanburyu Denshozukuri Daiginjo

"Refreshing, slightly dry daiginjo"

This Daiginjo has won many awards in competitions for its refreshing and slightly dry taste.

Daiginjoshu	720ml
Alc.	15-16%
Rice type	Rice suitable for sake brewing from Iwate Prefecture
Sake meter value	+4.0
Acidity	1.3

Junmai Daiginjo Migakiyonwari Gokujo

"Trademark of Daiginjo"

A top-quality sake brewed with the best sake rice, Yamadanishiki and spring water from Daijishimizu. It is characterized by a rich aroma and a light taste.

Junmai Daiginjoshu	720ml
Alc.	16-17%
Rice type	Yamadanishiki
Sake meter value	+1.0
Acidity	1.3

Azumamine Shuzoten

Water source	Ou Mountains, Mt.Azumane subsoil water
Water quality	Medium soft

5 Aza-uchikawa, Tsuchidate, Shiwa-cho, Shiwa-gun, Iwate 028-3453　TEL.019-673-7221
E-mail: info@azumamine.com　http://www.azumamine.com

Sake made with local products, truly unique to Iwate

Nanbu-style sake brewing inherited over a long history.

Azumamine Shuzoten was founded in 1781, with its roots in the Gonbee Zakaya of Omi merchant Murai Gonbee who created the so-called Sumizake of the Kamigata school. It is not only the oldest sake brewery in Iwate, but also the birthplace of the Nanbu Toji (master brewer) method, and continues to brew Iwate-style sake by following it. The brewery is located in the inland part of Iwate Prefecture, where the climate is harsh, with temperatures dropping to below 10 degrees Celsius in winter. The brewery uses subsoil water from Mount Azumane for its brewing water, and named the brand Azumamine after the mountain. The brewery produces only Junmaishu and Junmaiginjo-shu. The sake rice is carefully selected through contract cultivation with local sake farmers, and the brewery strives to produce sake that is a perfect marriage for the rich food of Iwate.

純米吟醸酒
Azumamime Junmaiginjo Miyamanishiki Nama

"The standard of Azumamine"

This junmaiginjo-shu has a slightly sweet and gentle taste with a crisp finish, and is good for drinking during meals. Serve cold or at room temperature.

Junmai Ginjoshu	1,800ml/720ml
Alc.	15-16%
Rice type	Miyamanishiki
Sake meter value	-1.0
Acidity	1.7

純米酒
Yuraku Junmai Ginotome

"The taste of Iwate's nature"

This sake has a slightly sweet taste with a moderate acidity and a full-bodied flavor. As you drink, you will notice its rich flavor, but it never interferes with the taste of food. Serve cold, at room temperature, or lukewarm.

Junmaishu	1,800ml/720ml
Alc.	15-16%
Rice type	Ginotome
Sake meter value	-1.0
Acidity	1.6

| Water source | Ou Mountains |
| Water quality | Soft |

Iwate Meijo

13 Aza-shinmachi, Maezawa, Oshu-shi, Iwate 029-4208 TEL.0197-56-3131
E-mail: info@iwatemeijyo.jp https://www.iwate-meijo.com/

Sake brewed for local food

A local sake brewer with deep roots in the region, producing sake that goes well with local ingredients.

The brewery was founded in 1858 at the end of the Edo period (1603-1868). In 1955, the current Iwate Meijo Co., Ltd. was established jointly with Yoshida Shuzo, which was also located in the former Maezawa Town. The brand name Iwate Homare, which is still popular today, was created at the time of the establishment of the company. Under the motto of "local production, local trade," Iwate Meijo uses local sake rice and subsoil water from the Tanzawa alluvial fan to produce sake that is rooted in the local community, using the techniques of Nanbu Toji. Oshu City is also famous for its luxury Maezawa beef. The city pursues brewing of sake that matches local ingredients, seeking to create a sake with a pleasant taste that matches the characteristics of each ingredient. Reflecting their efforts, Daiginjo Yume Fubuki and Junmai Daiginjo Oshunoryu both won the prefectural governor's prize at the Iwate Prefecture Appraisal Contest in 2020.

大吟醸酒
Iwatehomare Daiginjo Yumefubuki

"A satisfying fruity aroma"
This sake has a splendid aroma with a well-balanced flavor that goes down easily. It has won many gold awards at the National New Sake Competition.

Daiginjoshu	1,800ml/720ml
Alc.	16%
Rice type	Yuinokaori
Sake meter value	+4.0
Acidity	1.3

純米大吟醸酒
Junmai Daiginjo Oushunoryu

"Like a dragon rising toward the crescent moon"
This elegant sake is made from highly refined Yuinoka rice grown in Oshu and brewed slowly at low temperatures to give it a fruity aroma. At lukewarm temperatures, it goes well with meat dishes.

Junmai Daiginjoshu	720ml
Alc.	16%
Rice type	Yuinokaori
Sake meter value	-1.0
Acidity	1.3

Nanbu Bijin

Water source	Orizume Bansen-kyo subsoil water
Water quality	Medium-hard

13 Aza-kamimachi, Fukuoka, Ninohe-shi, Iwate 028-6101 TEL.0195-23-3133
E-mail: sake@nanbubijin.co.jp https://www.nanbubijin.co.jp/

Bringing the pride and dreams of Iwate to the world

Building a World-Class City on the Northern Edge of Iwate Prefecture.

Ninohe City is located in the northern outskirts of Iwate Prefecture, bordering Aomori Prefecture.
Nambu Bijin operates in this small town with a total population of 26,000 people. Since its establishment in 1902, the company has been working together with the local community to become a sake brewery that serves the community. Ninohe City is the largest producer of raw lacquer in Japan and the culture of lacquer is still very much present. The city's lacquer is used as an adhesive for gold ornaments at Kinkakuji Temple in Kyoto and for repairs at Nikko Toshogu Shrine. The brewery obtained international certification as kosher in 2013 and as the world's first completely vegetarian brewery in 2019. Taking advantage of this, the brewery and its colleagues in the local food industry announced the "Ninohe Food Diversity Declaration" in 2020, becoming the first in Japan to commit to building a town that is global inclusive.

Nanbu Bijin
Tokubetsu Junmaishu

"A true local sake brewed with care"

This is the ultimate sake made from "Ginotome," Iwate Prefecture's original rice suitable for sake brewing, carefully brewed using traditional Nanbu Toji techniques.

Special Junmaishu	720ml
Alc.	15-16%
Rice type	Ginotome
Sake meter value	+3.0
Acidity	1.5

Nanbu Bijin
Junmai Daiginjo

"The winner of Japan's Best Junmai Daiginjoshu award"

Nanbu Bijin's flagship Junmai Daiginjoshu won the top prize in Japan in the Junmai Daiginjoshu category at the 2018 Sake Competition.

Junmai Daiginjoshu	720ml
Alc.	16-17%
Rice type	Yamadanishiki
Sake meter value	+1.0
Acidity	1.2

Water source	Mt.Azumane subsoil water
Water quality	Soft

Hirota Shuzoten

2-4 Aza-izumiyashiki, Miyade, Shiwa-cho, Shiwa-gun, Iwate 028-3447 TEL.019-673-7706
E-mail: hiroki@tj8.so-net.ne.jp http://hirotashuzoten.net/

Sake that makes everyone happy

The sake brewed using the traditional "Sanki Amazake Moto method handed down since the Edo period.

This small sake brewery is located in Shiwa-cho, Shiwa-gun, Iwate Prefecture, which is known as the "birthplace of Nanbu Toji". The brewery was founded in 1903. Kiheiji Hirota, the first generation, took over a sake brewery that was well known in the area at that time. The company has continued to brew sake since the end of the war under the name "Hirota Shuzo". The "Hiroki" brand was created to be "a sake that will be enjoyed by a wide range of people" at various occasions including festivals and celebrations in Shiwa Town. The flagship brand still remains the locals' favorite sake. Currently, the brewery is led by Ono Toji, the first female Nanbu Toji master. In order to deliver sake that can be enjoyed by a wider range of people, the brewery focuses on cultivating umami in rice. Since the 2017 brewing year, the entire Hiroki brand has been brewed using the "Sanki Amazake Moto" method.

Junmai Daiginjo Hiroki : 40% polished

"Made from the highest-quality sake rice Yuinokaori"

This Junmai Daiginjoshu is characterized by a gentle aroma that harmonizes the soft, fluffy sweetness and flavor of rice- goes down smoothly.

Junmai Daiginjoshu	720ml
Alc.	16%
Rice type	Yuinokaori
Sake meter value	-2.0
Acidity	unpublished

Tokubetsu Junmai Hiroki : 60% polished

"Sake during the meal, great at any temperature"

This special Junmaishu was brewed especially for locals. Can be enjoyed at any temperature.

Special Junmaishu	720ml
Alc.	14%
Rice type	Iwaterice
Sake meter value	+3.0
Acidity	unpublished

Washinoo

22-158 Ohbuke, Hachimantai-shi, Iwate 028-7111 TEL.0195-76-3211
E-mail: sake@washinoo.co.jp   http://www.washinoo.co.jp

Water source	Mt.Iwate groundwater
Water quality	Medium-hard

Sake brewing rooted in the local community

Enjoy the taste of sake as well as the sake container. The brewery that passes on Japanese sake culture to future generations.

The brewery was founded in 1829 at the foot of Mount Iwate. It continues to brew sake in the same building built when it was founded. Of the approximately 2,000 koku (1 koku = 180 milliliters) of sake produced, 99.9% is shipped within Iwate Prefecture. The name "Washinoo" (eagle tail) is said to have come from the fact that the sake is brewed with fresh water from the foot of Mt. Iwate, where a large eagle used to live. Others say that the name derives from the remaining snow on top of Mount Iwate that appears clearly in spring in the shape of a large eagle with its wings spread out. The brewery's goal is to produce sake with "enduring" quality that is delicious even without refrigeration. Since 2010, the brewery has been collaborating with local craftspeople to hold events called "sake and snack containers," with the idea that if you want to enjoy local sake, using locally-made sake cups and containers is the best.

普通酒
Washinoo Kinjirushi

"The iconic brand of Washinoo"

This brand has been a local favorite since it was classified as "second-level" sake. It is a blend of three kinds of moromi (mash), one of which is Yamahai yeast mash.

Futsushu	1,800ml
Alc.	15%
Rice type	Domestic rice

純米吟醸酒
Washinoo Yuinoka

"Iwate's finest sake"

This sake uses "Yuinoka" sake rice produced in Iwate. It has a mild aroma, a refined and light mouthfeel, and a clean taste like light snow.

Junmai Ginjoshu	720ml
Alc.	16%
Rice type	Yuinokaori
Sake meter value	-2.0
Acidity	1.25

| Water source | Own well water (underground water from Ou Mountains) |
| Water quality | Soft |

Ichinokura

14 Aza-okeyaki, Matsuyamasengoku, Osaki-shi, Miyagi 987-1393 TEL.0229-55-3322
E-mail: sake@ichinokura.co.jp https://ichinokura.co.jp/

A brewery in Miyagi that continues to propose new types of sake

Maintaining tradition while developing new products. Contributing to the region through rice cultivation and sake brewing.

Ichinokura is famous throughout Japan as one of the leading brands of local sake in Tohoku. Founded in 1973, Ichinokura was formed by four breweries in Osaki City, Miyagi Prefecture, which is known for its rice production and is blessed with a rich but harsh natural environment. The four breweries collaborate together with the aim of preserving traditional sake brewing and promoting regional development. Since its founding, the brewery has stubbornly maintained its traditional techniques and produced high quality sake with care and attention to detail. It has also developed new products like sparkling sake "Suzune" utilizing brewing femerntation technology – the brewery relies on both traditional skills and state-of-art technology. In 2004, Ichinokura established its agricultural division, Ichinokura Nosha, and the entire Ichinokura Group is contributing to environmental conservation and regional development with the goal of realizing what it calls the "Ichinokura-style sixth-order industry".

Ichinokura Tokubetsu Junmaishu Dry

"Dry sake that brings out the flavor of food"

With a calm and elegant aroma, this sake has the soft flavor of rice and a refreshing bitterness. This is a Junmaishu with a well-balanced blend of the soft flavor of rice and a refreshing bitterness, with a calm and elegant aroma.

Special Junmaishu	...1,800ml/720ml/300ml
Alc.	15%
Rice type	Sasanishiki/Kuranohana
Sake meter value	+1.0 – +3.0
Acidity	1.5 – 1.7

Ichinokura Sparkling Sake Suzune

"A pioneer in sparkling sake"

Sparkling sake made by applying the fermentation technology of Japanese sake to trap carbon dioxide gas. Refreshing with a light mouthfeel.

Sparkling Sake	300ml
Alc.	5%
Rice type	Toyonishiki
Sake meter value	-90.0 – -70.0
Acidity	3.0 – 4.0

Kakuboshi

Water source	Shishiori River system
Water quality	Medium-hard

245 Kamihigashigawane, Kesennuma-shi, Miyagi, 988-0801   TEL. 0226-22-0001
E-mail: center@kakuboshi.co.jp   https://kakuboshi.co.jp/

Delicious from the first cup to the last drop

Sake from a brewery loved by the people of Kesennuma for bringing the best out of fresh seafood.

The history of Kakuboshi dates back to 1906 when Sakubee Saito, the 14th-generation head of Saito-ya, began brewing and selling nigori-sake, or unfiltered sake. The first brewery was located in the village of Orikabe, Iwate Prefecture, about 20 km west of the current location. At that time, sake was brewed in Rikuchu (Orikabe Village) and sold in Rikuzen (Kesennuma). Therefore, the sake was named "Ryokoku", meaning "two countries". The brewery was named "Kakuboshi", because the light from a sacred mirror shone brightly on a sake cup during a ceremony to pray for the production of good sake, and it was seen as a good omen. Since its establishment, the brewery's motto is "quality first." Its sake, which is light but full of flavor, brings the best out of seafood from the Sanriku Kesennuma region. You will never get tired of drinking it, from the first to the last drop.

大吟醸酒
Kinmon Ryougoku Kisho

"A great match with fresh seafood"

The brewery takes pride in this sake. It has a light ginjo aroma and a smooth taste that keeps you coming back for more.

Daiginjoshu	1,800ml/720ml
Alc.	15.5%
Rice type	Yamadanishiki etc
Sake meter value	+3.0
Acidity	1.4

純米大吟醸酒
Mizutoriki Junmai Diaginjo-shu Kuranohana Yonwari-Yonbu

"Kura Master Gold Award-winning Sake"

This is a mellow, umami-rich sake with a full yet crisp flavor. It is best served at room temperature or lightly chilled with sashimi.

Junmai Daiginjoshu	1,800ml/720ml
Alc.	16.5%
Rice type	Kuranohana
Sake meter value	±0
Acidity	1.5

| Water source | Hirose River system / Abukuma River system |
| Water quality | Soft |

Saura

2-19 Motomachi, Shiogama-shi, Miyagi 985-0052   TEL.022-362-4165
E-mail: info@urakasumi.com   https://www.urakasumi.com/

Sake brewery where Yeast No. 12 originated

Aiming for authentic sake with harmonious taste and aroma.

Shiogama City, the hometown of Urakasumi, is a port town facing the Pacific Ocean. Many tourists visit the city every year to see the Shiogama Shrine, which has a history of more than 1,000 years, and Matsushima, considered one of the three most scenic spots in Japan. With one of the best fishing grounds in the world, the city is dotted with many sushi restaurants, where many people enjoy fresh seafood. Urakasumi, which is blessed with a rich climate, is characterized by its high quality sake brewing and expression of regional characteristics. The sake is made from local rice such as Sasanishiki, harvested in Miyagi Prefecture, known for its high-quality rice, and a good match for seafood from the Sanriku coast. The brewery also produces sake-based liqueurs using ingredients produced in Miyagi Prefecture, with the aim of communicating the beauty of the local climate and various ways to enjoy sake.

純米吟醸酒
Junmai Ginjo Urakasumi

"Great taste preserved since 1973"
This is a well-balanced Junmai Ginjoshu with a moderate aroma and a soft taste. It is a long-selling product of Urakasumi that is ideal for enjoying meals.

Junmai Ginjoshu	720ml
Alc.	15-16%
Rice type	Yamadanishiki/toyonishiki
Sake meter value	+1.0-2.0
Acidity	1.3-1.4

純米吟醸酒
Junmai Ginjo Urakasumi No.12

"Kyokai No. 12 is back"
This is a revival of the sake brewed using No.12 yeast, which was separated from Urakasumi's unpasteurized ginjo sake around 1965. It is characterized by a pleasant aroma and acidity.

Junmai Ginjoshu	1,800ml/720ml
Alc.	15-16%
Rice type	Kuranohana
Sake meter value	+2.0-3.0
Acidity	1.6-1.7

Niizawa Jozoten

Water source	Zao Mountains subsoil water
Water quality	Medium soft

63 Aza-kitamachi, Sanbongi, Osaki-shi, Miyagi 989-6321 TEL.0229-52-3002
E-mail: info@niizawa-brewery.co.jp http://niizawa-brewery.co.jp/

Loved by the locals
The ultimate sake for food

Named after the legend of the famous horse that ascended to heaven, the brewery puts its soul into this sake.

This sake brewery has been in business since 1873 in Sanbongi, located in the south of Osaki City, Miyagi Prefecture, the birthplace of famous rice brands Sasanishiki and Hitomebore. More than half of the employees at this brewery are women, making it a good place for both men and women to work. 2003 saw the launch of "Hakurakusei" brand under the concept of "the ultimate sake for food. The sake is popular both within and outside of the prefecture. In the Sanbongi region, there is a legend about a horse connoisseur named Hakuraku who raised a famous horse that ascended to heaven. This legend is where the name of the sake comes from. The Niizawa Brewery is aiming for highest quality in rice polishing by using a rare flat rice polisher, and a state-of-the-art diamond rice polisher. The attention to detail in ingredients gives its sake sharpness, making it great sake to be enjoyed during a meal. The brewery continues to further improve the quality of its sake.

純米吟醸酒

Hakurakusei Junmai Ginjo

"Ginjo aroma like banana and melon"

The concept of this sake is "the ultimate sake for food." It has a pleasant fruity taste and refreshing acidity. Serve chilled in a thin glass.

Junmai Ginjoshu	1,800ml/720ml
Alc.	15%
Rice type	Kuranohana
Sake meter value	+3.0
Acidity	1.6

特別純米酒

Hakurakusei Tokubetsu Junmai

"Sake with rich umami of the rice brings the most out of food"

It has a subtle aroma and sharp mouth-feel, with a good balance of rich rice flavor and clear acidity.

Special Junmaishu	1,800ml/720ml
Alc.	15%
Rice type	Yamadanishiki
Sake meter value	+3.0
Acidity	1.6

Water source	Reidozawa Natural Water
Water quality	Soft

Hagino Shuzo

52 Arikabeshinmachi, Kannari, Kurihara-shi, Miyagi, 989-4806 TEL.0228-44-2214
E-mail: info@hagino-shuzou.co.jp http://www.hagino-shuzou.co.jp/

There are many kinds of sake – just enjoy the one you think is delicious

Once you have our sake, you'll never forget its taste. We only brew good sake in small amounts.

Founded in 1840, the brewery has a long history of 182 years in Arikabe, which flourished as a post town on the old Oshu Highway. The brewery produces handmade sake with locally grown rice and water in the cold climate and rich nature of northern Miyagi Prefecture. The local brewers put care and effort into sake brewing. The brewers have strived to produce sake that they themselves think is truly delicious, rather than accommodating every customer's preference. They aim to brew unique sake that brings the best out of dishes and remains in the memory of those who drink it. Kurihara City, where the brewery is located, is blessed with vegetables, wild plants, and mushrooms. In order to make the most of the delicate flavors of the mountain vegetables, the taste and the aroma of its sake isn't too strong, and has a soft and modest quality to it.

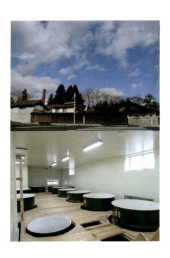

純米大吟醸酒

Haginotsuru Junmai Daiginjo

"Delicate aroma produced with care"

Daiginjo usually comes with flowery aroma, but this sake is purposefully made to be more purely elegant in its taste, bringing the most out of the delicate taste of Miyamanishiki rice. Serve cold or at room temperature.

Junmai Daiginjoshu	1,800ml/720ml
Alc.	16%
Rice type	Miyamanishiki
Sake meter value	+1.0
Acidity	1.6

純米酒

Hiwata Kimoto Junmaishu

"A good match with home cooking from the mountainous countryside"

It has a depth in its flavor – typical of sake made in the traditional "Yamahai" way of brewing – and a fresh mouthfeel. The rich taste is a favorite of everyone.

Junmaishu	1,800ml/720ml
Alc.	16%
Rice type	Gohyakumangoku
Sake meter value	+4.0
Acidity	1.8

Yamawa Shuzoten

Water source	Funagata Mountains subsoil water
Water quality	Soft

109-1 Aza-minamimachi, Kami-machi, Kami-gun, Miyagi 981-4241 TEL.0229-63-3017
E-mail: yamawa@nona.dti.ne.jp

Always delicious, because it's simple

A popular sake brewery loved by the local people in the snowy area on the border between Yamagata and Miyagi.

It was founded in 1896 by Mr. Wahei Ito, the first generation of the brewery, who closed down his family's apothecary business and started brewing sake. The brewery is located in a rural area covered with virgin beech forests and blessed with subsoil water from the Funagata mountain range. The region is designated as a heavy snowfall area during the brewing season. The main brand "Washigakuni" has a wide range of products from ordinary sake to ginjo-shu, and is very popular within the prefecture, especially among local people. Other brands include "Meiso-sui" ginjo, which is exported overseas, and "Yamawa" which only has a lineup of junmai or higher-refined sake with limited distribution. The concept behind Yamawa's sake brewing is "simple is best". The brewery believes that truly good sake is one that's loved for a long time regardless of trends.

特別純米酒
Washigakuni Tokubetsu Junmai

"Brewed with Miyagi's sake rice Kuranohana"

The sake is brewed using Miyagi yeast and "Kuranohana" rice produced in Miyagi Prefecture.

Special Junmaishu	1,800ml/720ml/300ml
Alc.	15%
Rice type	Kuranohana
Sake meter value	+3.0
Acidity	1.5

純米大吟醸酒
Yamawa Junmai Daiginjo

"daiginjo with an elegant taste"

This sake is made by polishing Yamadanishiki to 40% and fermenting it slowly and carefully at low temperatures in small batches.

Junmai Daiginjoshu	1,800ml/720ml
Alc.	16-17%
Rice type	Yamadanishiki
Sake meter value	+3.0
Acidity	1.8

Water source	Subterranean water of the Omono
Water quality	Soft

Akita Shurui Seizo

4-12 Kawamotomutsumimachi, Akita-shi, Akita 010-0934 TEL.018-864-7331
https://www.takashimizu.co.jp/

Akita's rice, water, and high-quality spring water

Sake made with the utmost care, using carefully selected Akita rice, water, and techniques.

In 1944, twelve sake brewers in Akita City and its suburbs, who had been making sake from the Edo period (1603-1868) to the early Showa period (1926-1989), merged to form Akita Shurui Seizo. Kikusui, the oldest of the breweries, was founded in 1655, and Iidagawa, the youngest, was founded in 1921. At the time of the merger, the name of the brewery's flagship brand, "Takashimizu", was chosen from 5,037 entries solicited in the local newspaper. It was named after a sacred spring that still provides water in the Teranai Ooji area of Akita City (known as "Sakura Ooji"). Takashimizu is the most widely enjoyed sake in Akita today, made with high quality rice from the prefecture and clean underground water from the Ou Mountains. It is brewed with a great passion and care by the Yamauchi Toji (master brewer) in keeping with tradition.

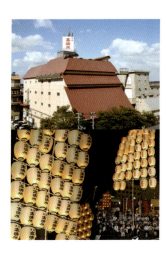

本醸造酒
Takashimizu Honjozo

"Sake brewed in snowy Akita: delicious even when served hot"

A full-bodied, tasty sake with an elegant mouthfeel and a rich aroma, brewed in the traditional Akita style of cold brewing.

Honjozoshu	… 1,800ml/720ml/300ml/180ml
Alc.	15.5%
Rice type	Akita rice
Sake meter value	+1.0
Acidity	1.9

純米酒
Sakenokuni Junmaishu

"Mild and elegant quality""
This Junmaishu has a gentle, full aroma that is not overpowering, as well as a full, rich taste typical of Akita, with just the right amount of acidity to make it crisp.

Junmaishu	… 1,800ml/720ml/300ml
Alc.	15.5%
Rice type	Akita rice
Sake meter value	+1.0
Acidity	1.7

Akita Meijo

Water source	Subsoil water from the Omono River
Water quality	Soft

4-23 Daikumachi, Yuzawa-shi, Akita 012-0814 TEL.0183-73-3161
E-mail: ranman@ranman.co.jp https://www.ranman.co.jp/ranman/

Like a beautiful woman in full bloom

A sake brewery born to spread the mellow and deep beauty of Akita's sake throughout Japan.

Akita's sake is made from high-quality rice and abundant water. In 1922, Akita Meijo Co., Ltd. was established in Yuzawa City, which is well suited for sake brewing, by a group of leading sake brewers and political and business leaders in the prefecture. Its flagship brand is "Ranman". Since its establishment, the company has been committed to quality first, sparing no time or effort in its pursuit of making delicious sake. The brewery developed a new kind of sake rice "Ginsan" in collaboration with the Akita Prefectural government. Efforts are being made to collaborate with local agricultural producers and to use the new variety at breweries in the prefecture. The brewery has two branches: the Ontake Brewery, where sake is produced by computerized scientific analysis, and the Ogatsu Brewery, where sake is produced by hand using traditional methods based on experience and intuition.

純米吟醸酒
Kaori-ranman Junmai Ginjo

"The revolutionary aroma that changes the concept of Japanese sake"

This is a new type of sake created through technological innovation. It has a splendid aroma like apple and melon, and a taste that makes the most of the flavor of the rice.

Junmai Ginjoshu	1,800ml/720ml
Alc.	15-16%
Rice type	Sakekomachi
Sake meter value	+3.0
Acidity	1.4

特別純米酒
**Ranman
Tokubetsu Junmaishu**

"Enjoy this sake in a wine glass"

This is a Special Junmai made from rice developed jointly with the prefecture and brewed with a special yeast developed by the prefecture. It has a fresh, splendid aroma and an elegant taste.

Special Junmaishu	1,800ml/720ml
Alc.	15%
Rice type	Ginsan
Sake meter value	+3.0
Acidity	1.6

Water source	Subsoil water originating from the Dewa Hills and the Yomono River
Water quality	Medium-hard

Kariho Shuzo

275 Aza-jinguji, Jinguji, Daisen-shi, Akita 019-1701 TEL.0187-72-2311
E-mail: kariho@oregano.ocn.ne.jp https://www.igeta.jp/

A brewery that organized a sake rice cultivation association with local farmers

Pouring passion into all steps, beginning with sake rice cultivation.

The historic brewery, built in 1853, is located on the shores of the Omono River, the largest river in Akita.

The Kariho brewery was built in 1850, three years before the arrival of Perry's Black Ships. Built on the shore of the Omono River, the largest river in Akita Prefecture, it served as a base for water transportation. In 1913, Kyonosuke Ito and others who ran a sake brewery in a neighboring village took over this historic brewery and started a business, thus beginning the history of Kariho Brewery. The Senboku Plain, one of the largest grain-growing areas in Akita Prefecture, has a climate with four distinct seasons. Warm and humid in summer and autumn, the region is suitable for growing sake rice. Cold and snowy in winter, it is ideal for sake brewing. The brewery has organized a sake rice cultivation association with the local farmers, and it is committed to all processes of producing sake including the cultivation of rice. The brewery's sake is sold through Akita Seishu Co., Ltd.

Kariho Ichihozumi
Junmai Ginjo

"The lone warrior of sake, Ichihozumi"

It is brewed with a unique low-temperature saccharified sake yeast evolved from the Yamahai brewing method. It is an ace of the new generation of brewers that will be registered as a new breed in fiscal 2020.

Junmai Ginjoshu	720ml
Alc.	17%
Rice type	Ichihozumi
Sake meter value	+3.0
Acidity	1.7

Kariho Hyakuden
Junmai Ginjo

"Unique brewing method: low-temperature saccharification"

New-age sake rice "Hyakuden" is used. This Junmai Ginjoshu is characterized by a soft and expansive aroma.

Junmai Ginjoshu	720ml
Alc.	17%
Rice type	Hyakuden
Sake meter value	-0.4
Acidity	1.9

Dewatsuru Shuzo

Water source	Natural groundwater from the Dewa Mountains
Water quality	Soft

81 Aza-akutono, Nangai, Daisen-shi, Akita 019-1901 TEL.0187-63-1224
E-mail: y-ito@igeta.jp https://www.igeta.jp/

The motto of our brewery is "Harmony makes good sake"

Brewing sake from traditional rice, rooted in the local Akita climate.

In 1865, at the end of the Tokugawa shogunate era when the Meiji Restoration was just around the corner, Dewatsuru brewery was founded by Shigeshiro Ito, the 12th-generation head of the Ito family, as the Yamato Sake Brewery. Shigeshiro was a pioneer farmer who made his living cultivating land and growing rice, and even today the brewery's policy is to brew sake starting with rice production. The brewery has established an in-house agricultural division, which cultivates rice for sake from spring to autumn in the fields near the brewery. In 1994, the brewery launched a new brand, "Yamato Shizuku," named after the company's original name, in cooperation with local voluntary sake dealers. The company's goal is to make only Junmaishu.

純米吟醸酒

Dewatsuru awa sake Ashita-e

"The new frontier of sake made possible by the brewery's techniques"

This is a sparkling sake with fine bubbles. It can be served chilled with Western food, fruit, or dessert.

Junmai Ginjoshu	720ml
Alc.	13%
Rice type	Akitasakekomachi
Sake meter value	-1.0
Acidity	1.3

純米酒

Dewatsuru Shizenmai-shu Matsukura

"Sake blessed with the power of nature"

It is Junmaishu with a deep, harmonious taste of acidity and sweetness. It is made entirely from specially cultivated organic rice, making it smooth and full of flavor.

Junmaishu	720ml
Alc.	15%
Rice type	Special cultivation rice(Akitakomachi)
Sake meter value	+2.0
Acidity	1.8

| Water source | Chokai mountain subsoil water |
| Water quality | Soft |

Tenju Shuzo

117 Aza-hachimorishita, Jonai, Yasimamachi, Yurihonjo-shi, Akita 015-0411 TEL.0184-55-3165
E-mail: info@tenju.co.jp   https://www.tenju.co.jp

192 years since establishment
Sake produced by traditional artisans

Aiming for the best sake in Akita, the kingdom of fine sake. The best sake that can be made in this region.

Tenju Shuzo was founded in 1830 in the village of Yajima, at the northern end of the trailhead for Mt. Chokai, a beautiful mountain in the Tohoku region. The company's goal is to produce "the best sake possible here". In the Koyoshigawa River basin, where the highest-quality rice is produced, the Tenju Sake Rice Research Association was established to ensure the best sake rice, based on the idea that sake brewing starts with rice production. The brewery was one of the first in Japan to start contract cultivation of sake rice together with local farmers and brewers. Used in this sake is the soft subsoil water of Mount Chokai, which takes permeate. The long-standing history of brewery and its famous sake brands are made possible by rice grown with a continuous passion, an abundance of water in the region, and traditional craftsmanship.

純米大吟醸酒
Junmai Daiginjo Chokaisan

"The winner of many competitions"
This is the work of the seventh-generation chief, who is an agricultural university graduate, brewed with Miyamanishiki rice produced by the Tenju Sake Rice Research Group, and ND-4, a flower yeast developed by an agriculture university.

Junmai Daiginjoshu	1,800ml/720ml
Alc.	15%
Rice type	Miyamanishiki
Sake meter value	+1.5
Acidity	1.6

純米酒
Junmaishu Tenju

"Aims to locally produce the best sake possible"
Junmaishu with acidity, produced from special rice – one will never get tired of this sake. It is delicious served cold, at room temperature, lukewarm, or hot.

Junmaishu	1,800ml/720ml
Alc.	15%
Rice type	Miyamanishiki
Sake meter value	+3.0
Acidity	1.6

Fukurokuju Shuzo

Water source	Own groundwater
Water quality	Medium-hard

48 Aza-shitamachi, Gojome-machi, Minamiakita-gun, Akita, 018-1706 TEL. 018-852-4130
E-mail: info@fukurokuju.jp   https://www.fukurokuju.jp

Local "rice", "water", and "people" Sake brewing with a hint of culture

Located in Oga Peninsula, famous for Namahage and near the site of a morning market with a 500-year history.

The brewery was founded in the first year of the Genroku era (1688). It was founded by Watanabe Hikobee, the first-generation head of the brewery, beginning in the town of Gojome in Akita Prefecture. The town is located east of Hachirogata Lagoon on the Oga Peninsula, famous for Namahage, and has many industries that use Akita cedar as a material. In front of the warehouse is a morning market that has been going on for 500 years. The area produces wild vegetables in spring and mushrooms in fall. It first developed as a place where food from the mountains and the sea were exchanged. The Fukurokuju Shuzo is located in a mountainous farming village with a wide variety of terrain, from steep mountains to fertile rice paddies. The brewing water is groundwater from the brewery's premises, and the rice is grown under contract with farmers who belong to the local "Gokome Town Sake Rice Study Group", which was established in 2008 to produce high-quality sake rice. Its representative brand is "Ippaku Suisei." The brewery's motto is sake with a hint of culture, produced by locals, using locally-grown rice and local water.

特別純米酒
Ippaku Suisei Ryoshin

"A comet in Akita, the kingdom of sake"
This sake is brewed with local rice and local water, and by local people, in co-operation with local farmers. It has an elegant flavor and excellent sharpness.

Special Junmaishu	1,800ml
Alc.	16%
Rice type	Ginnosei/Sakekomachi
Sake meter value	+2.0
Acidity	1.3

純米大吟醸酒
Ippaku Suisei Premium

"The true spirit of Ippaku Suisei"
This sake is made from rice carefully selected by the Gojome Town Sake Rice Research Association. It is a limited-quantity product filled with the soul of the brewer.

Junmai Daiginjoshu	720ml
Alc.	16%
Rice type	Sakekomachi
Sake meter value	+2.0
Acidity	1.4

Water source	Ou Mountains
Water quality	Soft

Maizuru Shuzo

184 Aza-asamai, Asamai, Hirakamachi, Yokote-shi, Akita 013-0105 TEL.0182-24-1128
E-mail: asanomai@poplar.ocn.ne.jp

An elaborate sake brand with Yamahai Junmai

A happy and relaxing sake to drink.

The Maizuru shuzo was founded in 1918 by a group of local volunteers. When the brewery was founded, cranes flew into the spring water pond near the brewery every morning and soared through the sky.

Thus, the name of the brewery was changed to "Asanomai" (dance in the morning). The Asamai district of Hiraga-cho, Yokote City, where the brewery is located, is in the center of the Yokote Basin in the southern part of Akita Prefecture. It is in the heart of a vast area of rice fields with the Ou mountain range. The terrain is at the end of the Minasegawa fan-shaped plateau. The brewery uses the abundant spring water from the subsoil as brewing water, and brews only Junmaishu which has a refreshing, pleasant acidity and a full-bodied flavor at its core.

In addition, the snowfall in the Yokote Basin makes it difficult for us to get enough snow.

In addition, to make the most of the large amount of snowfall in the Yokote Basin, the "Kamakura Snow Storage" system was introduced a few years ago. The low-temperature storage improves and stabilizes the quality of the sake.

純米酒
Tabito

"Sake during the meal that matches well with Akita's special dishes"

Tabito comes in two kinds: fast-brewed and Yamahai-brewed. Both are aged for three years or more. It has a unique taste and is best when served super hot or super cold.

Junmaishu	1,800ml/720ml
Alc.	15%
Rice type	Domestic rice
Sake meter value	+5.0
Acidity	1.8

純米吟醸酒
Gekka-no Mai

"The flavor remains unchanged whether heated or cooled"

This Junmai Ginjoshu has a modest aroma and a delicious flavor. It has a pleasantly dry taste. We recommend drinking it lukewarm or hot.

Junmai Ginjoshu	1,800ml/720ml
Alc.	15%
Rice type	Ginnosei
Sake meter value	+3.5
Acidity	2.2

Eau de Vie Shonai

Water source	Mt.Chokai subsoil water
Water quality	Soft

Otsu123, Hamanaka, Sakata-shi, Yamagata 998-0112 TEL.0234-92-2046
E-mail: satou@kiyoizumigawa.jp http://kiyoizumigawa.com

A step ahead of the rest
The taste of tradition

Goes well with all kinds of food. Aiming to be the "ultimate sake for meals".

The brewery was founded in 1875 in Sakata City, Yamagata Prefecture, blessed with the natural beauty of the Sea of Japan and Mt. Chokai. As a small sake brewery located on the seashore of the Sea of Japan in the Tohoku region, the brewery aims to produce "the ultimate sake for meals". The brewery's ideal is sake that goes well with all kinds of food. Since its establishment, the brewery has marketed "Kiyoizumigawa", and has continued to put effort and care into brewing sake with local ingredients. For sake rice, it uses "Dewasansan", "Dewanosato", and "Yukimegami" from Yamagata Prefecture. For water, it uses underground water (subsoil water from Mt. Chokaisan. The company's name, "Eau de Vie", which means "water of life" in French, was chosen because of its particular focus on water. Currently, the most popular products are "Kin no Kura" and "Gin no Kura" The former focuses on the balance of aroma and taste, and the latter is characterized by local ingredients used in its brewing.

純米吟醸酒
Kiyomizugawa
Junmai Ginjo Gin no Kura

"Sake made from "Dewasansan" rice produced in Yamagata. Special sake for meals"

Characterized by a fruity aroma, neat taste, and refreshing and crisp mouthfeel, this sake goes well with squid and sweet shrimp sashimi, as well as vegetable tempura.

Junmai Ginjoshu	720ml
Alc.	15%
Rice type	Dewasansan
Sake meter value	+4.0
Acidity	1.4

純米吟醸酒
Kiyomizugawa
Junmai Ginjo Sumirebana

"Junmai Ginjoshu with a revolutionary flavor: Shirokoji brewing method"

This sake has a white wine-like color, refreshing citrus aroma, and sweetness with a hint of acidity from the white rice malt.

Junmai Ginjoshu	720ml
Alc.	13%
Rice type	Yamagata rice
Sake meter value	-7.0
Acidity	3.2

*This is a limited edition, not currently available at Eau de Vie Shonai. It was made available through crowdfunding in the summer of 2020.

Water source	Yamagata Prefecture Gassan water system
Water quality	Soft

Koikawa Shuzo

42 Aza-koya, Amarume, Shonai-machi, Higashitagawa-gun, Yamagata 999-7781   TEL.0234-43-2005

Sake brewery depends on agriculture

A brewery that is determined to brew locally, and cultivates its own rice.

Founded in 1725, Koikawa Shuzo is located in Shonai-machi in the Shonai region of Yamagata Prefecture. Surrounded by rice paddies, the brewery uses local rice and water. The brewers at this brewery are all locals. One of the major accomplishments of Koikawa Shuzo is the revival of Kameno-o, sake rice which had not been in production for a while. The brewery received only a handful of Kameno-o rice seeds, which had been discontinued, and spent many years carefully cultivating the rice until it could be used to make sake. A sake brewery in Niigata Prefecture revived the rice a year earlier, and in the manga "Natsuko no sake" (Natsuko's Sake), which is set there, the rice appears as "Ryu Nishiki," a fantastic sake rice. The brewer says, "Sake brewing cannot be done by one person. We rely on agriculture." This is why the master brewer grows rice himself.

Koikawa Junmai Ginjo
Kameji-kojitsu Kameno-o 100%

"Mild aroma and soft mouthfeel"

This Junmai Ginjoshu is well-balanced, with clean, crisp flavor that is characteristic of sake rice. We recommend drinking it lukewarm.

Junmai Ginjoshu	1,800ml/720ml
Alc.	15-16%
Rice type	Kamenoo
Sake meter value	+6.0
Acidity	1.5

Koikawa Junmai Daiginjo
Beppin Yukimegami 100%

"Gorgeous ginjo aroma and mellow flavor"

This sake is made from Yamagata Prefecture's "Yukigami" sake rice and is suitable for chilling. It has an elegant aroma and a clear, dry taste.

Junmai Daiginjoshu	1,800ml/720ml
Alc.	16-17%
Rice type	Yukimegami
Sake meter value	+1.0
Acidity	1.4

Kozaka Shuzo

Water source	Iide Mountains subsoil water
Water quality	Hard

7-3-10 Chuo, Yonezawa-shi, Yamagata 992-0045 TEL.0238-23-3355
E-mail: ko-bai@abeam.ocn.ne.jp   http://www.ko-bai.sakura.ne.jp/

Severe cold in Yonezawa
Handcrafted in the severe cold of Yonezawa

We brew sake as lovingly as if it were our own child. Sake made by "seeing and touching".

The brewery was established in 1923 in Yonezawa, a castle town of Uesugi with deep snow. The main brand, "Kobai", is brewed in the cold, taking advantage of the location of Yonezawa where the temperature can reach -10°C in winter. The brewery also insists on doing everything by hand, from washing the brewing rice with bare hands to putting the label on the sake bottle. However, that does not mean that the brewery sticks to traditional techniques. The sake storage tanks are temperature-controlled by computers, and the latest technology is flexibly incorporated to ensure stable quality. The motto of the brewery is "competing with quality!!" In addition to traditional sake brewing, the brewery also produces rather unique sake, such as one made from glutinous rice fermented with wine yeast and super dry sake known as the highest-grade in Yamagata.

純米大吟醸酒
Junmai Daiginjo Kobai

"A limited edition made from rice washed by hand"

This sake is made with 100% Yamadanishiki. It has a full-bodied aroma that spreads to the palate, but is not too floral. It is dry and easy to drink. Goes well with Yonezawa beef sukiyaki.

Junmai Daiginjoshu	720ml
Alc.	15%
Rice type	Yamadanishiki
Sake meter value	+3.5
Acidity	1.4

純米吟醸酒
Junmai Ginjo Kobai

"Yamaga's flavor with a reasonable price"

The rice used for this sake is "Dewa sansan", which has a soft flavor. It is dry and refreshing, and goes well with Yamagata's famous potato stew.

Junmai Ginjoshu	720ml
Alc.	15%
Rice type	Dewasansan
Sake meter value	+4.0
Acidity	1.4

| Water source | Mogami River System |
| Water quality | Medium soft |

Goto Shuzoten

1462 Oaza-nukanome, Takahata-machi, Higashiokitama-gun, Yamagata 999-2176   TEL.0238-57-3136
E-mail: gotobenten@gmail.com   https://www.benten-goto.com/

Sake blessed by the climate of severe cold

Traditional sake brewing techniques nurtured by Yamagata's harsh winters and water from the Mogami River.

Since its establishment in 1788, this local sake brewery in the Okitama region of southern Yamagata Prefecture has focused on making high-quality sake in small quantities by hand. The brewery produces Junmaishu and Junmai Ginjoshu using mainly rice grown on contract in the brewers' own fields, and also brews sake using mainly Yamagata's own brewing-grade rice such as "Dewasansan", "Dewanosato", and "Yukigami" delivering delicious sake made in Yamagata's rich nature to consumers. With the motto of "quality-driven sake brewing", the brewing process is basically done once a year, from October to the end of March. In recent years, the company has achieved good results at various sake competitions, with four of its brands winning gold medals at the 2020 U.S. National Sake Appraisal. Its "Dewasansan Junmai Daiginjo Genshu" won the Platinum Award at the 2020 Kura Master.

Benten Yamadanishiki Gokuju Daiginjo Genshu

"Enjoy the mellow and delicate aroma"

This sake was specially brewed to showcase at a competition. It is characterized by a mellow aroma and a refreshing taste. Serve at room temperature or lightly chilled.

Daiginjoshu	1,800ml/720ml
Alc.	18%
Rice type	Yamadanishiki
Sake meter value	+1.0
Acidity	1.3

Benten Dewasansan Junmai Daiginjo Genshu

"Carefully, locally crafted flavor of raw sake"

This sake is characterized by a gentle aroma and a soft, broad flavor. It goes well with strong-flavored dishes. Serve at room temperature or lightly chilled.

Junmai Daiginjoshu	1,800ml/720ml
Alc.	17%
Rice type	Dewasansan
Sake meter value	±0
Acidity	1.4

Koya Shuzo

Water source	Gassan system
Water quality	Soft

2591 Shimizu, Okura-mura, Mogami-gun, Yamagata 996-0212 TEL.0233-75-2001
E-mail: info@hanauyo.co.jp https://hanauyo.co.jp

Sake loved by hot-spring bathers

Founded in 1593, the oldest sake brewery in Yamagata Prefecture, nurtured by its hometown.

The Mogami River, whose source is on Mt. Nishiazuma on the border of Yamagata and Fukushima prefectures, is the most important river for the industry, economy, and culture of Yamagata Prefecture. Kiyomizu, Okura Village, where Koya Brewing is located, is in the Shinjo Basin where the Mogami River turns west. From the ancient Warring States period to recent years, this area was a key distribution point for the Mogami River, and was granted the right to transfer people and goods to Shimizu boats. From the beginning of the Edo period until the Meiji period (1868-1912), it was home to village headmen, wholesalers, and the main office for various feudal lords, as well as a contractor for the hand boats of Lord Uesugi of the Yonezawa domain. The village of Okura is also home to the Hijiori hot spring village (opened in 807), which has long prospered as a lodgings for Mt. Gassan climbers, and it is said that the sake produced by the brewery has been loved by those who visit the village to take the waters and cure their illnesses.

大吟醸酒
Daiginjo Kinu

"An artwork brewed with only the best of Yamadanishiki"

It is characterized by its fruity ginjo aroma and silky soft taste. Good to served well chilled with salmon meuniere.

Daiginjoshu	720ml
Alc.	17%
Rice type	Yamadanishiki
Sake meter value	+2.0
Acidity	1.2

純米吟醸酒
Hanauyou Dewasansan Junmai Ginjo

"Rice, water, koji, yeast, and people—everything is local"

This Junmai Ginjoshu is slowly fermented at a low temperature using traditional techniques. It is characterized by its mellow taste, richness, rich aroma and sharpness.

Junmai Ginjoshu	720ml
Alc.	15%
Rice type	Dewasansan
Sake meter value	+3.0
Acidity	1.3

Water source	Gassan water system
Water quality	Soft

Sato Sajiemon

255 Aza-machi, Amarume, Shonai-machi, Higashitagawa-gun, Yamagata 999-7781   TEL.0234-42-3013
E-mail: yamatozakura@beige.plala.or.jp

Sake made from rice grown in rice paddies where killifish live

Hand-made, gentle-flavored sake is like the gentleness of the Shonai people.

Sato Sajiemon was founded in 1890. The brewery is located in the middle of the Shonai Plain in Yamagata Prefecture and is surrounded by rice fields. It is a truly local sake brewery that cherishes the local rice, water, and air of Shonai. The brewers, all from the Shonai area, uses Yamagata's original sake rice and yeast. They aim to produce sake that is soft and rich, which can be served before or during meals. The brewery's flagship brand is "Yamatozakura". In recent years, the killifish living in the rice paddies of the area almost went extinct due to infrastructure development, but local residents and elementary school students helped to temporarily evacuate the tiny fish to a conservation pond. The rice grown in the environmentally-friendly rice paddies, where killifish were able to live again, is used to make another brand of sake called "Junmaishu Killifish Rice" "Junmaishu Medaka Rice".

Daiginjo Yamatozakura Gold Label

"Daiginjo with elegant and gorgeous aroma"

Daiginjoshu goes well with meals—its aroma and taste make it ideal as a sake to drink with meals. Serve cold or at room temperature with grouper dengaku or sashimi.

Daiginjoshu	720ml
Alc.	16.7%
Rice type	Yamadanishiki
Sake meter value	±0
Acidity	1.3

Junmai Ginjo Yamatozakura Dewasansan

"Rice, koji, and yeast are all from Yamagata"

All of the ingredients used for this sake are from Yamagata. Goes well with cod roe soup or natto soup. Serve cold or at room temperature.

Junmai Ginjoshu	720ml
Alc.	15.5%
Rice type	Dewasansan
Sake meter value	±0
Acidity	1.4

Shindo Shuzoten

Water source	Azuma Mountains subsoil water
Water quality	Soft

1331 Oaza-takei, Yonezawa-shi, Yamagata 992-0116 TEL.0238-28-3403
E-mail: info-sake@kurouzaemon.com http://www.kurouzaemon.com/

A brewery evolving with tradition and free ideas

Award-winning sake brewed with the blessings of nature is a sign of innovation.

The Shindo Sake Brewery is located in the eastern corner of Yonezawa City, which is famous for its high-grade Japanese beef. Making use of the farmland inherited from the brewery's first generation, small lots of high quality sake is produced with refined traditional techniques. The brewery's main goal is to realize the full potential of local sake. The fact that top-grade sake such as ginjo-shu has won numerous awards both in Japan and abroad proves that the brewery is constantly refining its techniques to provide sake that meets the needs of the times. In order to brew sake of today's quality, it is necessary not only to rely on the master brewer's intuition, but also to use one's eyes, ears, nose, and tongue, as well as the latest technology to quantify the results of analysis. With this in mind, the company continues to be an innovator in sake brewing technology despite it being a small, local brewery. Shindo leads the way in defining the taste of cutting-edge sake.

純米大吟醸酒
Gasanryu Gokugetsu

"Delicate and elegant taste"
This luxurious sake is brewed with Dewasansan rice grown by the company and polished to 40%. Then the mash is put in a bag and only the drops that come out of it are bottled.

Junmai Daiginjosh	1,800ml/720ml
Alc.	16.2%
Rice type	Dewasansan (in-house cultivation)
Sake meter value	+1.0
Acidity	1.4

大吟醸酒
Gasanryu Kisaragi

"Moderate aroma. Light, soft flavor"
The rice, brewing water, and yeast are all from Yamagata. It has a well-balanced aroma with a hint of rice flavor and is easy to drink.

Daiginjoshu	1,800ml/720ml
Alc.	14.2%
Rice type	Dewasansan (in-house cultivation)
Sake meter value	+3.0
Acidity	1.2

Water source	Subsoil water from the Okitama River
Water quality	Soft

Suzuki Shuzoten Nagaikura

1-2-21 Yotsuya, Nagai-shi, Yamagata 993-0015 TEL.0238-88-2224

The life of the brewery comes from the well water that springs in the "City of Water, Greenery and Flowers."

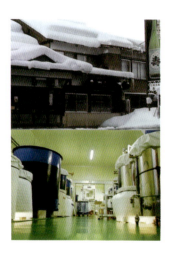

Founded in 1931 as Toyo Shuzo, it was renamed Suzuki Shuzo Nagai Brewery in 2011.

The company was founded in 1931 as Toyo Shuzo by influential people in Koide Village (now the southern part of Nagai City), which prospered thanks to the Mogami River shipping industry. The brand names of sake sold when the brewery was established were "Toyo no Homare" and "Kiku Toyo." Later, "Isshou Kofuku" became the flagship brand. In 2011, the company name was changed to Suzuki Shuzo Nagaikura. Nagai City, known as a city of water, greenery, and flowers. The Shirakawa River, with its source in the Iide mountain range, and the Okiamino River, with its source in the Asahi mountain range, flow into the mother Mogami River. The source of this water is snow, which flows through the scattered villages in the rice paddy area. The well water in the area is the life of the brewery. "Kotobuki," one of the sake brands at the brewery, has an elegant flavor that anyone can enjoy. "Isshou Kofuku" is made from rice grown in Yamagata Prefecture, and is easy to drink and has a local flavor.

Iwaki Kotobuki Yamahai Junmai Daiginjo

"Unique sake with friendliness"

This sake aims for a fine but imposing flavor as well as a fine aftertaste, with a hint of vintage sake.

Junmai Daiginjoshu ...	1,800ml/720ml
Alc.	16%
Rice type	Yamadanishiki/Omachi
Sake meter value	±0
Acidity	1.4

Daiginjo Isshou Kofuku

"Perfect as a gift"

This daiginjo has a beautiful and full aroma with a smooth mouthfeel and lingering aftertaste.

Daiginjoshu	1,800ml/720ml
Alc.	17%
Rice type	Yamadanishiki
Sake meter value	+4.0
Acidity	1.0

Takagi Shuzo

Water source	Natural water from Mt. Hayama
Water quality	Soft

1826, Oaza-tominami, Murayama-shi, Yamagata 995-0208　TEL.0237-57-2131

Artistic sake where people and nature are in harmony

A brewery in Yamagata with a 400-year history that continues to hone its techniques.

Founded in 1615, the brewery boasts 400 years of history and is one of the oldest in Yamagata Prefecture. The brewery has been using subterranean water from Mt. Hayama in the Dewa mountain range as its water source for generations, and continues to brew artistic sake in harmony with nature and people. The current head of the brewery is Tatsugoro Takagi, the 14th generation head of the family. The representative brand "Jyuyondai" is a popular brand brewed by making full use of both traditional and modern techniques, not merely relying on the brewery's long-standing history. It is now so well known that anyone who loves sake knows this brand. For a while, there was a boom in the sake world for "light and dry" sake, but Jyuyondai has replaced it with "mellow and umami" sake. While it has a robust rice flavor that is not found in tanrei-style sake, the aftertaste is more refreshing.

特別本醸造酒
Jyuyondai Honmaru

"Honjozoshu with high-quality sweetness"

The sweet aroma and flavor are truly mellow. It is best served cold or lukewarm.

Special Honjozoshu	1,800ml
Alc.	15%
Rice type	Yamadanishiki/Aiyama

純米吟醸酒
Jyuyondai Tatsu no Otoshigo

"Youthful and rich in flavor"

This Jyuyondai is made with "Tatsu no Otoshigo", a kind of rice suitable for sake brewing developed in-house. It is characterized by its fresh youthfulness.

Junmai Ginjoshu	1,800ml
Alc.	15%
Rice type	Tatsunootoshigo

Water source	Tazawa River
Water quality	Soft

Tatenokawa Shuzo

27 Aza-shimizuda, Yamadate, Sakata-shi, Yamagata 999-6724 TEL.0234-52-2323
E-mail: info@tatenokawa.com https://www.tatenokawa.com

Our goal is to make sake that can compete globally
TATENOKAWA

The first sake brewery in the Ginjo Kingdom of Yamagata to produce only Junmai Daiginjo.

Tatenokawa Shuzo was founded in 1832. A vassal of the Uesugi clan who visited Shonai at the time was surprised by the quality of the water and recommended that Heishiro Sato brew sake, and thus Heishiro started a sake brewery. In 1854, when Lord Sakai of the Shonai domain visited the brewery, he offered the sake made their to the lord, who was so pleased that he named the brewery "Tatenokawa". Today, Tatenokawa Shuzo brews only Junmai Daiginjoshu as part of its attempt to achieve the "100-year vision of Tatenokawa," which aims to communicate the excellence of Japanese traditional culture, sake, both in Japan and abroad. In addition to traditional sake brewing, the brewery is also engaged in various businesses, such as the production of "Komei" with 1% rice polishing and collaboration sake with the American band "Foo Fighters".

Tatenokawa Junmai Daiginjo Seiryu

"Light, soft sake recommended for beginners"

It is a brand that is representative of the brewery. As its name implies, it has the clarity of a clear stream flowing at the foot of Mt. Chokai.

Junmai Daiginjoshu	1,800ml/720ml
Alc.	14%
Rice type	Dewasansan
Sake meter value	-2.0
Acidity	1.4

Tatenokawa Junmai Daiginjo Jyu-hachi(18)

"This bottle should be called a gem of sake"

The ace of the Tatenomekawa Brewery. Only the best "nakatori" part of the pressing is bottled for this fragrant, splendid sake.

Junmai Daiginjoshu	720ml
Alc.	15%
Rice type	Yamadanishiki
Sake meter value	-1.0
Acidity	1.4

Dewazakura shuzo

Water source	Ou Mountains
Water quality	Soft

1-4-6 Hitoichimachi, Tendo-shi, Yamagata 994-0044 TEL.023-653-5121
https://www.dewazakura.co.jp

Bringing "Ginjo" to the world from Yamagata

A pioneer of Ginjoshu that has won numerous awards in and outside of Japan.

Founded in 1892, the Dewazakura Sake Brewery is a local sake brewery rooted in Yamagata that strives to produce local sake loved by the people of Yamagata. The company policy of "quality first" led to the commercialization of Oka Ginjoshu in 1980. Since then, Dewazakura became the most iconic brand of ginjo-shu. Everyone who is fond of ginjo-shu knows this brand. The brewery is definitely one of the most well-known in the Tohoku region, and it has been exporting its products since 1997. With the goal of making ginjo-shu more popular around the world, Dewazakura is exported to 100 cities in 35 countries, including embassies and other diplomatic missions of the Ministry of Foreign Affairs. Dewazakura has been consistently rated as one of the best in domestic and international competitions. "Ichiro" brand and "Dewazakura" brand won the highest award in the sake category of the International Wine Challenge (IWC) in 2008 and 2016, respectively.

吟醸酒
Dewazakura Oka Ginjoshu

"Light and full flavored"
The oldest wine merchant in the UK, BB&R, chose it as the first sake it marketed. Locally, it is indispensable for cherry blossom viewing parties.

Ginjoshu	1,800ml/720ml/300ml
Alc.	15%
Rice type	Domestic rice
Sake meter value	+5.0
Acidity	1.2

純米大吟醸酒
Junmai Daiginjo Ichiro

"Elegantly fragrant and elegantly sweet"
In 2008, it won the highest award in the sake category at the IWC, the "Sake Champion" award. Recommended for drinkers not familiar with sake.

Junmai Daiginjoshu	720ml
Alc.	15%
Rice type	Yamadanishiki
Sake meter value	+4.0
Acidity	1.3

Water source	Private well
Water quality	Soft

Kaitou Otokoyama Shuzo

785 Nakaaraikuboi, Minamiaizu-machi, Minamiaizu-gun, Fukushima 967-0005 TEL.0241-62-0023
E-mail: kuramoto@otokoyama.jp https://otokoyama.jp

Power and skill handed down from generation to generation

Making the most of the cold climate for unique sake brewing.

The town of Minami-Aizu, in Fukushima Prefecture's Minami-Aizu District, nurtured its own culture as a territory under the direct control of the Shogunate during the Edo period. Tajima Gion Festival is famous as a representative example of its traditional culture. The Kaitou Otokoyama Shuzo, which operates a sake brewery in this area, was founded in the first year of Kyoho (1716). The name "Kaitou" derives from , Harumasa Watanabe, the third-generation head (the Chinese characters 開当 can be read both "Harumasa" and "Kaitou"). Since then, the brewery has been rooted in this cold region for 300 years over 14 generations. All the brewers at the company are from the prefecture. They work day and night to make sake with great care. In recent years, the brewery has won many gold awards at the National New Sake Competition. The brewery's motto is "to expand the range of choices for the customers and provide them with more sake to their liking".

Kaitou Otokoyama Junmaishu

"Delicious taste that makes you feel relaxed"

Brewed with pure water in severe coldness. The aroma is full and the taste is mellow.

Junmaishu	1,800ml
Alc.	15%
Rice type	Yumenokaori
Sake meter value	+3.0
Acidity	1.3

Kaitou Otokoyama Junmai Ginjo

"The more you drink it, the more delicious it tastes"

The mild aroma and the soft taste. The more you drink it, the more delicious it tastes – a good junmai ginjo.

Junmai Ginjoshu	1,800ml
Alc.	15%
Rice type	Yamadanishiki
Sake meter value	+4.0
Acidity	1.3

Kitanohana Shuzojo

Water source	Iide Mountains subsoil water
Water quality	Soft

4924 Maeda, Kitakata-shi, Fukushima 966-0862 TEL.0241-22-0268
E-mail: info@kitano87.jp   http://www.kitano87.jp

A young brewer in Kitakata

The youngest of the 11 local breweries, but one that you will want to keep an eye on.

Founded in 1919 in Kitakata, Fukushima Prefecture, a town of ramen noodles and breweries, the Kitanohana Sake Shuzo uses only local rice, water, and yeast, and the brewer himself carefully brews small quantities in small tanks. The main brand name of the brewery, "Kitanohana," has the meaning of aiming to be the best in the sake town of Kitakata, and the wish that there will be many happy and wonderful things (hana) for the customers. The brewing water used is fresh water that gushes from the underground bedrock veins of the Iide Mountains. Water is underground for an estimated 500 years before it emerges as a spring. The Kitanohana brewery depends on this soft, delicious water to make its sake. The brewery also works with a sake rice production group of local farmers to reduce the use of pesticides. The group times rice cultivation to suit the sake brewing process.

純米酒
Karakuchi Junmai Kuradaiko +10

"Flavor that retains a bit of sweetness and crispness"

This junmaishu has a clean, dry taste that retains sweetness, but is also crisp. It can be enjoyed cold for its refreshing taste and warmed for its soft sweetness.

Junmaishu	1,800ml/720ml
Alc.	15-16%
Rice type	Domestic rice
Sake meter value	+10.0
Acidity	1.2-1.4

純米大吟醸酒
Junmai Daiginjoshu Kitanohana

"A splendid sake that is best served cold"

The soft, stunning sweetness is as gentle and beautiful as a peony flowers, and it goes down well. Brewery yeast for this brand is produced in Fukushima.

Junmai Daiginjoshu	1,800ml/720ml
Alc.	16%
Rice type	Yamadanishiki
Sake meter value	−6.0
Acidity	1.3

| Water source | Namie Town water supply (well water) |
| Water quality | Soft |

Suzuki Shuzoten

40 Aza-chimeiji, Oaza-kiyohashi, Namie-machi, Futaba-gun, Fukushima 979-1513 TEL.0240-35-2337
E-mail: namie@iw-kotobuki.co.jp http://www.iw-kotobuki.co.jp

The sake for men of the sea to share their joy

Sake brewing that aims to express the time that has passed for the local people.

The brewery was founded during the Tempo period (around 1840) in the Edo period (1603-1868), when the Soma clan allowed the brewery to produce unfiltered sake. The brewery was located across the embankment from the sea, and brewed sake for the men of the sea. At the time of the Great East Japan Earthquake, all the buildings were washed away and the brewery was forced to close down, but was rebuilt in Nagai City, Yamagata Prefecture. In March 2021, the new Suzuki Shuzo brewery started operation in Namie Town and the company made a fresh start. The brewing process takes place over all four seasons. Suzuki Shuzo uses only contract-grown rice as the raw material, and does its own milling. The brewery aims to express the time that has passed for the local people, and brews sake that is close to their hearts, both in the past and on into the future. At the brewery building, which is attached to the "Namie no Waza, Nariwai-kan" (Roadside Station Namie), you can observe the sake brewing process, and enjoy tasting and eating Ohbori Soma-yaki dishes.

Iwakikotobuki Tairyoiwai Kompeki

"Sake that is part of the local living style"

This is a favorite Namie sake back in production after a 10-year halt. It has a smooth flavor that can be paired with any meal, and served either cold or warm.

Junmai Ginjoshu	720ml
Alc.	15.5%
Rice type	Koshihikari

Iwakikotobuki Kijoawazake

"Let it cool completely and enjoy the taste of sparkling sake"

Sparkling sake made from rice grown in Nagai City and brewed in Namie. The elegant sweetness and refreshing acidity make it easy to drink.

Junmaishu	720ml
Alc.	7%
Rice type	Domestic rice

Takahashi Shosaku Shuzoten

Water source	Oto Mountains subsoil water
Water quality	Soft

755 Murahigashi, Oaza-ichinoseki, Monden-machi, AizuwAkamatsu-shi, Fukushima 965-0844 TEL.0242-27-0108
E-mail: sakeshou@nifty.com   http://aizumusume.a.la9.jp/

Sake brewing starts from rice cultivation

A brewery committed to "locally-made with local techniques", a tradition since its founding. The brewery is located about 6 km south of the city of Aizuwakamatsu.

The Takahashi Shosaku Shuzoten, known for its brand "Aizumusume", is located in the midst of a quiet paddie area in Mondenmachi Ichinoseki, Aizu's grain-growing district. The brewery is said to have been founded in the early Meiji period, and its name can be found in brewing records from 1877. It seems that the family were originally wealthy farmers who brewed sake using their own rice. Later, the business was closed down due to the Wartime Enterprise Development Order, but was eventually revived after much effort. In the 1960's, the company stopped making so-called sanzoshu and adopted the Dosan-Doho method of sake brewing, restoring the brewery's oldest traditions. Currently, Takahashi sells a limited number of sweets made from sake dregs. The name of the sake, "Aizumusume," is named after the beauty of the Aizu girls, who are compassionate and quiet, but have a strong core.

純米酒
Aizumusume Junmaishu

"Sake with a full flavor of rice"
A simple, unpretentious sake with a full flavor of rice that you will never get tired of drinking. Its flavor changes when served cold or warmed.

Junmaishu	1,800ml
Alc.	15%
Rice type	Gohyakumangoku
Sake meter value	+2.0 - -3.0
Acidity	1.4-1.6

純米酒
Aizumusume Muishin

"Genuine local sake made from organic rice"
Junmaishu with a comfortable flavor. Made from organic sake rice, including "Gohyakumangoku", which is grown by the brewer and the brew master.

Junmaishu	1,800ml
Alc.	15%
Rice type	Gohyakumangoku(organic rice)
Sake meter value	+2.0~-3.0
Acidity	1.4

| Water source | Natural spring water from the foot of Mt. Bandai and Mt. Iide. |
| Water quality | Soft |

Tatsuizumi Shuzo

5-26 Uwamachi, Aizuwakamatsu-shi, Fukushima 965-0034　TEL.0242-22-0504
E-mail: info@tatsuizumi.co.jp　http://tatsuizumi.co.jp

A revival of a fantastic rice for sake brewing
Kyonohana

Sake brewing using rice and water from Aizu.

In 1877, Ryuzo Shinjo, the first-generation head of the Shinjo family, established a sake brewery in Bakuro-machi, Aizuwakamatsu City, independent of the original Shinjo family. Since then, he has avoided mass production and mass sales, and has maintained the quality of handcrafted sake. The brewery's brewing process starts by growing good rice with local farmers, including Kyonohana, a rare variety of rice suitable for sake brewing. The brewing water is natural spring water from the foot of Mt. Bandai. The brewing process is limited to December through March, the coldest months of the year, and the brewery is very particular about the way it brews sake. All of this is done in order to create a sake that will impress the drinker with the fragrant aroma, mellow flavor, and umami sake was meant to have.

Junmai Daiginjo Kyonohana

"Sake made with "Kyonohana", a rare variety of sake rice"

Using only "Kyonohana" rice, this sake has a unique mellowness and deep flavor.

Junmai Daiginjoshu	1,800ml/720ml
Alc.	16%-17%
Rice type	Kyonohana
Sake meter value	±0
Acidity	1.5

Junmai Ginjo Kyonohana

"Junmai Ginjoshu that goes well with food"

Kyonohana makes up 60% of the rice used. This sake is slightly dry with a deep flavor. It is Junmai Ginjoshu with a mild aroma that does not interfere with food.

Junmai Ginjoshu	1,800ml/720ml
Alc.	15%-16%
Rice type	Kyonohana/Yumenokaori/etc
Sake meter value	+2.0
Acidity	1.4

Tsurunoe Shuzo

| Water quality | Soft |

2-46 Nanoka-machi, Aizuwakamatsu-shi, Fukushima 965-0044 TEL.0242-27-0139
E-mail: tsurunoe@nifty.com https://www.tsurunoe.com/

Aizu, the land of flowers, is famous for sake

Times have changed, but the commitment to taste has not.

The Hayashi family of "Tsurunoe" is part of the Eiboya clan, who served the head of the Aizu domain. The brewery was founded in 1794 by a branch of the Eihoya family, and was called Eihoya back then. The head of the family took the name "Heihachiro" for generations and brewed the brands "Nanayo Masamune" and "Hosen". In the early Meiji period, the name of the brewery was changed to "Tsurunoe," which represents Tsurugajo Castle, the symbol of Aizu, and Lake Inawashiro. "Aizuchujo" is a brand named after an official rank of the feudal lord Hoshina (younger brother of Iemitsu Tokugawa), and is now the the best known sake of this brewery. As a Haiku goes -- "Aizu, the land of flowers in Michinoku, is the land of sake" -- the good rice, good water, and severe winter in Aizu are the most favorable conditions for sake brewing. The brewery's traditional handmade sake tailored to the local climate has won many awards. Yuri, brewed by mother and daughter master brewers, is also popular among women and is recommended as a sake for meals.

純米大吟醸酒
Junmai Daiginjo Yuri

"The sake served at the Japan-U.S. Summit Dinner in 2009"
A refreshing, dry sake brewed by mother and daughter master brewers. It is named after the daughter of the 7th generation brewer.

Junmai Daiginjoshu	720ml
Alc.	15%
Rice type	Gohyakumangoku
Sake meter value	±5.0
Acidity	1.4

純米大吟醸酒
Aizuchujo Junmai Daiginjo Tokujyoshu

"Winner of the 2008 Tohoku Sake Competition Grand Prize"
Using only Yamadanishiki, this sake has an aroma that spreads in the mouth and has a full, elegant flavor.

Junmai Daiginjoshu	1,800ml/720ml
Alc.	16%
Rice type	Yamadanishiki
Sake meter value	±0
Acidity	1.3

| Water source | Abukuma Mountains |
| Water quality | Medium-hard water (well water) / Soft water (spring water) |

Niida Honke

139 Aza-Takayashiki, Kanezawa, Tamura-machi, Koriyama-shi, Fukushima 963-1151   TEL.024-955-2222
E-mail: info@1711.jp   https://1711.jp/

Committed to natural rice The sake brewery that protects rice farming

We hope that local people will be happy to drink Niida Honke.

The first generation of the Niida family started brewing sake in 1711 in Kanazawa, Tamura-cho, southeast of Koriyama City. In 1967, the company began brewing and selling "Kinpo", a pioneering brand of natural sake, using only natural rice grown without the use of pesticides or chemical fertilizers under contract with local farmers. In 2011, when the company celebrated its 300th anniversary, it achieved its long-held dream of brewing Junmaishu using only natural rice. The company's next dream is to turn all 60 hectares of rice fields in Kanazawa into natural rice fields and become a sake brewery that protects Japan's rice fields.

Niida-Shizenshu Junmai Genshu

"Sweet sake with umami"

This sake is made with a four-stage brewing process using traditional yeast and no added yeast (instead using the natural yeast in the air or living in the brewery) to bring out the sweetness and flavor of the natural rice.

| Junmaigenshu ... 1,800ml/720ml/300ml |
| Alc. .... 16% |
| Rice type ... Toyonishiki/Chiyonishiki/Koshihikari |

Odayaka Junmal Ginjo

"Elegant aroma like a melon"

This Junmai Ginjoshu has a good harmony "f "sweet, spicy, sour, bitter, and astring"nt" and has the original flavor of sake, which goes well with all kinds of food.

| Junmai Ginjoshu ... 1,800ml/720ml/160ml |
| Alc. .... 15% |
| Rice type .... Miyamanishiki |

Bandai Shuzo

Water source	subsoil water of Mt. Bandai
Water quality	Soft

2568 Kanagamidan, Bandai, Bandai-machi, Yama-gun, Fukushima 969-3301　TEL.0242-73-2002
E-mail: info@bandaishuzou.com　https://www.bandaishuzou.com/

Bandaisan Sake
Sake brewing rooted in the region with Mt. Bandai

The sake brewery that serves local consumers at the foot of Mt. Bandai.

Located northeast of Aizu Wakamatsu City, between Lake Inawashiro and the Aizu Basin, Bandai Shuzo was founded in 1890 in Bandai, a plateau town gently spreading at the foot of Mt. Bandai The fifth generation and head brewer, President Kuwabara, says, "Our company motto is 'Wajo Ryoshu' (Harmony makes good sake). The brewer and brewmaster work together to make sake. The sake is brewed using subsoil water from the Bandai west foothills spring group, which has been selected as one of the 100 best waters in Japan, and is made with rice from Bandai and by brewers from Bandai. The brewery is always ready to take on new challenges. It produced "Akai-sake Aizuzakura" made with black rice, an ancient rice produced at the request of a local farmer, and the "Jotanbo Special Junmai" named after the head of the sect at Enichiji Temple, the birthplace of local Buddhist culture.

普通酒
Shirogane Bandaisan

"Futsushu produced with techniques of stubborn Aizu Toji"

It has a natural aroma and a mellow flavor that makes it a good match for food. It is ideal as sake during the meals. Pair it with Japanese or Western-style dishes.

Futsushu	1,800ml/720ml
Alc.	15.3%
Rice type	Domestic rice
Sake meter value	±0
Acidity	1.4

純米大吟醸酒
Jotanbo Junmai Daiginjo

"Sake named after a high priest of Aizu Buddhsm"

You can taste the splendid ginjo aroma, elegant sweetness, and strong rice flavor. The acidity and sharpness are excellent, and tastes best cold.

Junmai Daiginjoshu	1,800ml/720ml
Alc.	16.4%
Rice type	Gohyakumangoku
Sake meter value	-1.0
Acidity	1.3

Water source	Subterranean water from Mt. Adatara
Water quality	Soft

East Japan Shuzo Cooperative Association

167 Yasumiishi, Nihonmatsu-shi, Fukushima 964-0082 TEL.0243-3117
https://higashinihonsyuzo.jp/

A brewery committed to unprocessed sake full of flavor

A brewery dedicated to producing high-quality sake for customers, with love.

East Japan Shuzo Cooperative Association is a relatively new brewery, established in 1974. Its parent company Okunomatsu Shuzo was established in 1716. Higashinihon Shuzo started as a cooperative association established by Okunomatsu Shuzo with several breweries in the neighboring area. All the sake brewed here is basically undiluted sake. The brewery has its own rice mill and polishes all of the rice it uses. For brewing water, they use subsoil water from the famous Mt. Adatara in Fukushima Prefecture, and strive to produce high quality sake with the motto "sake quality first". Okunomatsu Shuzo's sake "Adatara Ginjo," made based on unprocessed sake produced by Higashinihon Shuzo, was crowned the world's best sake at the IWC (International Wine Challenge) 2018.

大吟醸酒
Genshu Daiginjo

"An exquisite Ginjo flavor"
It has a rich and full aroma of ginjo and a perfect balance of flavors.

Daiginjoshu	1,800ml
Alc.	17.5%
Rice type	Yamadanishiki
Sake meter value	+1.0
Acidity	1.3

純米大吟醸酒
Genshu Junmai Daiginjo

"A sake that impresses with its aroma and flavor"
This Junmai Daiginjoshu is made with the finest traditional techniques. The delicate, mellow aroma and fullness are exquisite.

Junmai Daiginjoshu	1,800ml
Alc.	15.6%
Rice type	Yamadanishiki
Sake meter value	−6.0
Acidity	1.4

Minenoyuki Shuzojo

Water source	Iide mountain system groundwater
Water quality	Soft

1-17 Aza-Sakuragaoka, Kitakata-shi, Fukushima 966-0802　TEL.0241-22-0431
E-mail: info@minenoyuki.com　https://www.minenoyuki.com

Young sake brewer who bears the future of Kitakata

Sake brewed slowly by Aizu Toji using pure underground water from Mt. Iide.

This brewery was established in 1942 as a branch of the original Yamatonishiki Shuzo. It is the youngest brewery in Kitakata City, where sake brewing has been thriving since ancient times under the Iide mountain range. In addition to "Minenoyuki," the brand carried by the brewery since its establishment, the main product is "Yamatoya Zennai," which was inherited from the original brewery and reissued a few years ago. Young brewers are also working hard to produce a wide range of sake, from traditional, such as yamahai-brewed sake and doburoku, to rare varieties. Of particular note is the brewing of mead (honey wine), which is said to be the oldest sake in the world. In other countries, mead is brewed with wine yeast by adding grains and other ingredients, but the first purely domestic mead in Japan, "Miroku no Mori," was brewed with sake yeast using only local honey from Aizu.

*When including breweries that succeeded in brewing mead using imported ingredients, Minenoyuki Shuzojo was the third in Japan to successfully brew mead as of 2009.

特別純米酒
Yamatoya Zennai Junmai Namazume

"A flagship brand inherited from the original family for generations"

This is the standard version of the revived "Yamatoya Zennai" series. It is characterized by an elegant aroma, rich sweetness, and a strong, tight flavor.

Special Junmaishu	720ml
Alc.	15-15.9%
Rice type	Gohyakumangoku
Sake meter value	-2.0
Acidity	1.4

純米酒
Hatsuyukiso Clear

"Rice wine rather than sake"

The fresh, crisp flavor of newly-cultivated rice and just the right amount of sweetness. A new type of sake that goes well with both Japanese and Western food.

Junmaishu	720ml
Alc.	13%
Rice type	Gohyakumangoku
Sake meter value	-3.0
Acidity	1.5

| Water source | Subsoil water of Mt. Iide |
| Water quality | Soft |

Yamatogawa Shuzoten

4761 Aza-Teramachi, Kitakata-shi, Fukushima 966-0861 TEL.0241-22-2233
E-mail: sake@yauemon.co.jp http://www.yauemon.co.jp/

Living with Kitakata

A brewery committed to rice cultivation as the first step of sake brewing.

Since its establishment in 1790, the brewery has produced sake for nine generations. Aizu Kitakata faces Mt. Iide, which is perennially covered in snow on its northwest side, and industries that make use of its abundant subsoil water are thriving. Yamatogawa sake is brewed with this pure water. The single-minded spirit of successive generations of toji (master brewers) has produced famous sake brands, such as Yauemonshu. In recent years, the brewery has introduced new technologies with modern facilities, but at the same time, it has continued to preserve traditional techniques used since its establishment. In 2007, the brewery established an agricultural corporation called "Yamatogawa Farm" to create a cyclical soil system using rice bran and sake lees from the sake brewery as organic fertilizer. The brewery aims to produce high-quality sake that is in harmony with the Kitakata climate.

Junmai Kasumochi Genshu Yauemon-shu

"Kasumochi unprocessed sake with a sweet flavor"

The recipe of this tasty, sweet sake is a family secret. It is brewed with almost twice the amount of koji and glutinous rice as usual, and has a rich sweetness. Serve it on the rocks if you like.

Junmaigenshu	1,800ml/720ml
Alc.	17%
Rice type	Yumenokaori
Sake meter value	-20
Acidity	2.1

Junmai Karakuchi Yauemonshu

"An impeccable balance of umami and sharpness"

Junmai dry sake made from Fukushima Prefecture's "Yumenokaori" sake rice grown in the company's own fields. This sake is loved as a daily drink in Kitakata.

Junmaishu	1,800ml/720ml
Alc.	15%
Rice type	Yumenokaori
Sake meter value	+7.0

Guide to Sake Breweries and Famous Sake

Kanto Region

Ibaraki, Tochigi, Gunma, Saitama, Chiba, Tokyo, Kanagawa

Food Culture in the Kanto Region

Ibaraki: Faces the Pacific Ocean to the east and lakes and marshes of Kasumigaura to the south along the Tone River. In addition to "anko-nabe" hot pot dish and "kenchin-jiru" soup made from monkfish and other seafood caught in the Pacific Ocean, mushroom harvesting and rice farming are widespread in the mountains.

Tochigi: A land-locked prefecture, Tochigi has a large difference in temperature between day and night, and grows crops, such as strawberries and barley, as well as river fish cooked in sweetened soy sauce or grilled on a skewer and covered with dengaku miso. It is also famous as a production area for preserved food, such as kampyo and yuba.

Gunma: Much of the land contains volcanic ash. Field farming is more common than rice farming, and the prefecture is known as a producer of wheat. There are many dishes made from wheat flour, such as yakimanju, udon, and "okkirikomi". Fish is cooked in a sweet sauce, and highland cabbage and Shimonita green onions are grown in the area.

Saitama: Surrounded by the Chichibu Mountains and the Sayama Hills, the area is characterized by hot and humid summers and dry winters. The prefecture is known for vegetables such as leeks, spinach, taro, and kuwai.

Chiba: Sardine fishing is popular, and the Kuroshio Current flowing off the Boso Peninsula is good for growing lobsters, resulting in large catches. It is also known for the production of peanuts and vegetables such as turnips.

Tokyo: Warm climate with the Musashino Plateau and Tokyo Bay. Tokyo sushi and tempura dishes, known as Edomae, is made from fish caught in Tokyo Bay, such as sea eel, clams, and octopus.

Kanagawa: Facing Tokyo Bay and Sagami Bay, the Misaki area in the prefecture is famous for its tuna catches. It is also famous for horse mackerel sushi, shirasu (baby sardines), and kamaboko (fish paste). The plains surrounded by the Tama Hills and Tanzawa Mountains produce vegetables, tea, and tangerines.

Water source	Private well
Water source	Soft

Urazato Shuzoten

982 Yoshinuma, Tsukuba-shi, Ibaraki 300-2617 TEL.029-865-0032
E-mail: info@kiritsukuba.co.jp https://www.kiritsukuba.co.jp

Sake brewed with heart by the Nanbu Toji

The brand name produced by the Nanbu Toji is "Kiritsukuba," The painting on the label is a registered trademark.

The Urasato Sake Brewery, which brews "Kiritsukuba", is located in Yoshinuma, Tsukuba City, with Mt. Tsukuba. Founded in 1877, the brewery uses only carefully-selected rice that is suitable for sake brewing. It uses Ogawa yeast, produced in the prefecture and discovered by Chikara Ogawa, and its own water from the Ogai River system. Sake is made with care by Nanbu Toji Sasaki Keihachi and the brewers. All "Kiritsukuba" sake are designated as ginjo sake. The label features the art work "Kiritsukuba" by the late Shoichiro Hattori, a Western-style painter who was a member of the Art Academy of Japan, and the name of the work is used as a trademark. Hattori is from Ibaraki, where the brewery is located, and is one of Japan's leading landscape painters, who expressively captures the nature of Ibaraki, including Kasumigaura, Suigo, and Mt. Tsukuba.

Kiritsukuba
Tokubesu Honjozo

"Clear, sharp, and dry sake"
This sake, with a high sake content, is refreshing and dry. It is a special brew that can be enjoyed cold or warm.

Special Honjozoshu ... 1,800ml/720ml
Alc. 15-16%
Rice type .... Gohyakumangoku
Sake meter value .... +5.0

Kiritsukuba
Tokubetsu Junmaishu

"The most popular brand at the brewery"
Brewed with Gohyakumangoku rice from Toyama Prefecture and stored at under 15°C. It is a young and refreshing Special Junmaishu.

Special Junmaishu ... 1,800ml/720ml
Alc. 15-16%
Rice type .... Gohyakumangoku
Sake meter value .... +3.0

Okabe Gomei-gaisha

Water source	Satogawa river subsoil water
Water source	Soft

2335 Ozawa-cho, Hitachiota-shi, Ibaraki 313-0038 TEL.0294-74-2171
https://www.matsuzakari.co.jp/

Sake brewery with spirit With good rice, water, and friends who share hearts

Sake made by hand, with all our heart and soul.

Okabe Gomei-gaisha was founded in 1875. The city of Hitachiota City in Ibaraki Prefecture, where the brewery is located, has rice paddies around the Satogawa River in the southern part of the city, making it an ideal environment for sake brewing. It is blessed with water and rice. Since its establishment, the brewery has been focused on how to pass on to future generations the good qualities of local sake, which is a product of the local climate. Its motto is to brew sake with all its heart and soul, using locally grown rice and water, and making it by hand with like-minded people. They have continued to hold this belief and aim to brew sake that conveys their "heart" through their products. In recent years, they have organized "Hitachiota Jizake Project," in which they plant Miyamanishiki rice and brew sake together with the general public, an event that attracts participants not only from within the city but also from outside the prefecture.

純米吟醸酒
Matsuzakari Junmai Ginjo unfiltered, unpasteurized, undiluted sake

"Local sake with elaborate terroir"

This is the most popular brand, with a splendid aroma and a clean attack. You can enjoy both coolness and juiciness at the same time.

Junmai Ginjoshu	720ml
Alc.	16%
Rice type	Miyamanishiki
Sake meter value	+1.0
Acidity	1.4

純米酒
Goendane Junmai

"Hand-made sake"

Made from sake rice cultivated together with the participants of the local sake project. You can enjoy a juicy mouthfeel and a crisp aftertaste.

Junmaishu	720ml
Alc.	15%
Rice type	Gohyakumangoku
Sake meter value	+3.0
Acidity	1.5

Water source	Mt. Tsukuba water system
Water source	Soft

Hirose Shoten

880 Takahama, Ishioka-shi, Ibaraki 315-0045 TEL.0299-26-4131
E-mail: hirose@shiragiku-sake.jp https://shiragiku-sake.jp

Local sake that you'll never get tired of drinking

Consistent philosophy for sake-making and profound flavors loved by local people.

Hitachi Takahama, where the view of Mt. Tsukuba is said to be the most beautiful, once prospered as a base for water transportation in Kasumigaura. It is said that the Hirose family operated a sake brewery in this area about 200 years ago, and when power was still undeveloped, they introduced a steam engine and highly polished rice to produce high-quality sake. Based on this tradition, Hirose Shoten is committed to maintaining the delicious taste of local sake and uses its own revived rice for sake brewing. In addition to its flagship brand "Shiragiku," the brewery also produces Daiginjoshu "Shiragiku," Junmai Ginjoshu "Tsukuba no Kohbai Ichirin," Junmaishu "Kasuminosato," and honjozo "Tsukuba no Hakubai". The name "Shiragiku" means "white chrysanthemum," a typical Japanese flower that blooms in autumn when sake matures.

純米吟醸酒
Junmai Ginjo
Tsukuba no Kobai Ichirin

"Brewed with Miyamani-shiki rice polished to 50%"
A light, dry sake with a balanced aroma and taste. Recommended for drinking cold.

Junmai Ginjoshu	1,800ml
Alc.	15%
Rice type	Miyamanishiki
Sake meter value	+5.0
Acidity	1.0-1.1

大吟醸酒
Shiragiku Daiginjo

"Sake that you want a sip after another"
A crisp daiginjo with a splendid aroma, and a well-balanced flavor that is well-integrated into the clean quality of the sake.

Daiginjoshu	720ml
Alc.	16-17%
Rice type	Yamadanishiki
Sake meter value	+4.0-+5.0
Acidity	1.3

Hagiwara Shuzo

Water source	Tone River system
Water source	Soft

565-1 Sakai-machi, Sashima-gun, Ibaraki 306-0433 TEL.0280-87-0746
E-mail: info@tokumasamune.com https://www.tokumasamune.com

The most beautiful sake under heaven
Tokumasamune

With a history of more than 150 years, taking advantage of the blessed waters of the Tone River.

Sakaimachi, where the Hagiwara Shuzo is located, faces Sekiyado Town (now Noda City) in Chiba Prefecture across the Tone River, and has prospered as a transportation hub since the Muromachi period (1333-1573), taking advantage of the abundant water resources of the Tone River basin. Toemon started brewing sake in 1855, and the current brewer is the sixth-generation. The name of the sake, Tokumasamune, comes from a Chinese poem that says, "When you are happy, drink sake; when you are sad, make up for it with sake; together with the sorrow and joy of life, there is sake, and that is called the virtue of sake". Although the times have changed and the way we drink sake has also changed, this is a passionate sake brewery that keeps the tradition alive and never forgets the spirit of challenge. Since the beginning of the Heisei era (1989), the brewery has won awards for excellence and gold medals in many sake competitions.

純米吟醸酒
Tokumasamune Junmai Ginjo Minori

"Umami of rice peculiar to Junmai"
This is additive-free sake made by long-term fermentation at low temperatures using only rice and rice malt. The label features a motif of "fruitful fields".

Junmai Ginjoshu	1,800ml
Alc.	15%
Rice type	Miyamanishiki
Sake meter value	+4.0
Acidity	1.4-1.5

大吟醸酒
Tokumasamune Daiginjo

"Clear and flesh flavor"
This is excellent sake made by a skilled master brewer during the cold season. It has an elegant aroma and a smooth taste.

Daiginjoshu	1,800ml
Alc.	16%
Rice type	Yamadanishiki
Sake meter value	+5.0-+6.0
Acidity	1.4-1.5

| Water source | Kinugawa River system subsoil water |
| Water source | Soft |

Buyu

144 Yuki, Yuki-shi, Ibaraki 307-0001 TEL.0296-33-3343
E-mail: sakagura@buyu.jp   http://www.buyu.jp

Good sake rice, good water, Good technique

We continue to brew "true sake" by carefully preparing each bottle.

The company was founded during the Keio era (1603-1868) in the late Edo period (1603-1868) by Yukichi Hosaka, the first generation, in the castle town of Yuki in the northern Kanto region. From the Edo period to the Heisei period, the brewery has gone through major changes, and is now led by the fifth-generation head, Yoshio Hosaka. The brewery followed the traditions of Echigo Toji brewing for generations, but since 1996, it has been brewing sake with the hands of the local Yuki Toji. Rather than brewing to increase production, they brew in the so-called Sanki-jozo procedure (three-season brewing) to improve the quality of the sake, and try to manage the fermentation of each bottle to the best of their ability. In order to bring out the characteristics of the raw rice in the sake, the natural color of the rice is left without carbonation. The unique sake is made possible by using good rice, good water, and good brewing techniques.

本醸造酒
Buyu White Label

"Unique flavor and refreshing aftertaste"
This sake is brewed mainly for the local market. It is characterized by its unique taste and refreshing aftertaste, thanks to the use of rice suitable for sake brewing.

Honjozoshu	1,800ml
Alc.	15-16%
Rice type	Syuzo Suitable rice
Sake meter value	+1.0
Acidity	1.2

純米酒
Buyu Karakuchi Junmaishu

"Imposing flavor"
This sake exemplifies Junmaishu, with its well-balanced sharpness and umami. It has natural color and taste thanks to maturation.

Junmaishu	1,800ml
Alc.	15%
Rice type	Yamadanishiki/Gohyakumangoku
Sake meter value	+3.0
Acidity	1.3

Yamanaka Shuzoten

Water source	Well water in the brewery
Water source	Soft

187 Shinishige, Joso-shi, Ibaraki 300-2706 TEL.0297-42-2004
E-mail: info@hitorimusume.co.jp https://hitorimusume.co.jp

Brewed with fresh water from the Kinugawa River
A love of the dry taste

Despite the effects of the earthquake, a brewery that has maintained its traditions since the Edo period.

The Yamanaka Shuzoten is located in the middle of the Kanto Plain, with Mount Tsukuba to the east and the Kinu River to the west. Although the exact date of the brewery's establishment is unknown due to the damage of historical documents, the brewery is said to have been founded in 1805, the second year of the Bunka era. The Edo Shogunate's major renovation of the Tone, Kinugawa, and Watarase Rivers led to the expansion of rice cultivation areas, and the brewery began to focus on sake production, using the Kinugawa River adjacent to the brewery to transport sake to Edo (Tokyo). After painstaking research on brewing water, the brewery devised its own two-stage brewing method. Since then, for 217 years, the brewery has been producing dry sake that is soft on the palate. The brewery's sake is not only appreciated by the local community, but has also been highly evaluated overseas, winning the Monde Selection High Quality Trophy and the IWSC Gold Award.

Hitorimusume
Junmai Daiginjo

"Sake with a smooth and full aroma"

This sake has a quality like fresh water. Best served at room temperature or chilled with seafood and poultry dishes.

Junmai Daiginjoshu ... 1,800ml/720ml	
Alc.	16-17%
Rice type	Yamadanishiki
Sake meter value	+5.0
Acidity	1.3

Hitorimusume
Tokubetsu Junmai Cho-Karakuchi

"Harmony of various flavors and depth of taste"

Similar to a daiginjo, this sake was made with the aim of having a smooth mouthfeel. It can be served lukewarm, at room temperature, or slightly chilled.

Special Junmaishu ... 1,800ml/720ml	
Alc.	15.5%
Rice type	Chiyonishiki
Sake meter value	+8.0
Acidity	1.5

Water source: Kasahara water source
Water source: Super soft

Yoshikubo Shuzo

3-9-5, Honcho, Mito-shi, Ibaraki 310-0815 TEL.029-224-4111
E-mail: info@ippin.co.jp https://www.ippin.co.jp

Brewing sake with water loved by Mito Komon

Preserving Tradition and Reaching New Heights with High Quality Hitachi Rice and Rich Water.

Tokugawa Mitsukuni, the grandson of Tokugawa Ieyasu and known as "Mito Komon," was born in Hitachi Province (present-day Ibaraki Prefecture). He ordered the establishment of the Kasahara Waterworks, with its source in the Kasahara Fudo Valley, when he was the second feudal lord of Hitachi. In 1790, he changed his business from a rice merchant to a sake brewer in order to make sake from this abundant water and the rice of Hitachi, one of the largest granaries in the Kanto region. The 12th generation brewer, Hiroyuki Yoshikubo, brews sake with young brewers using tradition, innovation, and imagination. Toji Tadayuki Suzuki has been working at the brewery since he was 18 years old, learning the Nanbu Toji technique from senior generations of Toji brewers. He won the gold medal at the 2015 New Sake Competition. In recent years, special sake specifically for autumn salmon that run up the Naka River and mackerel, the local fish of Ibaraki Prefecture, has been gaining popularity.

清酒
SALMON de SHU

"A good match with salmon"
This sake was launched on November 11, 2020 (Salmon Day) to "make salmon taste better". It is a blend of Junmai-shu that brings out the best in the taste of salmon.

Seishu	1,800ml/720ml/300ml
Alc.	15%
Rice type	Domestic rice
Sake meter value	private
Acidity	private

清酒
SABA de SHU

"A good match with mackerel"
This sake was developed before SALMON de SHU, specifically for mackerel. It has a high acidity and amino acid content, which removes excess fat from mackerel and increases its flavor.

Seisyu	1,800ml/720ml/300ml
Alc.	15%
Rice type	Domestic rice
Sake meter value	private
Acidity	private

Raifuku Shuzo

Water source	Mt. Tsukuba
Water source	Soft

1626 Murata, Chikusei-shi, Ibaraki 300-4546 TEL.0296-52-2448
E-mail: info@raifukushuzo.co.jp http://www.raifukushuzo.co.jp

300 years since establishment Committed to quality

With the hope that drinking "Raifuku" will bring you good fortune.

Raifuku Shuzo was founded in 1716 by an Omi merchant in a place with good water at the foot of Mt. Tsukuba. The brand name, "Raifuku," which has been in use since the company's founding, is derived from the haiku poem which goes, "Happiness will come to those who drink sake and laugh a lot". If you drink "Raifuku," good fortune will come to you. This is the wish of the brewery. The brewery uses about 10 types of rice suitable for sake brewing and home-cultivated natural flower yeast. In the 300 years since its establishment, the sake has been made with a single focus on quality, and is sold only through special dealers, not on the Internet. Raifuku also makes a new brand of sake called Raifuku X, the specifications of which are kept secret.

純米大吟醸酒
Junmai Daiginjo Raifuku

"The best-tasting sake"
This is a limited edition Junmai Daiginjoshu made with Aizan, a rare variety of rice. The rice is polished to 40%.

Junmai Daiginjoshu	720ml
Alc.	16%

純米吟醸酒
Junmai Ginjo Makko Shobu

"Head-on fight with true quality"
This is a Junmai Ginjoshu with a graceful, splendid aroma and a deep, well-balanced flavor.

Junmai Ginjoshu	720ml
Alc.	15%

Water source	Shimono Kinugawa subsoil water
Water source	Soft

Inoue Seikichi Shoten

1901-1 Shirasawa-cho, Utsunomiya-shi, Tochigi 329-1102 TEL.028-673-2350
E-mail: sawahime-hiroshi@bz01.plala.or.jp http://sawahime.co.jp

Aiming for creation of a true "local sake"
Back to the basics

Made only with Tochigi-grown rice. This is the taste of Tochigi" to the world!

The company was founded in 1868. Sawahime is brewed in Shirasawa-juku, a post town on the Oshu Highway, using the famous underground water of the Kinugawa River. Hiroshi Inoue, the first Shimono Toji, is a graduate of a university program specializing in brewing. His catch-phrase is "True Local Sake Declaration". In 2010, the company's flagship brand, "Sawahime Daiginjo Shin Jizake-sengen" (Declaring intention to make a true local sake), won the highest award in the SAKE category, Champion Sake, at the International Wine Challenge (IWC), the world's largest international liquor competition. Currently, the company exports its products to several countries, including the United States and Hong Kong

Sawahime Daiginjo Shin Jizake-sengen

"A masterpiece with a splendid aroma"

This is a gem brewed by young brewers. It has a mellow, slightly dry taste with a splendid aroma. It won the "Champion Sake" award.

Daiginjoshu	1,800ml/720ml
Alc.	17%
Rice type	Hitogocochi
Sake meter value	+3.0
Acidity	1.3

Sawahime Junmai Daiginjo Shin Jizake-sengen

"A pioneer of local sake"

This sake is brewed using a new variety of sake rice "Yumesasara", and is a culmination of the brewery's techniques and passion. It has mellow umami with a refreshing aroma.

Junmai Daiginjoshu	1,800ml/720ml
Alc.	16%
Rice type	Yumesasara
Sake meter value	+1.0
Acidity	1.3

Utsunomiya Shuzo

Water source	Kinugawa River
Water source	Soft

248 Yanagita-machi, Utsunomiya-shi, Tochigi 321-0902 TEL.028-661-0880
E-mail: infoshikisa248@shikisakura.co.jp https://www.shikisakura.co.jp

Sake is a blessing from heaven and earth

It is the spirit of brewing good sake that determines the taste of sake.

Sake is brewed with the blessings of nature and the earth, and its taste is determined by the spirit of the brewers who want to brew good sake – that is the philosophy of the brewery. Shikisakura is brewed with subsoil water from the Kinugawa River, and is made from "Yamadanishiki", "Gohyakumangoku", and "Miyamanishiki", which are all home-polished rice varieties suitable for sake brewing, and "Tochigi no Hoshi", which is non-sake rice . It has won the gold medal at the National New Sake Competition for six consecutive years and is highly evaluated for its quality. In particular, Shikizakura Junmai Daiginjo Imai Shohei has an exceptional taste made possible by the motto left by the previous generation, "The taste of even a small cup of sake is infinite, if it contains the heart of the brewer.

純米酒
Shikisakura
Tochigi no Hoshi Junmaishu

"Enjoy pairing it with food"
Slightly dry Junmaishu brewed with "Tochigi no Hoshi", the same rice used in the "Daijo-sai" harvesting festival, to pair with food. Moist taste and mellow aroma.

Junmaishu	720ml
Alc.	15%
Rice type	Tochiginohoshi
Sake meter value	+2.0
Acidity	1.3

純米大吟醸酒
Shikisakura
Junmai Daiginjo Imai Shohei

"An exquisite taste, great when served hot"
This is a sake made with the passion passed onto the current generation from the brewery's last chief. A mellow, dry Junmai Daiginjoshu that blossoms with flavor and aroma when served lukewarm.

Junmai Daiginjoshu	500ml
Alc.	15%
Rice type	Gohyakumangoku
Sake meter value	+4.0
Acidity	1.3

| Water source | Nikko Mountains subsoil water |
| Water source | Soft |

Kobayashi Shuzo

743-1 Soshima, Oyama-shi, Tochigi 323-0061　TEL.0285-37-0005
h.kinsyo@tvoyama.ne.jp

We want to share the terroir of Tochigi with the world

The nationally popular brand "Ohobiden" has a splendid ginjo aroma and mellow taste.

Kobayashi Brewery, founded in 1872, is located in Oyama City in the southern part of Tochigi Prefecture, surrounded by rice paddies fed by subsoil water from the Nikko Mountains. The brewery's products are based on the theme of water from Nikko, a World Heritage Site. The brand "Ohobiden," which has gained nationwide popularity, is an elegant and fresh work of art that can only be expressed in the harsh but rich terroir of this region, where the climate, agriculture, and brewery are all united. The impact of the splendid ginjo aroma and mellow flavor will make sake fans swoon. Masaki Kobayashi, a fifth-generation sake brewer, hopes that the values of his hometown, Tochigi, will be shared throughout the world.

Ohobiden Akaban
Junmai Daiginjoshu

"A sake that conveys the thoughts of the brewer"

Yamadanishiki, the king of sake rice, is polished to 40% for this sake. It is a very popular Junmai Daiginjoshu in the higher class of "Ohobiden" brand

Junmai Daiginjoshu	1,800ml/720ml
Alc.	16%
Rice type	Yamadanishiki
Sake meter value	±0
Acidity	1.6-1.8

Ohobiden NIKKO

"Brewed with traditional yeast mash"

It is characterized by rich acidity, mellow taste, muscat-like ginjo aroma, silky touch and clean and refreshing aftertaste.

Junmai Ginjoshu	1,800ml/720ml
Alc.	16%
Rice type	Yumesarara
Sake meter value	±0
Acidity	1.6-1.8

Sagara Shuzo

Water source	Subsoil water from Nikko Mountains
Water source	Soft

3624 Shizuka, Iwafune-machi, Tochigi-shi, Tochigi 329-4307　TEL.0282-55-2013
E-mail: info@asahisakae.com　http://asahisakae.com

190 years of history with the motto "harmony makes good sake"

Brewing sake that bonds people.

In the northern part of the Kanto Plain, the Ashio Mountains and the Nikko Mountain Range can be seen to the north, and mountains such as Mikamo, Iwafune, and Ohira begin to rise. Sagara Shuzo is located in Iwafune-cho, Tochigi City, surrounded by nature. Since its establishment in 1831, the brewery has been dedicated to brewing sake using subterranean water from the Nikko Mountain Range and Tochigi-grown yeast and rice. The brewery is rooted in the local community. It uses Asahinoyume, a locally produced variety of non-sake rice, for its Junmaishu and Honjozo. Junmai Ginjoshu and Ginjoshu are made from sake rice produced in the prefecture and are good for matching with food. Typical dishes in this region are on the salty side, so the brewery makes sake that has a clear, delicate taste and a balance of moderate acidity and sharpness to match the salt and soy sauce seasoning.

特別純米酒
Asahisakae Tokubetsu junmai

"Dry tokubetsu junmai with sharpness"
Dry sake with a pleasant crispness. Serve it chilled, but if you prefer, you can drink it warm to accentuate its sharpness.

Special Junmaishu	1,800ml/720ml
Alc.	16%
Rice type	Asahinoyume
Sake meter value	+2.5
Acidity	1.8

純米吟醸酒
Asahisakae Junmai Ginjo

"Fruit-like, fresh aroma"
The taste is clear and crisp, and the finish is smooth and dignified. This Junmai Ginjoshu is best served chilled.

Junmai Ginjoshu	1,800ml/720ml
Alc.	16%
Rice type	Gohyakumangoku
Sake meter value	+2.5
Acidity	1.8

| Water source | Nakagawa River subterranean water |
| Water source | Medium-hard |

Shimazaki Shuzo

1-11-18, Chuo, Nasukarasuyama-shi, Tochigi 321-0621 TEL. 0287-83-1221
E-mail: uroko@azumarikishi.co.jp http://www.azumarikishi.co.jp/

One of the best in Japan
One of the best long-term matured sake in Japan

Unique local sake to accompany food culture of Nasu.

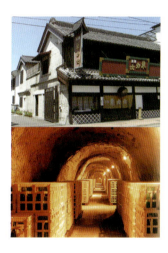

Shimazaki Shuzo was founded by Hikobee Shimazaki in 1849. The second-generation family head, Kumakichi, moved the brewery to its current location in Karasuyama, Nasu. Kumakichi was an unparalleled sumo enthusiast, and named his representative brand "Higashi Rikishi". The taste of this sake is rich and super sweet. Surrounded by the Yamizo Mountains, this region has long been home to many people engaged in agriculture and forestry, and people like to eat food with a strong flavor. In order to provide what the local people want, Shimazaki's sake has been made to match the food, making it unique local sake that is unlike any other in Tochigi Prefecture, or even in Japan. Another feature of the brewery is its long aging process in a huge cave storage facility. The brewery is a pioneer in the production of long-term matured sake, and its "Uroko" is one of the best in Japan.

Uroko Daiginjo
Hizo 10 years

"Daiginjo matured in a cave for 10 years"

It has a sweet, apricot seed-like aroma and a deep flavor. It can be served cold for a smooth taste or lukewarm for a softer taste.

Daiginjoshu	720ml/300ml
Alc.	17%
Rice type	Yamadanishiki
Sake meter value	+2.0
Acidity	1.2

Uroko
Yamahai Junmai Genshu

"Matured twice in the brewery and a cave"

A perfect balance of fragrant, aged aroma and smooth flavor that complements food. Serve it cold at 10-15 °C or medium warm.

Junmaishu	1,800ml/720ml
Alc.	17%
Rice type	Domestic rice
Sake meter value	+2.5
Acidity	1.9

Tentaka Shuzo

Water source	Nasu Mountains
Water source	Medium-hard

2166 Hiruhata, Otawara City, Tochigi Prefecture, 324-0411 TEL: 0287-98-2107
E-mail: tentaka@tentaka.co.jp https://www.tentaka.co.jp

Delicious, safe, and enjoyable

Sake brewing that is friendly to people, nature, and the future. The brewery strives to reduce the burden on the environment.

Tentaka Shuzo was founded in 1914 and is located in a rural area between two rivers on the southern edge of Nasu Plateau. The brewery has an affiliated company that specializes in organic farming, and its employees also grow the rice used as ingredients. The brewery is certified organic in the US and EU, as well as in Japan, and practices sustainable agriculture and environmentally friendly brewing. The company uses reusable bottles as much as possible, and the brewing process is based on hot water disinfection with minimal use of chemicals. Machines, large equipment, and buildings that cannot be disinfected with hot water are washed with water and then sanitized with alcohol spray. About 30% of the electricity used is generated by solar panels on the roof, and the hot water from steaming the sake is used to heat the store and office, reducing the environmental impact. The brewery also produces organic food products, such as organic Amasake drink and organic rice koji.

純米酒
Organic Junmaishu Tentaka

"Precious Vegan Society Certified Sake"
This dry junmaishu is made from only organic rice grown by contract farmers, and has a strong rice flavor and gentle acidity. It can be served cold or warm.

Junmaishu	1,800ml/720ml
Alc.	15%
Rice type	Organic Hyakumangoku/organic Asahinoyume
Sake meter value	+5.0
Acidity	1.9

??酒
Kyubi

"Mysterious sake with different ingredients each time"
Like the "nine-tailed fox" of Nasu, the rice, polishing ratio, yeast, and brewing method are different each time. Every bottle has a different taste.

??shu	1,800ml/720ml
Alc.	?%
Rice type	?
Sake meter value	?
Acidity	?

Water source	Shojinzawa spring
Water source	Soft

Tomikawa Shuzoten

998 Otsuki, Yaita-shi, Tochigi 329-1575 TEL.0287-48-1510
E-mail: chuai@peach.ocn.ne.jp http://www.kubun.jp

Brewing satisfying local sake with some of the most delicious water in Japan

"Local Sake is the Local Food Culture". Inheritance of local traditions and culture are deepply engraved in our heart.

The brewery was founded in 1913. At the time, the spirit of loyalty and patriotism was at its height, and the brewery named its sake "Chuai" (loyalty and deep love). Sake is made from water from the Shojinzawa Spring, a tributary of the Arakawa River. The water in the spring was chosen as one of the 100 best waters in Japan. The brewery produces sake with care in this rich natural environment. Its goal is to make the best use of the original flavor of the mellow rice, and to create satisfying local sake. The representative brands are "Chuai", or "Nakadori" sake sold since 2012, and "Tomikawa" which was rebranded in 2020. Chuai is characterized by a flavor that makes the most of the original flavor and aroma of the rice. "Tomikawa," which is made using traditional techniques, has a refreshing taste and goes well with all kinds of food.

Chuai Nakatori
Junmai Daiginjo Banshu Aiyama

"Quintessential taste of "Nakatori"

Sake brewed with Aiyama, a rare variety of rice in the Banshu area. Junmai Daiginjoshu with a splendid, juicy flavor.

Junmai Daiginjoshu	1,800ml
Alc.	17-18%
Rice type	Aiyama
Sake meter value	−3.0
Acidity	1.4

Tomikawa
Shiboritate Namagenshu

"Richness and umami that you can't get from Junmai"

This is our most popular unfiltered, unpasteurized sakes, made with the same methods used since the brewery's founding.

Namagenshu	1,800ml
Alc.	19-20%
Rice type	Asahinoyume
Sake meter value	+3.0
Acidity	1.1

Matsui Shuzoten

Water source	Private spring at the foot of Mt.Takahara
Water source	Ultra-soft

3683 Funao, Shioya-cho, Shioya-gun, Tochigi 329-2441 TEL.0287-47-0008
E-mail: info@matsunokotobuki.jp   http://www.matsunokotobuki.jp

The brewer himself produces sake as Toji

The name of the brewery derives from the strength of an old pine tree.

You can feel the brewer's love and passion in every drop of sake. It is said that the first brewer, Kuroji Matsui, moved from Niigata to this area where high-quality water is in abundance, and established the brewery during the Keio era. The ultra-soft spring water from the cedar forest behind the brewery is used for brewing. The brand name "Matsunokotobuki" (Longevity of Pine Trees) was named after an old pine tree, which is a symbol of happiness and strength. The brewing process at Matsui Shuzoten follows traditional methods, sparing no effort and putting love into every drop of sake. Nobutaka Matsui, the president of the brewery, serves as the toji (master brewer). On November 21, 2006, he was certified as the first Toji of Shimono, and continues to explore his own tastes.

Matsunokotobuki Honjozo Otoko no Yujo

"Feel the softness of the brewing water"

This sake was born out of an exchange between the brewer and Toru Funamura, a composer from the same hometown. It has a crisp acidity and a mouthfeel that goes down gently.

Honjozoshu	1,800ml
Alc.	15.4%
Rice type	Asahinoyume
Sake meter value	+6.0
Acidity	1.4

Matsu no Kotobuki Daiginjo

"A splendid aroma from Yamadanishiki rice"

A soft, fragrant and refreshing daiginjo made with Yamadanishiki, the best rice for sake brewing.

Daiginjoshu	1,800ml
Alc.	15.4%
Rice type	Yamadanishiki
Sake meter value	+4.5
Acidity	1.25

| Water source | Mt.Takahara, Shojinzawa Spring |
| Water source | Soft |

Morito Shuzo

645 Higashiizumi, Yaita-shi, Tochigi 329-2512 TEL.0287-43-0411

Doing business under the name of "Jyuichiya"
What is the motto hidden in the name?

Not too sweet, not too dry. Delicious sake you will never get tired of drinking.

The company was founded in 1874 by Kiyohei Morito. The brewery is located in a rich natural environment: in April, when the brewing season ends, frogs can be heard singing in chorus. Fireflies flutter about in early summer. Abundant underground water gushes forth. Mt. Takahara has been selected as one of the 100 best forests for "shinrinyoku" (forest bathing), and the water from the "Shojinzawa Spring" has been selected as one of the 100 best waters in Japan. In such an ideal natural environment, the sake "Jyuichi Masamune" is brewed using Shojinzawa spring water. As the name "Jyuichiya" suggests, the motto of the brewery is "quality-driven sake that is neither (-) sweet nor (+) spicy and has a delicious taste that will never leave you feeling tired of drinking.

"Tennen Ginka"
Jyuichi Masamune Sakura Genshu

"On the rocks or mix with hot water"

This sake is brewed with natural ginkga yeast cultivated from cherry blossoms, and has a mellow mouthfeel and a sharp taste.

Genshu	1,800ml/720ml
Alc.	18%
Rice type	Tochigishu14
Sake meter value	+3.0-+5.0
Acidity	1.0-1.3

Jyuichi Masamune
Shojinzawa Yusui Jikomi

"Using spring water from Shojinzawa, one of the 100 selected water of Japan"

This sake is with a large amount of malted rice, and brewed slowly at low temperatures in the coldest season. It is a traditional handmade gem with a rich taste and aroma.

Junmai Daiginjoshu	1,800ml/720ml
Alc.	15%
Rice type	Yamadanishiki/Hattannishiki
Sake meter value	+3.0
Acidity	1.5

Watanabe Shuzo

Water source	Naka River system
Water source	Soft

797-1 Susagi, Otawara-shi, Tochigi 324-0212 TEL.0287-57-0107
E-mail: kyokko@sake-kyokko.co.jp

As long as we call ourselves a local sake brewery, we continue to serve the local community

A brewery that produces casks for sake brewing in order to preserve the vanishing forestry culture.

Watanabe Brewery, founded in 1892, is located in Kuroha Town, sandwiched between the foot of Mt. Nasu and Mt. Yamizo, using the clear water flowing from the mountains as brewing water and relying on the cold wind blowing from the mountains to slowly brew sake. The lumber produced in the Yamizo Mountains, called Yamizosugi, is one of the best lumbers in Tochigi Prefecture. In recent years, however, the demand for lumber has diminished and the forestry industry has fallen into decline. Watanabe Shuzo is concerned about the loss of the forestry culture, so they make casks for sake brewing using Yamizo cedar, and trying to brew sake based on a traditional method using casks. The company's motto is "Local sake is sake that is loved by the local community." It strives to contribute to the local economy by cultivating sake rice and developing yeast.

純米酒
Kyokko Junmai Daiginjo Yumesasara

"Sake brewed from a new breed of rice "Yumesasara"
This Junmai Daiginjoshu is made with "Yumesasara", a new breed of rice developed in Tochigi Prefecture that is ideal for sake brewing. The yeast used is ND-3, and it was developed in-house.

Junmaishu	720ml
Alc.	16%
Rice type	Ginfu
Sake meter value	+3.0
Acidity	1.5

純米酒
Kyokko Karakuchi Junmai Ginjo Yamaoroshi Haishimoto

"Brewed in casks for sake"
Junmai Ginjoshu brewed using casks made from Yamizo cedar, a local specialty.

Junmaishu	720ml
Alc.	16%
Rice type	Ginfu
Sake meter value	+10.0
Acidity	1.5

Water source	Subsoil water from Mt.Asama range
Water source	Soft

Asama Shuzo

1392-10, Oaza-Naganohara, Naganohara-machi, Agatsuma-gun, Gunma 377-1304 TEL.0279-82-2045
E-mail: info@asama-sakagura.co.jp https://asama-sakagura.co.jp/

Brewed by a local master brewer Sake as a local specialty

Sake that has become a local specialty. The brewery also considers the preservation of the local environment.

Located in the highlands of Gunma Prefecture, Asama Shuzo , named after Mt. Asama, was founded in 1872, 150 years ago. In 2004, the brewery decided that it was necessary to train and establish local Asama Toji, instead of having a group of Echigo Toji brewers. In addition to sake production, the brewery also operates a tourism center, which offers products and services unique to the region to tourists visiting nearby sightseeing spots, such as Kusatsu Onsen and Karuizawa. As part of this effort, the brewery rents abandoned rice fields to grow its own rice to produce sake, a local specialty, while preserving the local environment. Currently, the company is working with farmers throughout Agatsuma County, and is expanding into a specialty product not only for Naganohara, but for the entire county.

純米大吟醸酒

Higen Junmai Daiginjo Rei

"The new flagship of the brewery"

Junmai Daiginjoshu with a delicate flavor brewed using "Kairyo Shinko", a breed of rice developed by the brewery and polished to 20%.

Junmai Daiginjoshu	720ml
Alc.	18%
Rice type	Kairyoushinkou
Sake meter value	private
Acidity	private

大吟醸酒

Higen Premium Daiginjo

"An aroma that resembles sweet pear"

Daiginjo with a splendid aroma and full flavor that is both mellow and crisp.

Daiginjoshu	720ml
Alc.	18%
Rice type	Yamadanishiki
Sake meter value	private
Acidity	private

Ohtone Shuzo

| Water source | Oze water system |
| Water source | Soft |

1306-2 Takadaira, Shirasawa-machi, Numata-shi, Gunma 378-0121 TEL.0278-53-2334
E-mail: info@sadaijin.co.jp   http://www.sadaijin.co.jp

A drop of the abundance of Oze

A gem of handmade local sake brewed with traditional methods and nurtured by nature.

Ohtone Shuzo is located at the foot of Mt. Oze, surrounded by the mountains of Okutone. Although the company was founded in 1902, the stone shrine dedicated to Matsuo-sama, the god of sake brewing, on the premises of the brewery bears the inscription "Genbun 4 nen" (1739), indicating that sake brewing began here as far back as the mid-Edo period. Traditional methods of brewing Oze's local sake have been handed down over generations and generations. The brewery contends that only local breweries can pass on the spirit of sake brewery entrusted by previous generations and rooted in the traditions and cultures of the region. Clear air. Green mountains. A pure stream. Sake is made from a single drop of the abundance of nature. With these words in mind, the brewery strives to make local sake that cannot be produced by big businesses.

純米酒
Sadaijin Junmaishu

"Warm it up and enjoy the pleasant aftertaste"

A rich, deep, full-bodied sake with a pleasant aroma and a moderate sweetness, made with the aim of producing sake that the brewer himself would want to drink.

Junmaishu	1,800ml/720ml
Alc.	15-16%
Rice type	Prefectural rice
Sake meter value	−1.0
Acidity	1.4

純米吟醸酒
Junmai Ginjo Hanaichimonme

"Delivering the spirit of sake on a flower petal"

This Junmai Ginjoshu has a mellow aroma and a clean taste that makes the most of the deliciousness of the rice. It goes well with sashimi and soy sauce-based dishes.

Junmai Ginjoshu	1,800ml/720ml
Alc.	14-15%
Rice type	Wakamizu
Sake meter value	−2.0
Acidity	1.4

Water source	Mt. Akagi
Water source	Slightly hard

Kondo Shuzo

1002 Omama-cho, Midori-shi, Gunma-ken 376-0101 TEL.0277-72-2221
https://akagisan.com/

Using subsoil water from Mt. Akagi
Dry taste only

A brewery that continues to make sake that is loved by the locals for its crisp, dry taste that never gets boring.

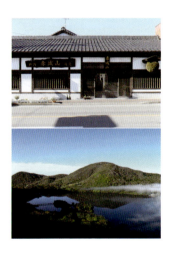

Founded in 1875 in the town of Omama, Gunma Prefecture, Kondo Shuzo celebrates its 147th anniversary in 2022. Initially, the sake was sold under the name "Akagi" because of the brewery's location at the foot of Mt. Akagi, and was later renamed "Akagi Masamune". Since its establishment, the brewery has focused on brewing dry sake. Since the third-generation chief was known for his love of the moon, the brewery changed its name to "Akagiyama" in reference to Kunisada Chuji, a gambler in the Edo period who was associated with Mt. Akagi. It was during the days of the 5th generation chief that "Otoko no Sake Akagyama Karakuchi" became a popular product. Its refreshing and crisp dry taste, which never gets boring, was appreciated by sake fans. The most important thing for Kondo Shuzo is teamwork, and everyone works together to make sake that is loved by the local community and is the best sake in the world.

大吟醸酒
Akagiyama Daiginjo

"The fruity and elegant aroma of ginjo"

This is the ultimate sake made entirely by hand by Iwate's Nanbu Toji. Made with Yamadanishiki rice polished to 35%.

Daiginjoshu	1,800ml
Alc.	17.5%
Rice type	Yamadanishiki
Sake meter value	+2.0
Acidity	1.3

純米大吟醸酒
**Akagiyama
Junmai Daiginjo Kurobi**

"The first centrifugally separated sake in Gunma"

This sake was made using a centrifugal separator. Brewed with Yamadanishiki rice from Hyogo Prefecture's A district polished to 35%.

Junmai Daiginjoshu	720ml
Alc.	16.5%
Rice type	Yamadanishiki
Sake meter value	±0
Acidity	1.2

Shimaoka Shuzo

Water source	Subsoil water of Mt.Akagi range
Water source	Hard

375-2, Yura-cho, Ota-shi, Gunma 373-0036　TEL.0276-31-2432

"Otokozake" nurtured by hard water

Sake awarded the highest "Yokozuna" prize in the honjozo category of Japanese sake.

Nittanosho-Takaraizumigo (now Ota City) is surrounded by the beautiful Jomo mountains and the clear streams of the Tone and Watarase rivers. This area was the center of ancient eastern culture and also the birthplace of the Nitta clan, loyal retainers of the Kemmu era. As the name "Takaraizumi" (treasure spring) suggests, the area has many historical sites and pure water resources. The well water used as brewing water at Shimaoka Shuzo is hard water with a high mineral content, and the brewery has been using the traditional method of Yamahai Shamo for many years, believing that it is the best way to make the most of the water's characteristics. Handmade koji and long-term, low-temperature fermentation using a small-scale brewing method produce sake with a smooth and delicious taste. In particular, the flagship brand "Gunma Izumi Yamahai Honjozo" was rated as a "Yokozuna" (grand Sumo champion) in the Honjozo category of the March 2010 issue of Shokuraku, a Japanese sake ranking magazine.

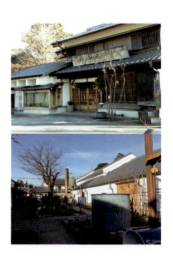

Gunmaizumi Yamahai Honjozo

"Natural, gentle, and healing sake"

Honjozo with full-bodied, matured taste. The strength unique to Yamahai can be enhanced when served hot.

Honjozoshu	1,800ml/720ml
Alc.	15-16%
Rice type	Wakamizu/Asahinoyume
Sake meter value	+3.0
Acidity	1.6

Gunmaizumi Yamahai Cho Tokusen Junmai

"An authentic, old-fashioned brew"

This sake is brewed using the traditional method of Yamahai. It has a refreshing acidity and a profound mature flavor.

Junmaishu	1,800ml
Alc.	15-16%
Rice type	Wakamizu
Sake meter value	+3.0
Acidity	1.7

| Water source | Subterranean water from Tanigawa Mountains and Mt.Hotaka range |
| Water source | Soft |

Nagai Honke

703 Shimohotchi-machi, Numata-shi, Gunma 378-0074 Japan TEL.0278-23-9118
https://www.nagaihonke.co.jp

With gratitude for the abundance of nature's that makes sake brewing possible

Our commitment to the famous sake "Tone Nishiki" - this is a delicious Gunma sake.

The Tambara Plateau is located in the mountainous northern part of Gunma Prefecture, surrounded by the Tanigawa mountain range and the Hotaka mountain range. Nagai Honke is a small sake brewery located at the entrance of the sacred Kashozan in the south of the plateau. The area is blessed with a natural environment of clear water and clean air, making it ideal for brewing. The brewery brews sake by hand, taking advantage of these favorable conditions. The water sources are Mt. Tanigawa and Mt. Hotaka, among others. The water that eventually flows into the Tone River is clean and abundant. Yamadanishiki, one of the highest-quality rice varieties is transformed into a fine, mellow sake with a great flavor. The clean air of Gunma, purified through the layers of snow clouds in the Joshinetsu mountains, nurtures the excellent microorganisms that are essential for sake brewing. This, combined with the skills of traditional brewers, results in the creation of delicious local sake from Gunma.

純米酒
Tonenishiki Junmaishu

"Special sake like a flower"
This hand-made sake is the work of a master brewer who has mastered his art craft. The flavor of Junmaishu blends into the refreshing taste.

Junmaishu	720ml
Alc.	15-16%
Rice type	Miyamanishiki/Wakamizu
Sake meter value	+3.0
Acidity	1.4

大吟醸酒
Tonenishiki Daiginjo

"Umami of Yamadanishiki, the king of sake rice"
This is an excellent product made with Yamadanishiki, the king of sake rice, and the famous water of the Tone River. It has a full-bodied taste that draws out the flavor of the rice.

Daiginjoshu	720ml
Alc.	17-18%
Rice type	Yamadanishiki
Sake meter value	+4.0
Acidity	1.3

Yanagisawa Shuzo

Water source	Subsoil water at the southern foot of Akagi
Water source	Soft

104-2 Fukatsu, Kasukawa-machi, Maebashi, Gunma 371-0215 TEL.027-285-2005
E-mail: yanagisawa@mbs.sphere.ne.jp

Traditional technique of mochi rice four-stage densho shikomi

Since its establishment in 1877, Yanagisawa Shuzo has been brewing one of the few "sweet" sake.

The brewery is located in Fukatsu, Kasukawa Town, at the southern foot of Mt. Akagi in Gunma Prefecture. Since the brewery's establishment in 1877, the wisdom passed down orally to the next generation is "sake brewing is all about preserving the original umami and sweetness of the rice", and Yanagisawa continues to focus on making one of the few sweet sake. In the past, the region was a place where sericulture and wheat farming flourished. Sweet sake was favored by people who worked hard under the blasts of the Akagi-oroshi wind to take away their fatigue. The brewery's signature sake, Katsuragawa Jyoshuichi, is made using the traditional mochi, four-stage brewing method, instead of the usual three-stage method.

Katsuragawa Jyoshuichi Honjozo

"Natural sweetness, without any added sugar"

This is sweet sake that has been loved by the local people since the brewery's establishment. It has been nurtured over the years by the climate and local diet.

Honjozoshu	1,800ml
Alc.	15-16%
Rice type	Domestic rice
Sake meter value	−7.0~−4.0
Acidity	1.4

Katsuragawa Tokusen Honjozo

"Fresh feel on your throat, good served cold"

This mellow, sweet sake is brewed using a traditional michi-rice brewing method. It has a distinctive, subtle aroma and a delicious rice flavor that spreads across the palate.

Honjozoshu	1,800ml
Alc.	15-16%
Rice type	Domestic rice
Sake meter value	−8.0~−5.0
Acidity	1.3

| Water source | Tone River system subsoil water |
| Water source | Soft |

Yamakawa Shuzo

185-3 Akaiwa, Chiyoda-machi, Ora-gun, Gunma 370-0503　TEL.0276-86-2182

Contemporary Yamahai technique with a unique twist

Brewed with underground water from the famous Tone River and the traditional Yamahai brewing method.

Yamakawa Shuzo, located in the "beak" at the eastern edge of "dancing crane-shaped" Gunma Prefecture, was founded in 1850 in Akaiwa, a post town that developed from early on. The first-generation of the brewery was Yagobee Yamakawa, who came to the village as a Niigata Toji. Since then, for six generations over 172 years, the brewery has produced sake using the traditional Yamahai-moto brewing method. Underground water from the Tone River, which is famous for its abundance of clean, clear water, has been used to produce a handcrafted product that is both delicious and soft. The brewing method used at this brewery is called "modern Yamahai", developed by the brewery based on the traditional Yamahai method. By keeping the temperature low and shortening the brewing period of the Yamahai moto, the brewery is able to ship the product after the summer, whereas the mainstream brewing methods do not bring flavor to its full potential until about a year after brewing.

本醸造酒
Koto Yamahaijikomi

"Delicious hot or cold"
A dry sake with the richness, fullness, and softness of modern Yamahai brewing that was developed based on traditional techniques.

Honjozoshu	1,800ml
Alc.	15-16%
Rice type	Domestic rice
Sake meter value	+3.0
Acidity	1.5

純米吟醸酒
Tonegawasodachi Junmai Ginjo

"Sake nurtured by local water"
This sake is handcrafted using subterranean water from the famous Tone River and rice suitable for sake brewing, and features a mild taste and aroma.

Junmai Ginjoshu	1,800ml
Alc.	15-16%
Rice type	Miyamanishiki
Sake meter value	+2.0
Acidity	1.5

Igarashi Shuzou

Water source	Chichibu mountain range subsoil water
Water source	Medium-hard

667-1 Kawadera, Hanno-shi, Saitama, 357-0044 TEL.042-973-7703
E-mail: iga_s@snw.co.jp https://www.snw.co.jp/~iga_s/

The taste and heart of Okumusashi
The sake brewery started by a Toji

The main brand name is "Tenranzan", a famous mountain in Hanno the emperor once climbed.

Hisazo, the first-generation head of the family who started the brewery, was born in Niigata Prefecture and worked as a toji (master brewer) at the Sawanoi Ozawa Shuzo in Oume. In 1989, he became an independent brewer and established his own business in Hanno. The main brand name is "Tenranzan", derived from the name of a famous mountain in Hanno that the Meiji Emperor climbed. The company is particular about every step of the production process, but its greatest care goes into washing the rice and absorbing water into the rice. If the rice is not washed well or the water is not absorbed well, the quality of the sake will be adversely affected. In particular, the temperature of the water is kept constant in a thermal tank with a cooling system, because the temperature of the well water is different every time it is pumped up. In addition, the nature of the rice and how well it has been polished are checked visually. Careful sake brewing is what makes the brewery's sake good.

純米吟醸酒
Tenranzan Junmai Ginjo

"The umami blossoms in your mouth"
Won the gold award in the premium heated-sake category of the national heated sake contest. We recommend drinking it cold or lukewarm.

Junmai Ginjoshu	720ml
Alc.	15%
Rice type	Ginginga/Miyamanishiki
Sake meter value	+2.0
Acidity	1.2

純米吟醸酒
Igarashi Junmai Ginjo

"Sake is only available at the brewery"
This sake is bottled immediately after pressing. Junmai Ginjoshu has a fresh, crisp, bubbly taste that only the direct pumping method can provide.

Junmai Ginjoshu	1,800ml
Alc.	16%
Rice type	Ginginga
Sake meter value	+2.0
Acidity	1.2

| Water source | Tone River subsoil water |
| Water source | Medium-hard |

Kamaya

1162 Kisai, Kazo-shi, Saitama 347-0105   TEL.0480-73-1234
E-mail: kamaya@rikishi.co.jp   https://www.rikishi.co.jp

Refining traditional techniques
Aiming for a new high by polishing traditional techniques

A sake brewery that has built its credibility by contributing to the local people of Kazo.

Kamaya Shinpachi, an Omi merchant, started a sake brewery in 1748 in Kazo, Saitama County, Musashi Province, where there was a post town on the Nakasendo Road. Shinpachi believed in the importance of "selling good quality products at low prices to gain the trust of customers", and this philosophy was passed down from generation to generation. The story goes that when the famine hit in 1886 and many people were displaced, the brewery built an opulent sake cellar that created a large number of jobs in the local community. "Kazo no Mai", which was launched in 2015, uses sake rice grown with care by farmers in Kazo, brewed by Kamaya, and sold at local sake stores. It is truly "made in Kazo", with the hope of revitalizing the region.

本醸造酒
Rikishi Kinsen Honjozo

"Sake for drinking every night"

Kamaya's standard taste. It is dry sake that you will never get tired of drinking. It can be served cold, at room temperature, or warmed.

Honjozoshu	1,800ml
Alc.	15.5%
Rice type	Miyamanishiki/Normal rice
Sake meter value	+3.0
Acidity	1.4

大吟醸酒
Rikishi Daiginjo

"Elegant taste. Serve cold"

Has a noble ginjo aroma and a crisp taste. The low-temperature aging process brings out the flavor of the rice. The taste is clean and free of any odd flavors. Ideal as an aperitif.

Daiginjoshu	1,800ml/720ml
Alc.	16.7%
Rice type	Yamadanishiki
Sake meter value	+3.8
Acidity	1.1

Koyama Honke Shuzo

Water source	Well water on the premises
Water source	Soft

1798, Oaza-Sashiogi, Nishi-ku, Saitama-shi, Saitama 331-0047　TEL.048-623-0013
https://www.koyamahonke.co.jp

Skills and the heart that has been handed down for 200 years

The brewery is eager to solve environmental issues to protect water, the key to good sake.

The brewery was founded in 1808 by Matabee Koyamaya, who was born in Hyogo Prefecture and learned the art of sake brewing in Nada and Fushimi. The brewery's stance of providing reasonably priced "good sake" that customers will never get tired of drinking, in order to make their day a "good time", is summed up in the catchphrase in the Koyama Honke brand logo: "Good day, good sake, good time". In recent years, Koyama Honke has been working on environmental issues, donating a portion of the proceeds from the sales of its main brand, Kinmon Sekai Taka, to the Saitama Prefecture NPO fund for the restoration of greenery and rivers. The brewery is well aware of the importance of water, which is the life blood of sake.

吟醸酒
Kinmon Sekaitaka Ginjo 50

"Sake born out of love for the local community"
This sake is characterized by its fruity taste and crisp aftertaste that can satisfy a wide range of people, from sake beginners to connoisseurs.

Ginjoshu	1,800ml/720ml
Alc.	15%
Rice type	Domestic rice
Sake meter valu	±0
Acidity	1.2

普通酒
Koyama Honke Kai

"A taste you can tell the difference after one sip"
A high-quality packed sake that provides a delicious, friendly, and comfortable feel as an evening drink.

Futsushu	2,000ml/900ml
Alc.	17%
Rice type	Domestic rice
Sake meter value	+1.0
Acidity	1.7

| Water source | Chichibu Mountains subsoil water |
| Water source | Hard |

Shinkame Shuzo

3-74 Magome, Hasuda-shi, Saitama 349-0114 TEL.048-768-0115
E-mail: shinkame1848@galexy.ocn.ne.jp https://www.shinkame.co.jp

Nothing to hide, nothing to add.

Belief that we will do our best to serve sake that you can drink anytime during your lifetime.

Shinkame Shuzo was founded in Hasuda City, Saitama Prefecture, in 1848, at the end of the Edo Period (1603-1868). At that time, the company's name was Iseya Honten. It is also called "Hasuda no Mori" by Shinkame lovers, referring to the area where the brewery is located. The name "Shinkame" comes from the "divine turtle" that used to live in Tenjin Pond behind the brewery. In 1987, the seventh-generation head of the brewery, Yoshiyuki Ogawara, converted all of its sake to junmai, becoming the first brewery in the post-war era to do so. Using only rice malt and water as ingredients, the brewery does not add anything or hide anything. It continues to make sake slowly and carefully with the help of craftsmanship, farmers, and time. The brewers strive to make the best sake in the world that is good, even when it's served hot.

純米酒
Shinkame Junmai

"The standard sake of Shinkame"

Junmaishu aged for two years or more. The harmony of plumpness, depth, richness, and sharpness that results from the aging process is a perfect match for food.

Junmaishu	1,800ml
Alc.	15-16%
Rice type	Domestic rice
Sake meter value	+6.0
Acidity	1.6

純米酒
Hikomago Junmai

"Junmaishu matured for three years, good served warm"

This sake is aged for three years or more. It is smooth on the palate, with a mild rice flavor and a sharp edge.

Junmaishu	1,800ml
Alc.	15-16%
Rice type	Yamadanishiki
Sake meter value	+6.0
Acidity	1.6

Bukou Shuzo

Water source	Underground water from Mt. Bukou
Water source	Medium-hard

21-27 Miyakawa-cho, Chichibu-shi, Saitama 368-0046 TEL.0494-22-0046
https://www.bukou.co.jp

A historical brewery that has grown together with Chichibu

Bukou Masamune, nurtured by tradition and the secret techniques of master brewers.

Buko Masamune is named after Chichibu's famous Mt. Buko. Yanagida Sohonten, the brewer of "Bukou Masamune", has witnessed the history of Chichibu since its establishment in 1753, the middle of the Edo period. The brewery is located in a place with a famous well system with excellent quality and quantity, as described in the Chichibu folk tale "Seven Wells". It was selected as one of the "100 Best Waters of the Heisei Era" and designated as "the subsoil water of Mt. Bukou". The well is open to the public during business hours, and you can take some home with you if you bring your own container. The store retains the appearance of one of the oldest buildings in the Chichibu Valley. In February 2004, it was designated as a nationally-registered tangible cultural property. Over the years, Bukou Masamune, sake brewed by traditional and secret techniques of a master brewer, has won numerous awards at product and appraisal competitions. Its quality is known as one of the best.

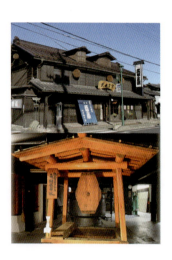

純米酒
Bukou Masamune Junmaishu

"Mellow sake with a wide range of flavors"

This traditional sake is made with only rice and rice malt, and has a crisp, clean finish. It is best served chilled or as warm as body temperature.

Junmaishu	1,800ml
Alc.	15-16%
Rice type	Miyamanishiki
Sake meter value	+2.0
Acidity	1.5

大吟醸酒
Bukou Masamune Daiginjo

"A fruity ginjo aroma"

This is a special limited edition of the brewery's signature work of art, the culmination of its brewing techniques. Serve it cold so that you can enjoy its aroma.

Daiginjoshu	720ml
Alc.	16%
Rice type	Yamadanishiki
Sake meter value	+3.0
Acidity	1.3

| Water source | Chichibu Mountains subsoil water |
| Water source | Slightly hard |

Kitanishi Shuzo

2-5-5 Kami-choi, Ageo-shi, Saitama 362-0037 TEL.048-771-0011
E-mail: furukawa@bunraku.net https://www.kitanishishuzo.co.jp

Science × Romance

A brewery that boasts a wide variety of sake brands to match various seasons and cuisines.

A sake brewery in Ageo City, Saitama Prefecture, founded by Kamekichi Kitanishi in 1894. Ageo is an old post town called Ageojuku along the Nakasendo, one of the five highways of Japan. During the Edo period (1603-1868), it was a busy transit point for feudal lords and the imperial family. It is also a historic sake brewing area known for high-quality subterranean water from Chichibu. Kitanishi Shuzo uses rich underground water for brewing, and while the brewing process is based on a scientific approach, the brewers also try to use their five senses. Although sake brewery has a long history and tradition, the brewers also aspire to try new things and be creative. In addition to the flagship "Bunraku" series -- which was named by the founder Kamekichi who was a great admirer of the performing art of Bunraku -- the brewery has produced a series of highly acclaimed sake products including the "AGEO" brand, which focuses on locally-produced ingredients, and the "Sara" brand, which features an unprecedented acidity.

純米吟醸酒

Bunraku Junmai Ginjo Ikkai Hiire

"A pleasant ginjo aroma that goes well with food"

This sake has a not-too-imposing, pleasant ginjo aroma and an exquisite balance of umami that spreads out in the mouth. Its clean, dry taste goes down the throat smoothly.

Junmai Ginjoshu	720ml
Alc.	15%
Rice type	private
Sake meter value	private
Acidity	private

純米吟醸酒

Junmai Ginjo Sara

"Wonderful, mysterious sake from the land of "Sainokuni""

This sake was designed over a period of five years. The harmony created by the elegant ginjo aroma, the sweetness of the rice, and the fine acidity is superb.

Junmai Ginjoshu	720ml
Alc.	16%
Rice type	private
Sake meter value	private
Acidity	private

Yao Honten

Water source	Arakawa River system
Water source	Soft

1432 Aza-kubonoiri, Bessho, Chichibu-shi, Saitama 368-0054　TEL.0494-23-8919
E-mail: honten@yao.co.jp　https://chichibunisiki.com/

Sophisticated sake from Chichibu

Inheriting tradition and living in the present in beautiful nature in Chichibu.

Yao Honten was founded in 1749 by Kihei Yao. A native of Omi Province, he leased a share in a sake brewery from the Matsumoto family, a local lord in Omiyago (present-day Chichibu City), and started the business under the name Masuya Rihei. The area is blessed with high quality water from the Arakawa River system, with its source on Mt. Kobushi, which straddles the three prefectures of Saitama, Yamanashi, and Nagano. The region where the brewery is located has clean air and a climate that varies greatly in temperature – the perfect condition for sake brewing. In Chichibu, where many Shinto rituals are held throughout the year, sake is an important part of the rituals. "Chichibu Nishiki", a local specialty sake born and beloved in Chichibu, is a gem nurtured in this deep mountain village over 270 years of brewery history.

大吟醸酒
Chichibunishiki Daiginjo Waza no Kiwami

"Sake made by Chichibu Toji with heart and soul"

Chichibu Toji put his whole heart and soul into this sake. It has a graceful and splendid aroma, a deep, mellow flavor, and a rich aftertaste. Produced in limited quantities.

Daiginjoshu	1,800ml/720ml
Alc.	17%
Rice type	Yamadanishiki
Sake meter value	+3.0
Acidity	1.3

特別純米酒
Chichibu Nishiki Tokubetsu Junmaishu

"The standard of Chichibunishiki"

The most popular line of Chichibu Nishiki's Junmaishu variations. The flavor is mellow and rich, but it feels good going down.

Special Junmaishu	1,800ml/720ml
Alc.	15-16%
Rice type	Miyamanishiki
Sake meter value	+2.0
Acidity	1.9

| Water source | Boso Hills subsoil water |
| Water source | Soft |

Inahana Shuzo

5841 Torami, Ichinomiya-machi, Chosei-gun, Chiba 299-4303   TEL.0475-42-3134
http://www.inahana-syuzou.com/

Sake brewing that has contributed to the local community since the Bunka-Bunsei Era

The brewery motto is "Be a local brewer for the local community" We want everyone to know how delicious Chiba's rice is.

Inahana Shuzo is said to have been brewing sake as early as the Bunka-Bunsei period of the Edo era (1603-1868), as historical documents show that the brewery submitted an application to the Tokugawa Shogunate for increasing the output of sake by 50 koku (1 koku = about 150 kg of rice). During this period, the wide swath of the Kujukurihama area saw record catches of fish. The Inahana family, who owned lands as well as fishing equipment in the region back then, is said to have made fertilizers from dried sardines and transported them to Edo. It is said that the family was also able to boost the production of sake during this era to celebrate the big catch. The family's motto is, "Be a local sake shop for the local community," and they place the highest priority on contributing to the local community. In recent years, the family has been working with local farmers to grow Yamadanishiki and other variations of rice suitable for sake brewing, and is taking on the challenge of making safe and reliable sake. The brewery will continue to brew good sake, hoping that more people will discover how delicious the rice grown in Chiba is.

大吟醸酒
Daiginjo
Kinryu Inahana Masamune

"Best when served cold"

Made from rice polished to 40% and Iwashimizu water, this is a gem that offers a great balance of ginjo aroma and mellow flavor.

Daiginjoshu	1,800ml/720ml
Alc.	16%
Rice type	Miyamanishiki
Sake meter value	+4.0
Acidity	1.3

純米酒
Junmai Kamoshizake

"Smooth as unpasteurized sake"

This Junmaishu stands out for its balance of gentle sweetness and acidity. The finish is crisp and clean, but you can feel the texture of sake in your mouth.

Junmaishu	1,800ml/720ml
Alc.	16%
Rice type	Miyamanishiki
Sake meter value	+4.0
Acidity	1.3

Iinuma Honke

Water source	Hokuso Plateau Groundwater
Water source	Medium-hard

106 Mabashi, Shisui-machi, Inba-gun, Chiba 285-0914　TEL.043-496-1111
E-mail: hokusou@iinumahonke.co.jp　https://iinumahonke.co.jp

The brewery that creates sake culture

Aiming to create a new value at the intersection of "people" and "sake".

The Iinuma family's ancestors first settled in Shisui about 400 years ago. They were originally engaged in farming and forestry, but in the Genroku era of Edo (1688-1703), they received permission from the shogunate to brew sake to be dedicated to shrines and temples. The town's name, Shisui, comes from "a sake well". When the sake brewery develops, Shisui develops, and when Shisui develops, the sake brewery develops, too. This mutual relationship is a priority for the brewery. In addition, Iinuma Honke is striving to engage in all processes of sake brewing, from the production of sake rice to the brewing and sales of sake. The brewery also operates a café and gallery, aiming to create a new experience for its customers by combining tourism and sake. The brewery is currently exporting its products to about 10 countries around the world. Its popular products include Junmai Ginjoshu, the Migaki Hachiwari brand, and Yuzu liqueur.

純米吟醸酒
Kinoene Junmai Ginjo

"A fresh taste like unpasteurized sake"

It is characterized by a fruity ginjo aroma, refreshing acidity, and a lingering sweetness. You can also enjoy the fizzy sensation from fermentation.

Junmai Ginjoshu	1,800ml/720ml
Alc.	15-16%
Rice type	Yamadanishiki/Gohyakumangoku
Sake meter value	±0
Acidity	1.6

純米吟醸酒
Junmai Ginjo Nama-shu Kinoene Apple

"Cute bottle. Easy to drink. Legit taste"

This sake has an elegant aroma of malic acid and a refreshing aftertaste, with a harmony of sweetness and acidity. Good served cold.

Junmai Ginjoshu	1,800ml/720ml
Alc.	15-16%
Rice type	Yamadanishiki/Gohyakumangoku
Sake meter value	−18.0
Acidity	2.7

| Water source | Iwashimizu water |
| Water source | Soft |

Kameda Shuzo

329 Naka, Kamogawa-shi, Chiba, 296-0111 TEL.04-7097-1116
E-mail: kameda@jumangame.com http://jumangame.com/

The roots of the brewery are in sake devoted to the gods

255 years of traditional sake brewing. Sake devoted to the harvest festival at Meiji Shrine.

Kameda Shuzo started brewing sake 30 years ago, using a cold storage warehouse at the southernmost tip of Chiba Prefecture. Its flagship brands include "Jumangame," one of the most famous sake brands in Boso. The brewery also produces and sells authentic shochu (distilled spirits), liqueurs, and other alcoholic beverages made from local specialty crops. The brewing of sake in this region began in the Horeki era, when mountain priests made white sake and offered it to the gods. Since then, traditional sake brewing has continued for 255 years and has been handed down to the present day. In 1871, when the Meiji Emperor ascended to the throne, a rice field near the brewery was selected as the site for the first rice festival of the Meiji era (1871). Kameda Shuzo, as a member of the local religious community, is in charge of brewing white sake from rice grown in the Saita rice field, which is dedicated as sacred wine at the Meiji Shrine's Niiname Harvest Festival every year.

Cho Tokusen Daiginjo Jumangame

"An elegant aroma and rich umami"
Made with pure water from the Mineoka mountain system and Yamadanishiki rice from Yoshikawa-cho, Miki City, Hyogo Prefecture. This is the ultimate daiginjo, polished to an extreme 35%.

Daiginjoshu	720ml
Alc.	17%
Rice type	Yamadanishiki
Sake meter value	+3.0
Acidity	1.3

Junmai Daiginjo Jumangame Aiyama

"Carefully fermented for a long time at low-temperatures"
Aiyama sake rice is said to be difficult to cultivate. For this reason, sake made from this rice is a rare find.

Daiginjoshu	1,800ml/720ml
Alc.	16%
Rice type	Aiyama
Sake meter value	+1.0
Acidity	1.4

Kidoizumi Shuzo

| Water source | Well water |
| Water source | Medium-hard |

7635-1 Ohara, Isumi-shi, Chiba 298-0004 TEL.0470-62-0013
E-mail: s1879@kidoizumi.jp   http://www.kidoizumi.jp

Brewing sacred sake for traditional rituals to be passed down to future generations

Proactive participation in local events is a source of pride for the local sake brewery.

The brewery was founded in 1879. In addition to sake brewing, fishing was one of the company's business activities until just after the war. As such, sake at Kidoizumi Shuzo is a perfect match for local seafood. The brewery introduced its own high-temperature Yamahai brewing method in 1956, and continues to brew natural sake that enriches people's daily lives. The brewery, known for its matured sake, has some aged over 50 years in its collection. The connection with the local community can be felt most when its sake is dedicated at the Ohara Hadaka Festival, held every year during the "Higan" period in September.

純米酒
Kidoizumi Junmai Daigo

"Richness and sharpness created by high-temperature Yamahai brewing method"
A strong flavor from Yamadanishiki and a crisp acidity from lactic acid bacteria. It tastes great lukewarm!

Junmaishu	1,800ml
Alc.	16%
Rice type	Yamadanishiki
Sake meter value	private
Acidity	private

純米酒
Junmai Afusu Nama

"White wine-like Junmaishu"
This is Kidoizumi's signature sake with sweetness and acidity, like white wine. It is made through the so-called Ichi-dan-jikomi (one-step) method.

Junmaishu	1,800ml
Alc.	13%
Rice type	Fusanomai
Sake meter value	private
Acidity	private

| Water source | Well water |
| Water source | Medium-hard |

Kubota Shuzo

685 Yamazaki, Noda-shi, Chiba 278-0022 TEL.04-7125-3331
E-mail: kubotasb@cello.ocn.ne.jp

Signature local sake in Noda made with high-quality water

A brewery that used to be a large wholesale store in a region with delicious water.

Kubota Shuzo, located in the northernmost part of Chiba Prefecture, was founded in 1872. Sokichi, the first-generation head of the brewery, chose this area because of the availability of good water -- the most important factor in sake brewing -- and the convenience of water transportation by canal. He not only sold his own sake, mirin, honnaoshi, and shochu to the local community, but also served as a major wholesaler, shipping them to Shinkawa, Tokyo, and bringing rice and Nada sake on the return boat to wholesalers and retailers. Today, the brewery's sake attracts fans for its use of carefully-selected sake rice from Chiba Prefecture and other regions. Its traditional method of sake brewing is also well-received by customers. Its main brand is "Katsushika". The brewery produces a variety of local sake representative of Noda, such as daiginjo made with Yamadanishiki, Junmaishu with a rich taste of rice, and dry honjozo undiluted sake.

大吟醸酒
Katsushika Daiginjoshu

"Smooth and elegant taste"
Daiginjoshu made from Yamadanishiki rice polished to 40% and fermented slowly at low temperatures.

Daiginjoshu	1,800ml
Alc.	16%
Rice type	Yamadanishiki
Sake meter value	+1.0
Acidity	1.7

純米吟醸酒
Katsushika Junmai Ginjo Hosen

"Delicious served cold"
Made from Yamadanishiki polished to 55%. It is rich and sweet, with the fullness and flavor that only Junmai Ginjoshu can provide.

Junmai Ginjoshu	1,800ml/720ml
Alc.	15%
Rice type	Yamadanishiki
Sake meter value	−3.0
Acidity	1.8

Nabedana

338 Honcho, Narita-shi, Chiba 286-0026 TEL.0476-22-1455
https://www.nabedana.co.jp

Water source	Tone River system
Water source	Hard

Sake between hearts

A brewery located at the gate of Naritasan Shinshoji temple, with 330 years of tradition.

A sake brewery with an eccentric name, Nabedana, stands in front of the gate of Naritasan Shinshoji temple. The brewery was founded in 1689, when the Sakura clan gave the brewery 1,050 shares of sake production (equivalent to today's manufacturing license) and started brewing right next to Naritasan. At first, the brewery was called Nabeya, and then the name was changed to Nabedana by adding the word "store". This is the origin of the name. After the Meiji era, three breweries were built. During the World War II, however, all of them merged together in Kozakimachi, known for its fermentation culture. This is where the brewery currently operates its business. The brewery, which has a history of 300 years, has maintained its traditional methods while actively introducing new technologies and exporting its products to countries around the world. Nabedana sake has won many awards in various contests. The most popular brands are "Jinyu", named after the family motto, "A man should live with virtue and courage", and "Fudo", named in honor of Narita Fudo.

純米大吟醸酒
Fudo Junmai Daiginjo

"High-quality sake that you can keep drinking"

This Junmai Daiginjoshu is made using "Sakekomachi", a variety of rice suitable for sake brewing, produced in Akita Prefecture. Pasteurized only once, it has a mellow taste.

Junmai Daiginjoshu	720ml
Alc.	15.8%
Rice type	Sakekomachi
Sake meter value	+1.0
Acidity	1.3

純米吟醸酒
Fudo Junmai Ginjo

"Junmai Ginjoshu with an emphasis on sharpness"

This Junmai Ginjoshu is made with "Sakekomachi," rice suitable for sake brewing produced in Akita Prefecture and polished to 55%. It is lighter than Junmai Daiginjoshu and has sharpness.

Junmai Ginjoshu	720ml
Alc.	15.8%
Rice type	Sakekomachi
Sake meter value	+3.0
Acidity	1.4

| Water source | Ground water in Sawara diluvial uplands |
| Water source | Medium-hard |

Baba Honten Shuzo

614-1, Sawarai, Katori-shi, Chiba 287-0003 TEL.0478-52-2227
E-mail: babahonten@jasmine.ocn.ne.jp

Embraced by the Tone River
A brewery in Sawara, a "little Edo" embraced by the Tone River

A brewery that has maintained traditional methods for more than 300 years, part of the history of Sawara.

The Tonegawa River, built up around the beginning of the Edo shogunate, was a key transportation route for goods, and the prosperous water town of Sawara was known for its rice production. At the same time, sake brewing was so popular that it was called the "Nada of Kanto". Baba Zenbee, the founder of the company, established his business in this area during the Tenna era (1681~1683). Baba first opened as a koji store, but the fifth-generation family head started brewing sake. Since then, the family has continued the production of sake and mirin using traditional production methods. The name of the main brand "Kozen" comes from the business name, "Kojiya Zenbee". The taste of mirin seasoning affects the taste of food, so the family has preserved its original mirin flavor. They use the finest Yamadanishiki rice, domestic glutinous rice, and other ingredients, as well as groundwater from the Sawara Nambu diluvial uplands, striving to make safe and reliable sake and mirin.

大吟醸酒
Daiginjo Kaishu Sanjin

"Sake named after Katsu Kaishu's visit to the brewery"
This Daiginjo is brewed with the highest-quality Yamadanishiki polished to perfection. It has a fruity aroma and a refreshing yet deep flavor.

Daiginjoshu	1,800ml/720ml
Alc.	15-16%
Rice type	Yamadanishiki

本醸造酒
Honjozo Kozen

"The oldest brand at Baba Honten"
This sake is has a fresh taste and is slightly dry, but it also has a delicious sweetness and a clean finish. It can be served cold or warm.

Honjozoshu	1,800ml/720ml
Alc.	15-16%
Rice type	Miyamanishiki/Mitsuhikari

Miyazaki Honke

Water source	Well water
Water source	Hard

1699 Node, Sosa-shi, Chiba 289-3181　TEL.0479-67-2005
E-mail: info@miyazaki-honke.com　http://www.miyazaki-honke.com

Sake is your unfailing, unparalleled friend

Sake brewed slowly with the best water and the rich blessings of nature.

This brewery was established in the Keio period (1865-1868) and benefits from the rich nature of the Kujukuri Plain. The name of the main brand, "Fujinotomo," is said to be a combination of "unfailing friend," "unparalleled confidant," and the sacred Mt. Fuji. The historical city of Sosa City (formerly Noei-machi, Sosa-gun), where the brewery is located, is in the northeastern part of Chiba Prefecture, a 30-minute drive from Narita Airport. The city is surrounded by the rich nature in the southern part of the city, which overlooks the vast Kujukurihama beach, and the northern part, which was selected as one of the 100 Satoyama" of Japan. Fujinotomo is sake brewed with the blessings of the best water and the rich nature of the city over a long period of time. The brewery uses its own well water for brewing and growing low-pesticide rice to make sake.

純米酒
Fujinotomo Junmaishu

"Rich taste with umami of rice"
Dry, crisp sake with a full-bodied flavor and a rich aroma. Delicious at any temperature – serve i hot, lukewarm, room temperature, or cold.

Junmaishu	1,800ml/720ml
Alc.	14%
Rice type	Low-pesticide rice cultivated by the brewery
Sake meter value	+3.0
Acidity	1.5

純米吟醸酒
Fujinotomo Junmai Ginjo

"Light aroma and throat feel"
A refreshing sake with a light aroma and a hint of umami. It is best served cold or lukewarm to enjoy the aroma.

Junmai Ginjoshu	720ml
Alc.	15%
Rice type	Yamadanishiki
Sake meter value	+4.0
Acidity	1.6

| Water source | Well water (150m underground) |
| Water source | Medium-hard |

Ishikawa Shuzo

1 Kumagawa, Fussa-shi, Tokyo 197-8623 TEL.042-553-0100
E-mail: kura@tamajiman.co.jp https://www.tamajiman.co.jp/

Sake made with the air and water of Tama

Pride of the Tama region. "Tamajiman" is traditional, made-in-the-winter sake.

Fussa City, where Ishikawa Shuzo is located, is in the western part of the Tama region of Tokyo where the Tama River flows and lots of greenery and famous sightseeing spots are. The U.S. Yokota Air Base gives the city an exotic feel. The brewery was established in 1863, renting a warehouse from Morita Shuzo in Ogawa Village (present-day Akiruno City) on the other side of Tama River. As a sister brand to Morita Shuzo's "Yaegiku", the founding brand of Ishikawa Shuzo was named "Yaezakura". The current main brand "Tamajiman" started in 1933. The brewery also produced beer during the Meiji era, resuming production in 1998 and naming it "Tama-nomegumi". The brewery itself has a long history, but the Ishikawa family's sacred zelkova tree on the premises is 700 years old. It has been designated as a natural monument by the city, and is known as the "Zelkova of the Ishikawa Family.

純米酒
Tamajiman Junmaishu

"Great flavor when served hot"

Junmaishu that brings out the best of the flavor and sweetness of the rice. It has a harmonious acidity that makes it a perfect match for miso and soy sauce.

Junmaishu	1,800ml/720ml
Alc.	14%
Rice type	Sakekomachi
Sake meter value	−10.0
Acidity	1.8

大吟醸酒
Tamajiman Daiginjo

"Daiginjoshu using only Yamadanishiki rice"

This is a Daiginjoshu made only from Yamadanishiki rice, and has a splendid ginjo aroma. The taste is clean and crisp, with a smooth finish.

Daiginjoshu	1,800ml/720ml
Alc.	15%
Rice type	Yamadanishiki
Sake meter value	−2.0
Acidity	1.2

Ozawa Shuzo

Water source	Own well
Water source	Soft

2-770 Sawai, Ome-shi, Tokyo 198-0172 TEL.0428-78-8215
E-mail: tokyo@sawanoi-sake.com http://www.sawanoi-sake.com

The Great Nature of Okutama Produces Abundant Sake

Using Tokyo's remaining natural environment to brew the finest local sake that suits the tastes of the Kanto region.

Ozawa Shuzo, founded in 1702, is located in Okutama, a deep mountain gorge full of nature, even though it is located in Tokyo. "Sawanoi", named after the famous water village of Sawai, has been a popular local sake brand in Okutama for 300 years. Many people visit the area around the brewery, where you can see the brewing water gushing from the side wells, eat tofu and yuba made with the brewing water, enjoy the natural garden, and visit the "Kushi-Kanzashi Museum" (comb museum) where you can see artifacts made with the best techniques from the Edo period. When the mountains and their lush greenery, the clear air of Okutama, the selection of domestic rice, natural brewing water, traditional techniques refined, careful work, and sincerity are combined, the result is sake that will continue to be loved for years to come.

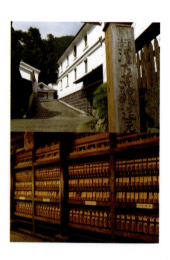

純米酒
Sawanoi
Junmai Ookarakuchi

"Fresh, authentic dry sake"

This is real dry sake that suits the food culture of the Kanto region, where soy sauce is often used. Its clean and crisp taste complements the richness of the rice.

Junmaishu	1,800ml
Alc.	15-16%
Rice type	Akebono
Sake meter value	+10.0
Acidity	1.7-1.9

特別純米酒
Sawanoi Tokubetsu Junmai

"Rich, moist taste"

This is a special Junmaishu brewed with rice polished to 60%. Elegant, slightly sweet, and full flavored.

Special Junmaishu	1,800ml
Alc.	15-16%
Rice type	Gohyakumangoku
Sake meter value	+1.0
Acidity	1.5-1.7

| Water source | Mt.Tanzawa range subsoil water |
| Water source | Medium-hard |

Kumazawa Shuzo

7-10-7 Kagawa, Chigasaki-shi, Kanagawa 253-0082 TEL.0467-52-6118
E-mail: kura@kumazawa.jp https://www.kumazawa.jp/

Brewing Shonan's climate into sake

Special dishes for special sake. A brewery that also runs Italian and Japanese restaurants.

Since its establishment in 1872, Kumazawa Shuzo has been brewing high-quality, handcrafted sake in small batches, and is the only brewery left in Shonan. The significance of the brewery's existence lies not only in its sake production, which is a source of pride for the local community, but also in its role in the local culture where people gather, drink sake together, and create something new. The brand name "Tensei" is taken from the Chinese expression that goes: "The color of blue that appears in between clouds when the rain stops and the sun starts to shine". The brewery aims to create a fresh and moist taste that is expressed in the Chinese sentence. The taste of the sake brewed with subsoil water from the Tanzawa mountain range is a gem of the Shonan climate. In the local Shonan area, the company operates four restaurants, including the MOKICHI TRATTORIA, where customers can enjoy their favorite Italian dishes and sake.

Senpou Tensei Junmai Ginjo

"Goes well with a wide range of dishes"

This Junmai Ginjoshu has a mild ginjo aroma, depth of flavor, and fine sweetness, which are characteristic of Yamadanishiki rice.

Junmai Ginjoshu	1,800ml/720ml
Alc.	16%
Rice type	Yamadanishiki
Sake meter value	+2.5-+3.5
Acidity	1.3-1.5

Nakazawa Shuzo

Water source	Subsoil water from Mt. Tanzawa range
Water source	Slightly soft

1875 Matsudasoryo, Matsuda-machi, Ashigarakami-gun, Kanagawa 258-0003 TEL0465-82-0024
E-mail: info@matsumidori.jp https://www.matsumidori.jp/

A young brewer's challenge to create a new kind of sake

"I want to brew unique sake." The 11th-generation head takes on the challenge of brewing a new kind of sake.

Nakazawa Shuzo was founded in 1825. At the time, the family was a merchant for the Odawara clan, supplying sake to Odawara Castle. The clan lord named the brewery's sake "Matsumidori", referring to the scenic beauty of the area around Matsuda where the brewery is located. The head of the family has been in charge of the brewery for generations. The brewer spends a long time brewing sake using subterranean water from the Tanzawa Mountains. Akira Kagewada, the 11th-generation head sake brewer, received training at Ichinokura, a long-established sake brewery in Miyagi Prefecture, and returned to his family's brewery in 2011. "Ryo", brewed with yeast that he discovered and isolated from the Kawazu cherry trees that bloom in the mountains behind the brewery, has become a specialty of Matsuda Town. Now, the whole community celebrates the blooming of the Kawazu cherry trees. The brewery has also released "Matsumidori Junmai Ginjo S.tokyo", which is made with the rare yeast discovered in 1909, to promote the appeal of Kanagawa's local sake both in Japan and abroad.

純米酒
Matsumidori Junmaishu

"Sake with both umami and sharpness"

This sake complements food, with a mild aroma and a delicious, sweet taste from the rice. The richness comes out even more when it's served hot.

Junmaishu	1,800ml/720ml
Alc.	15-16%
Rice type	Miyamanishiki
Sake meter value	+2.0-+3.0
Acidity	1.3-1.4

純米吟醸酒
Matsumidori Junmai Ginjo S.tokyo

"Sake brewed with yeast that is 111 years old"

This is a rich, smooth sake with a fresh, vinous aroma that spreads on the palate. The yeast used for this sake gives it great acidity.

Junmai Ginjoshu	720ml
Alc.	14-15%
Rice type	Domestic rice
Sake meter value	private
Acidity	private

Guide to Sake Breweries and Famous Sake

Koshinetsu Region

Yamanashi, Nagano, Niigata

Food Culture in the Koshinetsu Region

Yamanashi: Although there is no sea, the area is blessed with good-quality spring water flowing from the mountains, "wasabi zuke" (pickled wasabi) made from wasabi grown in the clear mountain streams is a popular dish in the prefecture. Many dishes are made with wheat flour, such as "azuki hoto" made with hoto noodles instead of rice cakes, as well as udon noodles. Grapes, peaches, and small plums are also grown in the area, and wine is made with the high-quality spring water.

Nagano: Much of the prefecture is covered with mountains, where lettuce and Chinese cabbage are grown in the highlands and mushrooms in the mountains. "Shinshu soba noodles" (made from wheat and buckwheat that grows well the cool climate) and "oyaki" buns (stir-fried vegetables seasoned with miso or soy sauce and wrapped in flour dough) are some of the prefecture's signature dishes. Preserved foods include "Shinshu miso" and "nozawana-zuke" pickles, which are made from rice malt and soybeans, and "sunkizuke", or turnip leaves and stems pickled using lactic acid fermentation.

Niigata: This is the region with the most snowfall in Japan, but the water from the melting snow moistens the rice fields that spread from the rivers to the plains, and the production of rice and sake flourishes. There are also many agricultural products such as edamame (green soybeans), sasa-zushi (sushi wrapped in bamboo leaves, sasa-dango (bamboo rice cake), and seafood from the Sea of Japan.

Ide Jyozoten

Water source	Subsoil water at the foot of Mt. Fuji
Water quality	Soft

8 Funatsu, Fujikawaguchiko Town, Minamitsuru-gun, Yamanashi 401-0301 TEL.0555-72-0006
E-mail: info@kainokaiun.jp https://www.kainokaiun.jp

The Only Brewery around the Fuji Five Lakes

The current owner is the 21st generation. Fuji's spring water is an important treasure.

Ide Jyozoten, located near the northern slope of Mt. Fuji and on the southern shore of Lake Kawaguchi, is the only sake brewery in the Fuji Five Lakes region that strives to produce sake loved by consumers. The forerunner of the brewery was soy sauce brewing, which started in the mid-Edo period (around 1700). At the end of the Edo period (around 1850), Yogoemon Ide, the 16th generation of the family, decided to make use of the cool climate of the northern foothills of Mt. Fuji at an altitude of 850 meters and the pure water that gushes forth in abundance: he started brewing sake in addition to traditional soy sauce. The main brand is "Kai no Kaiun", which is loved by the local people. The brewery also offers sake brewery tours by reservation, and conveys the appeal of sake to customers in Japan and abroad. (English tours are available).

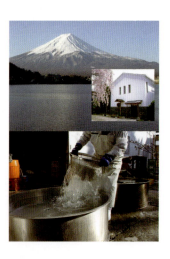

純米酒
Kai no Kaiun Junmaishu

"Umami produced by slow maturation-process"
The soft, mellow aroma brings out umami unique to junmai. It has a fine, smooth taste.

Junmaishu	1,800ml
Alc.	15-16%
Rice type	Yumesansui/Asahinoyume
Sake meter value	+3.0
Acidity	1.3

大吟醸酒
Kai no Kaiun Daiginjo

"A balanced, mild fruity flavor"
The acidity is low, and the taste is clean yet expansive.

Daiginjoshu	720ml
Alc.	15-16%
Rice type	Yamadanishiki
Sake meter value	+3.0
Acidity	1.0

Water source	Yatsugatake subsoil water
Water quality	Medium-hard

Takenoi Shuzo

1450 Minowa, Takane-cho, Hokuto-shi Yamanashi 408-0012 TEL.0551-47-2277
E-mail: sake-takenoi@nifty.com http://takenoishuzo.jp

The power of flower yeast originating at Tokyo University of Agriculture

Sake and shochu brewed with the blessings of nature, subterranean water from the Mt. Yatsugatake, and flower yeast.

The Takenoi Shuzo, which prides itself on sake brewed with subterranean water from Mt. Yatsugatake, is one of the leading sake breweries. It produces not only sake but also junmai shochu. It was founded in 1865, and the name "Takenoi", which is both the name of the company and the name of the sake, was chosen by combining "Take" of the founder (Takezaemon Shimizu) and the "I" meaning a well with good water. The sake produced in an environment blessed with nature has a rich taste that even impresses connoisseurs. It has been loved by the local community for centuries, and is valued at festivals and other events. It is also served as sacred sake at nearby shrines. The sake brewed by the managing brothers, who are graduates of Tokyo University of Agriculture, is worth trying. The older brother Koichiro, who is the Toji for the new brand, brews "Seiko," a junmai brand made with "Tsurubara yeast." The younger brother, Daisuke, who is the Toji for shochu, brews the "Blessings of the Sun and Earth" shochu series.

Seiko
Junmai Ginjo Omachi

"The best rice and yeast selected by Toji"

Sake made with 'Tsurubara yeast". Each class uses a different type of rice suitable for sake brewing from across Japan to produce a variety of flavors.

Junmai Ginjoshu	1,800ml/720ml
Alc.	15%
Rice type	Omachi
Sake meter value	+3.0
Acidity	1.7

Takenoi Junmai
Hitogokochi Certified

" "GI Yamanishi" sake"

This sake is made from only "Hitogokochi" rice grown in collaboration with local farmers in Hokuto City. Delicious in a wide range of temperatures.

Junmaishu	1,800ml/720ml/300ml
Alc.	15%
Rice type	Hitogokochi
Sake meter value	+2.0
Acidity	1.5

Tanizakura Shuzo

| Water source | Subsoil water at the southern foot of Mt. Yatsugatake |
| Water quality | Slightly soft |

2037 Yato, Oizumi-cho, Hokuto City, Yamanashi 409-1502 TEL.0551-38-2008
https://www.tanizakura.co.jp

A small sake brewery that started as a brewery offering sake to the gods

Tanizakura, sake with a genuine taste nurtured by the nature of Mt. Yatsugatake.

Tanizakura Shuzo was founded in 1848. At the time, a large amount of old coins were unearthed from the brewery's premises, which led to the business name "Kosenya" (meaning "old coin store"), and the brewery started out as a small sake brewery making sake to offer the gods. Since then, Yatsugatake Oizumi's magnificent nature and the blessings of its pure spring water have nurtured the spirit of the brewery's sake. It has since been committed to authenticity in pursuit of a taste favored by everyone in any era. The brewery is now run by the fourth generation. The brand name is "Tanizakura". The brewery uses high quality Yamadanishiki, Miyamanishiki, and other types of organic rice. The koji is carefully produced by the brewers, and subsoil water from the famous Yatsugatake South Foot Spring Group, which gushes into deep wells, plays a major role in the brewing. The brewers traditionally eat and sleep together during the brewing period from the end of October to April of the following year, and work as a team to make sake.

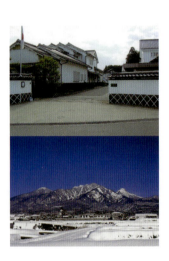

Namahai Zukuri
Honjozo Sakuramori

"A sake that can be enjoyed warm"

This sake has a mature aroma and a thick, strong body. It can be enjoyed warm, and goes well with hot pot dishes.

Honjozoshu	720ml
Alc.	14-15%
Rice type	Domestic rice
Sake meter value	+6.0
Acidity	1.9

Junmai Daiginjo Namashu
Kosenya no Sake

"Sake that bears the business name of the brewery"

This is a sake with a rich ginjo flavor, brewed in the tradition of Tanizakura. It has the roughness of Mt. Yatsugatake and a strong, sharp flavor.

Junmai Daiginjoshu	720ml
Alc.	17%
Rice type	Yamadanishiki
Sake meter value	+4.5
Acidity	1.6

| Water source | Well water |
| Water quality | Soft |

Yamanashi Meijo

2283 Taigahara, Hakushu-cho, Hokuto-shi, Yamanashi 408-0312 TEL.0551-35-2236
E-mail: info@sake-shichiken.co.jp https://www.sake-shichiken.co.jp/

Brewing with Famous Water for 300 years

The water used for brewing was selected as one of the 100 best waters in Japan. New ideas using "food".

In 1750, Mitsuyoshi Ihee Kitahara, the seventh-generation head of the Kitahara family who had been operating a sake brewery in Nagano Prefecture for generations, fell in love with the water of Hakushu and opened a branch of the family. Since then, the brewery has been producing sake using subsoil water from Mt. Kaikomagatake. The Ojira River flowing near the brewery has been selected as one of the 100 best waters in Japan, and the famous sake "Shichiken" has a 300-year history. What does it mean for employees of the company themselves to engage in brewing sake? While pursuing the answer to that question, the company is also proposing a new type of food culture. In recent years, the brewery has been developing products using amazake (sweet sake) and koji (fermented rice malt), and it strives to educate people on Japanese food culture. At the restaurant located next to the sake brewery, seasonal items on the menu are made with locally grown rice and vegetables. Among them, "salmon with malted rice" and "wasabi pickled in soy sauce" are gaining popularity.

Shichiken Junmai Ginjo Birodo no Aji

"Brewed with rice produced under contract cultivation"

This sake has a refreshing mouthfeel and a light ginjo aroma. Shichiken brand's standard lineup of junmai ginjo.

Junmai Ginjoshu	1,800ml
Alc.	15%
Rice type	Yumesansui

Kasuga Shuzo

Water source	Central Alps water system
Water quality	Soft

4878-1 Nishi-machi, Ina City, Nagano 396-0026 TEL.0265-78-2223
E-mail: kasuga-shuzo.co.jp

Ina Valley's famous sake brewed by two brothers

"Make good sake rather than make lots of sake". That's the motto of the brewery rooted in Ina, Shinshu.

Urushido Jozo, brewer of the famous "Inokashira" sake brand, is located in Ina City on the Tenryu River in Shinshu region's Ina Valley, surrounded by the Central and Southern Alps. The brewery was founded in October 1915 by Shuhei Urushido, the first generation of the brewery, in a rented store of a sake brewery "Isawa" in Inabu-juku port town, Ina City, whose business was temporarily closed. Inogashira" was named after the famous water in Tokyo's Inokashira Park, which was used as a water supply for 300 years by the Tokugawa family, and means a place with the best spring water. "Inokashira" has a well-balanced taste that will keep you coming back for more. Based on the idea of "making good sake rather than a lot of sake," the brewery uses Hitogochi and Miyamanishiki sake rice harvested in Ina, Shinshu. The two brothers (both graduates of a brewing faculty) are the only brewers on site.

純米吟醸酒
Inokashira Junmai Ginjo

"Fruit-like mellow aroma"

This Junmai Ginjoshu has a splended aroma and a slightly sweet taste. You can enjoy the aroma particularly when it is chilled. Best served with Shinshu salmon.

Junmai Ginjoshu	720ml
Alc.	15%
Rice type	Miyamanishiki from Nagano
Sake meter value	–2.0
Acidity	1.5

純米酒
Inokashira Junmai

"Acidity reminiscent of kiwi fruit"

A gentle Junmaishu with a mild aroma and a mellow feel in the throat. It is best drunk at room temperature to taste the umami.

Junmaishu	720ml
Alc.	15%
Rice type	Hitogokochi from Kamiina
Sake meter value	+2.0
Acidity	1.6

Water source	Subterranean water from the Northern Alps
Water quality	Medium-soft

Daishinshu Shuzo

2380 Shimadachi, Matsumoto-shi, Nagano 390-0852 TEL. 0263-47-0895
E-mail: info@daishinsyu.com http://www.DAISHINSYU.COM/

Made with gratitude to nature
Delicious sake blessed by the heavens

Magnificent nature, enthusiastic sake rice farmers, and the goal of genuine local sake.

The Daishinshu Shuzo is located in Matsumoto City, Nagano Prefecture. The brewery aims to produce a local sake that reflects the climate of Shinshu, aiming to produce delicious sake blessed by the heavens. The brewing water used is natural underground water that comes from the North Alps, which can be seen from the brewery. All of the rice used is from Nagano Prefecture and is suitable for sake brewing. The rice is grown with love and care by 10 contract farmers in the prefecture. The reason why the brewery does not brew throughout four seasons is because it does not want to manipulate nature with technology. Rather, it strives to make sake by appreciating the climate in the region. The phrase that expresses this philosophy is "love and appreciation" brewing. When the brewers begin to brew sake, they write these words out by hand and post them at their work stations.

純米大吟醸酒
Daishinshu NAC Kinmon Nishiki Junmai Daiginjo

"A sake with a smooth delicious taste"

All ingredients used for this sake are certified as high-quality by the Nagano Prefecture Designation of Origin Committee (N.A.C.).

Junmai Daiginjoshu	720ml
Alc.	16%
Rice type	Kinmonnishiki
Sake meter value	+5.0
Acidity	1.4

純米吟醸酒
DaIshInshu Telppal

"A great balance of rich flavor and aroma"

A sake that symbolizes the brewery's all-around (teippai) care of each step of the sake production process, from rice polishing to brewing and storage

Junmai Ginjoshu	1,800ml/720ml
Alc.	16%
Rice type	Kinmonnishiki
Sake meter value	+2.0
Acidity	1.5

Okazaki Shuzo

4-7-33 Chuo, Ueda-shi, Nagano 386-0012　TEL.0268-22-0149
http://www.ueda.ne.jp/~okazaki/

Water source	Sugadaira water system
Water quality	Soft

Sake brewery nurtured in Ueda, Shinshu

Passing on 350 years of traditional techniques to protect the local sake culture.

Okazaki Shuzo was founded in 1665, the 5th year of the Kanbun Era. The brewery' sake, made with traditional techniques that have been passed on over its history of 350 years, is loved by the local community. Ueda City, where the brewery is located, is a mortar-shaped basin stretching from the Sugadaira Plateau to the Utsukushigahara Plateau, and known as one of the biggest producers of rice, fruits, vegetables, and matsutake mushrooms in Japan. Ueda City is also famous for terraced rice paddies in Inakura. To preserve the rice paddies for future generations, Okazaki Shuzo has introduced a sake rice owner system and signed a partnership agreement with the owners of the terraced rice. In addition, five sake breweries in Ueda City have formed the "Sankeinishiki Project" to promote the use of Sankeishiki, a new type of rice developed in Nagano Prefecture.

純米大吟醸酒
Shinshu Kirei Miyamanishiki Junmai Daiginjo

"High-quality sake brewed with care and effort"

The elegant melon-like Ginjo aroma is harmonized with the sharp taste of Miyamanishiki. The delicate and transparent taste is a superb gem.

Junmai Daiginjoshu	1,800ml/720ml
Alc.	15%
Rice type	Miyamanishiki

純米酒
Shinshu Kirei Hitogokochi Junmai

"Junmai to be enjoyed during meals"

Once you sip it, a gentle taste spreads in your mouth. It has a crisp, clean finish with an excellent aftertaste.

Junmaishu	1,800ml/720ml
Alc.	15%
Rice type	Hitogokochi

Water source	Well water from Mt. Kirito
Water quality	Soft

Ono Shuzoten

992 Ono, Tatsuno-machi, Kamiina-gun, Nagano 399-0601 TEL.0266-46-2505
E-mail: info@yoakemae-ono.com https://yoakemae-ono.com/

Thoughts on the land of Ono

A brewery that has inherited the spirit of sake brewing that pursues the real thing even even if it costs them their lives.

Ono Shuzo was established in 1864. The main brand "Yoakemae" was launched by the fifth generation who respected Japanese poet Toson Shimazaki and wanted to create a brand named after him on the 100th anniversary of Toson's birth. Masaki Shimazaki, Toson's father and the model of the main character in Toson's novel Yoaekemae, had a close relationship with Ono, and the historical background of the story overlaps with the time when the brewery was founded. Considering these facts, Kusuo Shimazaki, the first chairman of the Toson Memorial Museum, gave his permission for the brewery to use the title of Toson's novel as the sake's brand name. At that time, Kusuo said, "I hope you will never forget the spirit of pursuing the real thing even if it costs you your life, and that you will stay focused on producing sake with a great taste throughout your life". The words of Kusuo became the motto of Ono Shuzo. The brand name "Yoakemae" reflects the brewery's commitment to respecting the local climate and brewing sake with care.

純米吟醸酒
Yoakemae "Namaippon"

"Once you start drinking, you want more and more"
An elegant Ginjo aroma spreads mellowly, while the aftertaste is refreshing. This is a representative sake of the "Yoakemae" (before dawn) lineups.

Junmai Ginjoshu	1,800ml/720ml
Alc.	16%
Rice type	Yamadanishiki
Sake meter value	±0

Kadoguchi Shuzoten

Water source	Nabekurayama spring water
Water quality	Soft

1147 Tokisato, Iiyama-shi, Nagano 389-2412 TEL.0269-65-2006
E-mail: info@kadoguchi.jp   http://www.kadoguchi.jp

Local Sake in Okushinano in the midst of deep snow

Sake brewing to contribute to the local community.

Kadoguchi Shuzo was founded in 1869 by Hikosaburo Muramatsu, who used to run a rice business under the name "Kokuya" at the current location of the brewery. It is the northernmost brewery in the prefecture, located in Iiyama City, Nagano Prefecture, one of the areas with the heaviest snowfall in Japan. Bordering Niigata Prefecture, the area gets covered with more than three meters of snow during the winter. The brewery uses plenty of clear, rich water from Mt. Nabekura, a beautiful mountain that embraces one of Japan's largest virgin beech forests. The rice used for its sake are highly polished "Kinmonnishiki", "Miyamanishiki", and "Hitogokochi" rice, all of which are suitable for brewing and grown in Nagano Prefecture.

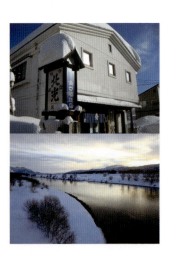

Hokko Masamune
Kinmonnishiki Tokubetsu Junmai

"Brewed with the same process as Daiginjo"

This sake is characterized by a versatile taste and sharpness that brings out the best of the contract-cultivated rice "Kinmonnishiki."

Special Junmaishu	720ml
Alc.	15%
Rice type	Kinmonnishiki
Sake meter value	+5.0
Acidity	1.6

Hokko Masamune
Kinmonnishiki Daiginjo

"A gem with a unique character"

The unique flavor of "Kinmonnishiki" rice spreads in your mouth as you feel its splendid, refreshing aroma and smoothness when it goes down your throat.

Daiginjoshu	720ml
Alc.	15%
Rice type	Kinmonnishiki
Sake meter value	+3.0
Acidity	1.0

Water source	Yatsugatake, Chikuma River subsoil water
Water quality	Soft

Kurosawa Shuzo

1400 Hozumi, Sakuho-machi, Minamisaku-gun, Nagano 384-0702 TEL.0267-88-2002
E-mail: info@kurosawa.biz   https://www.kurosawa.biz/

Deeply involved in agriculture
Contributing to the local community

The brewery is located at a high altitude, at the uppermost reaches of the Chikuma River.

Sakuho is a plateau town 800 meters above sea level at the upper reaches of the Chikuma River at the foot of Mt. Yatsugatake in Shinshu. Kurosawa Shuzo has been brewing sake with sincerity in this region since its establishment in 1858, taking advantage of the cool climate, clean air, and high-quality underground water of the Chikuma River. The pride of the brewery lies in the passion of the head brewer, Yohei Kurosawa, and the careful brewing of sake by all the brewers. The brewery uses only rice produced in Nagano Prefecture for its ingredients, and is deeply involved in agriculture by encouraging local farmers to plant sake rice, contracting with local farmers, and growing their own rice. The company also uses its own rice, and it invites the fans of the brewery's sake to events. It also operates a sake museum offering information on tourism and culture of the region.

Izutsucho Junmai Daiginjo Kurosawa

"The harmony of an elegant aroma and deep flavor"

It has a splendid and pleasant Ginjo aroma and a sweet, rich taste. A good balance of aromas and flavors. Serve well chilled.

Junmai Daiginjoshu	720ml
Alc.	15%
Rice type	Kinmonnishiki
Sake meter value	-2.0
Acidity	1.2

Kurosawa Kimoto Junmai Ginjo Vintage

"Kimoto undiluted sake matured for two summers"

This vintage sake has a well-balanced, deep flavor with a sharp snap and a lingering finish. Serve cold or lukewarm.

Junmai Ginjoshu	720ml
Alc.	17.5%
Rice type	Miyamanishiki
Sake meter value	+3.0
Acidity	2.0

Shusen Kurano

Water source	Chikuma River / Saikawa River
Water quality	Medium-hard

368-1 Imai, Kawanakajima-machi, Nagano-shi, Nagano 381-2226 TEL.026-284-4062
E-mail: info@shusen.jp http://www.shusen.jp

Sake that you can feel in your heart
Protecting sake brewed from the heart

A long-established sake brewery with 480 years of history. Evolving traditions backed by cutting-edge technology.

Shusen Kurano was founded in 1540. It is the oldest brewery in the Shinshu region, Nagano Prefecture. In Japan, it is the seventh oldest sake brewery, and there is an anecdote about Lord Takeda Shingen resting at the family temple of the Chino family, the chief brewer, and Chino's sake was served to him. The town of Kawanakajima, where the brewery is located, is a fan-shaped area surrounded by the Saigawa River, which originates in the Northern Alps, and the Chikuma River, which originates in Mt. Yatsugatake. Shusen Kurano uses rich subsoil water from the two rivers for brewing. They are also particular about using locally grown rice, which they contract with local farmers to grow. The variety is Miyamanishiki, a rice suitable for sake brewing, and the brewery makes sure it is cultivated in good soil. While preserving the traditional techniques and flavors of the past, Shusen Kurano's sake continues to evolve with the introduction of the latest technology and equipment.

Kawanakajima Tokubetsu Junmaishu

"A Junmai full of richness and sharpness"

Junmai is made entirely with Miyamanishiki rice from Nagano Prefecture. It has a rich and refreshing taste that you will never get tired of.

Special Junmaishu	1,800ml
Alc.	15-16%
Rice type	Miyamanishiki
Sake meter value	±0
Acidity	1.3

Genbu Junmai Giinjo [Unfiltered, undiluted sake]

"The taste of old-fashioned, handmade sake"

This is an excellent product that focuses on the flavor of the rice. It has a splendid aroma and a pleasantly refreshing acidity. Best served with tempura and pickled mackerel.

Junmai Ginjoshu	1,800ml
Alc.	16-17%
Rice type	Kinmonnishiki
Sake meter value	+3.0
Acidity	1.6

Water source	Kirigamine
Water quality	Soft

Sakenunoya Honkin Shuzo

2-8-21 Suwa, Suwa-shi, Nagano 392-0004 TEL.0266-58-0161
E-mail: info@honkin.net http://honkin.net

Brewing the authentic best
Brewing the best sake

Simple, but the flavor is deep. Local sake for drinking at night.

The brewery is located in Suwa City, Nagano Prefecture, a land with abundant water and greenery, where Lake Suwa, famous for "Omiwatari" (sacred cracks that form in the ice on the lake), is located. In Suwa City there are five sake breweries along a 500-meter stretch of the Koshu road. It was in 1756 that Isaburo Miyasaka, the founder of the company, took over the sake stocks and established the company under the business name "Shimonunoya". It is said that the Miyasaka family at that time devoted themselves to the development of Suwa's culture and commerce. It was engaged in many activities other than brewing sake, such as releasing river prawns into Lake Suwa, compiling a book called Suwa Kyuseki Shi (The History of Suwa), and serving as Shinto priests at shrines. It is a small sake brewery ran by a family of five. Most of the rice used for its sake brewing is purchased from local contract farmers. The brewery is deeply rooted in the local community.

普通酒
Honkin

"Sake that you would want to drink every day"

This yellow label is a long-selling product of Honkin, which aims to make sake that people want to drink every day. It is a refreshing sake with a good balance of sweet and spicy flavors, perfect for drinking every night.

Futsushu	1,800ml/720ml
Alc.	15-16%
Rice type	Prefectural rice

本醸造酒
Honkin Karakuchi Taichi

"The flagship sake of Honkin"

This sake is named after Mr. Taichi Kitahara, who worked as a Toji for more than 50 years at the brewery. It has a clear, dry taste with a refreshing acidity.

Honjozoshu	1,800ml/720ml
Alc.	15-16%
Rice type	Prefectural rice
Sake meter value	+7.0
Acidity	1.4

Shinshu Meijo

Water source	Yoda River subsoil water / Obsidian water
Water quality	Soft / Ultra soft

2999-1 Nagase, Ueda-shi, Nagano 386-040   TEL.0268-35-0046
E-mail: shinmei@ued.janis.or.jp   http://www.shinmei-net.com

Preserving and passing on tradition, while taking on a new challenge

Brewing quality sake by hand in the traditional way.

Shinshu Meijo was established about 180 years ago in the late Edo period (around 1834) in the town of Maruko(now Ueda City). The brewery's original name was "Masuya", but the name was changed to Shinshu Meijo in 1958. The brewery was initially established as a bottling company for four traditional sake breweries that had operated in the same area since the Edo and Meiji periods. Later, in 1973, they decided to collaborate in brewing sake as well, and the brewery's business took on its current form. The four breweries in Maruko, surrounded by beautiful mountains, all brought their own techniques and philosophies into making sake together – and the result is Shinshu Meijo's signature brand "Meiho Kikuzakari". The brewers, led by the Toji (chief brewer), quickly broke away from "Sanbai jozo" brewing style (increasing the volume by adding distilled alcohol, sugar, and other ingredients) that was popular during WWII as well as during the post-war period. The brewery is committed to preserving traditional techniques and knowledge based on experience and intuition, while introducing new technologies to make good sake.

普通酒
Meiho Kikuzakari

"Quintessential sake of Shinshu"
A full-bodied aroma and a clean, refreshing taste with a harmonious umami flavor. No sugar or other additives are used.

Futsushu	1,800ml/720ml
Alc.	15%
Rice type	Nagano rice
Sake meter value	-1.0-+1.0
Acidity	1.1-1.2

純米吟醸酒
Takizawa Junmai Ginjo

"Splendid sake with a hint of nobility"
A hint of fresh, fruity aroma, and you can taste the high-quality rice. It has a good balance of sharpness.

Junmai Ginjoshu	1,800ml/720ml
Alc.	16-17%
Rice type	Hitogokochi
Sake meter value	±0
Acidity	1.7

Water source: Subsoil water from the Southern Alps

Senjyo

2432 Kamiyamada, Takato-machi, Ina-shi, Nagano 396-0217 TEL.0265-94-2250
E-mail: info@senjyo.co.jp https://www.senjyo.co.jp

Passing on the culture of rice fermentation
Passing the rice fermentation culture onto the next generation

A brewery with a history of more than 150 years since the Edo period. Brewing sake for the younger generation.

Senjyo was established in 1866 at the end of the Edo period. It is located in Ina Valley between the Southern and Central Alps at an altitude of over 3,000 meters. The Ina Valley is blessed with abundant underground water and little precipitation, a large temperature difference between day and night due to the high altitude, and little damage from typhoons, all of which are ideal conditions for rice production. The brewery's philosophy is to "pass on the culture of rice fermentation to the future", and with the technology and know-how gained from over 150 years of working with rice and koji through sake brewing, the brewery hopes to popularize not only sake, but also amazake (sweet sake), shochu (distilled liquor), sake lees, and other fermented rice products to younger generations and people overseas who may not be familiar with these products. Currently, the brewery's sake is exported to more than fifteen countries and regions in North America, Europe, and Asia.

Kuromatsu Senjo Junmai Daiginjo Prototype

"Rich sweetness like a fruit"
Super-sweet Junmai Daiginjo. A translucent sweetness that makes the most of the characteristics of Hitogokochi sake rice.

Junmai Daiginjoshu	1,800ml/720ml
Alc.	16%
Rice type	Hitogokochi
Sake meter value	-15.0
Acidity	1.7

Kuromatsu Senjo Konna Yoruni Junmai Ginjo Yamame

"Made from only Miyamanishiki rice from Shinshu"
The label depicts yamame fish, a species found in many rivers in Nagano Prefecture. The sake rice and yeast are both produced in Nagano.

Junmai Ginjoshu	1,800ml/720ml
Alc.	16%
Rice type	Miyamanishiki
Sake meter value	private
Acidity	private

Daisekkei Shuzo

Water source	underground water from melting snow in the North Alps
Water quality	Soft

9642-2, Aizome, Ikeda-machi, Kitaazumi-gun, Nagano 399-8602 TEL.0261-62-3125
E-mail: info@jizake.co.jp   https://www.jizake.co.jp

"It tastes as good as I expected!" That's what we want to hear

Handmade local sake brewed in the beautiful nature of Azumino and with the passion of the brewers.

Daisekkei was founded in 1898 in the vast granary of Azumino, a region with an abundance of spring water under the silvery peaks of the Northern Alps, and has long been known for brands such as "Kikyo Masamune" and "Seikou Zakura". At that time, the company name was Ikeda Jyozo, but in 1949, after the war, the name was changed to Daisekkei Shuzo in honor of the great snow gorge of Mt. Hakuba. With the goal of "making sake loved by the locals even better," Daisekkei is striving to improve the quality of its regular sake. "It tastes as good as I expected!" are the words the brewery wants to hear from its customers. The brewery continues to contribute to the happiness of its customers, employees, the local community, and society as a whole, by providing its brand products and services. They also continue to take on challenges as a promoter of Japanese food culture, aiming to brew sake that is representative of Nagano as well as Japan.

純米吟醸酒
Daisekkei Jyosen

"The standard sake Daisekkei is proud of"

This is a standard sake that is faithful to the basics of Ginjo brewing. It's refreshingly dry, but with a hint of sweetness. It can be served cold or warm to your liking.

Junmai Ginjoshu	720ml
Alc.	16%
Rice type	Domestic rice
Sake meter value	+7.0
Acidity	1.5

純米吟醸酒
Daisekkei Junmai Ginjo

"Made from Miyamanishiki rice grown in Azumino"

Brewed by hand, this is a refreshing sake made with clean brewing water. It has a natural aroma and the flavor of the rice.

Junmai Ginjoshu	720ml
Alc.	15%
Rice type	Miyamanishiki
Sake meter value	+3.0
Acidity	1.7

| Water source | Mt. Asama system subterranean water |
| Water quality | Relatively hard |

Chikumanishiki Shuzo

1110 Nagatoro, Saku-shi, Nagano 385-0021 TEL.0267-67-3731
E-mail: info@chikumanishiki.com https://www.chikumanishiki.com

Saku in Shinshu is known for Chikumanishiki

Local sake made with the famous water of Shinshu, rice grown in Shinshu, and the passion of Shinshu brewers.

One of the ancestors of the family that established Chikumanishiki Shuzo is Toratane "Minonokami" Hara, one of the twenty-four generals of Takeda Shingen. Hara's descendant, Yahachiro Hara, started the brewery in 1681, while working as a landlord. The city of Saku, Nagano Prefecture, where the brewery is located, is a plateau 700 meters above sea level called Sakudaira. Mt. Asama is to the north, Yatsugatake to the south, and Chikuma River, the longest river in Japan, flows through the middle of the city. Surrounded by nature, where the temperature can drop to 10 degrees below zero in the early morning in midwinter, the area is blessed with the best environment for sake brewing. The brewing water comes from the brewery's own wells, two 13-meter shallow wells and two 60-meter deep wells, and the sake rice is milled in-house from Miyamanishiki produced in Nagano Prefecture. Chikumanishiki, a local sake with deep roots in Saku, continues to be the favorite of locals.

Junmai Ginjo Chikumanishiki

"Excellent sake brewed with famous water from the Asama mountain range"

The sake is made from highly-polished Miyamanishiki rice and brewed at low temperature for a long time. It was matured slowly in a low-temperature fridge.

Junmai Ginjoshu	1,800ml/720ml
Alc.	15%
Rice type	Miyamanishiki
Sake meter value	-3.0
Acidity	1.6

Kizan Sanban Junmai Ginjo Namazake

"Local sake that is committed to the basics of sake brewing"

An undiluted sake with the refreshing acidity and flavor of new sake made in spring. It is stored in bottles and kept at a stable temperature to maintain its flavor. A limited number of bottles are available.

Junmai Ginjoshu	1,800ml/720ml
Alc.	15%
Rice type	Miyamanishiki
Sake meter value	-15.0
Acidity	3.1

Tsuchiya Shuzoten

Water source	Chikuma River subsoil water / brewery well water
Water quality	Soft

1914-2 Nakagomi, Saku-shi, Nagano 385-0051 TEL.0267-62-0113
E-mail: info@kamenoumi.com https://kamenoumi.com/

Sake that reminds one of Saku, brewed with the heart

Making local sake using local sake rice, Tsuchiya Shuzo was founded in 1900.

The brewery is located in the city of Saku, Nagano Prefecture, in a highland basin at an altitude of more than 600 meters. The area has fertile soil and a warm climate that is ideal for rice farming and sake brewing. There are as many as thirteen sake breweries in the area. In this region, people traditionally enjoy sake with Saku carp, river fish, chicken, wild vegetables, and pickled vegetables (nozawana and white melon pickled in sake lees). The region was once known for sweet sake that matches wild vegetables. The Chikuma River, which is the longest river in Japan, runs through the center of the prefecture, which makes the land fertile. The brewery uses Miyamanishiki, Kinmonnishiki, Hitogokochi, and Yamaenishiki, all of which are grown with the abundant soft water and fertile fields in the prefecture. Recently, the brewery has been trying to use pesticide-free and organically-grown sake rice, and has been brewing sake with a high awareness of terroir.

Kamenoumi
Tokubetsu Junmai

"Cold and hot, whichever is enjoyable"

This sake has a well-balanced aroma with a harmonious combination of acidity and rice flavor. It is easy to drink thanks to the crisp aftertaste.

Special Junmaishu	1,800ml/720ml
Alc.	15%
Rice type	Hitogokochi
Sake meter value	+2.0
Acidity	1.8

Akanesasu
Tokubetsu Junmai

"The flavor of rice grown with attention to the soil"

This sake is brewed with a blend of "Kinmonishiki" and "Hitogokochi," both of which are grown without using pesticides. It is characterized by its elegant flavor and sweetness.

Special Junmaishu	1,800ml/720ml
Alc.	16%
Rice type	Kinmonishiki/Hitogokochi
Sake meter value	private
Acidity	1.6

| Water source | Subterranean water from Mt. Hachibushi |
| Water quality | Soft |

Toshimaya

3-9-1 Hon-cho, Okaya-shi, Nagano 394-0028 TEL.0266-23-1123
E-mail: shintaro@kk-toshimaya.co.jp https://jizake.miwatari.jp/

Sake brewed with people, rice, and water from Shinshu

More than eight types of sake rice are grown under contract in six areas of Nagano Prefecture.

Okaya City is located in the center of Nagano Prefecture, on the shore of Lake Suwa at an elevation of 759 meters above sea level. It is blessed with a beautiful natural environment, surrounded by the lake and mountains that change color with the seasons. The city has a beautiful natural environment, surrounded by a lake and mountains that shows different faces in different seasons. The slogan of the brewery is "elaborate local sake that makes people smile". All of the rice, which determines the flavor of sake, is produced in Nagano Prefecture, including Kinmonnishiki, Takananishiki, Sankeinishiki, Miyamanishiki, Hitogokochi, Shirakabanishiki, Yamadanishiki, and Yoneshiro. One or more of the eight different types of rice are used depending on the characteristics of the product. The source of water is the subsoil of Mt. Hachibushi, behind the brewery. If you want to drink sake made by the people, rice, and water of Shinshu, where the five tastes of sweetness, sourness, astringent taste, spiciness, and bitterness are in perfect harmony, look no further than the brand name "Miwatari".

純米吟醸酒

Miwatari Junmai Ginjo Petillant

"Sake that goes well with French cuisine"

This flavorful Junmai Ginjoshu can only be tasted at a sake brewery that retains the natural carbon dioxide gas produced during the fermentation of the mash.

Junmai Ginjoshu	720ml
Alc.	17%
Rice type	Hitogokochi
Sake meter value	-2.0
Acidity	1.3

特別純米酒

Miwatari Karakuchi Tokubetsu Junmai Sankeinishiki

"A new breed of sake rice, "Sankeinishiki""

This sake is brewed with only a new breed of sake rice called Sankeinishiki. It is dry special Junmaishu that was nurtured and brewed with great care by young brewers.

Special Junmaishu	20ml
Alc.	15-16%
Rice type	Sankeinishiki
Sake meter value	+3.5
Acidity	1.4

Totsuka Shuzo

Water source	Yatsugatake subsoil water
Water quality	Soft

752 Iwamurada, Saku-shi, Nagano 385-0022 TEL.0267-67-2105
E-mail: kanchiku@valley.ne.jp https://www.kanchiku.com/

Sake like the dignified appearance of bamboo that grows in the severe cold

A brewery that is committed to quality, nurtured by the magnificent nature of Mt. Yatsugatake.

Totsuka Shuzo was founded in 1653 by Hiraemon Totsuka, the first-generation head of the family to make Doburoku. After that, the brewery served the Iwamurada clan for several generations, and was entrusted with the role of town official. The name of the brewery's main brand, "Kanchiku," comes from an actual bamboo called Kanchiku, which is a rare type of that produces shoots in winter. The sake "Kanchiku" can only be produced at this time of the year in extremely cold in Saku. The area has abundant underground water and is known for its many sake breweries. Since its establishment, Totsuka Shuzo has been dedicated to handcrafting sake and focusing on quality. The brewery continues to produce sake with wisdom and care, nurtured by the magnificent nature of Saku, a place of longevity, good water and good rice.

大吟醸酒
Kanchiku Daiginjo

"Delicious sake with a distinguished aroma"

A refreshing and elegant taste and aroma. It has richness and umami, but also a sharp flavor. It is best served between 5°C and 15°C.

Daiginjoshu	720ml
Alc.	16-17%
Rice type	Yamadanishiki
Sake meter value	+5.0
Acidity	1.3

普通酒
Kanchiku

"The taste changes depending on the temperature – try it at room temperature or heated"

Sake with umami that can be enjoyed at different temperatures, from room temperature to warm.

Futsushu	1,800ml
Alc.	15%
Rice type	Miyamanishiki
Sake meter value	+2.0
Acidity	1.2

Water source	Mt. Yatsugatake water system
Water quality	Soft

Hanno Shuzo

123 Nozawa, Saku-shi, Nagano 385-0053 TEL.0267-62-0021
E-mail: sakagura@sawanohana.com http://www.sawanohana.com

The first cup is "delicious" The second "feels comfortable"

A sake brewery in Saku that aims to make simple and delicious sake.

Founded in 1901, Tomono Shuzo is a brewery located in Saku, Nagano Prefecture, a sake region surrounded by beautiful nature and cool weather. Utilizing high-quality sake rice grown in the Shinshu region, underground water from the Yatsugatake mountain range, and the cold climate of Saku in winter, Tomono Shuzo brews great sake fermented at low-temperatures. This local brewery has been loved by the local community for a long time, and is now expanding its sales channels throughout Japan and the world with the hope of offering a "healing" sake that reminds people of their hometown. The name of the main brand "Sawanohana " means a beautiful flower blooming in the clear streams in Saku. The brand's symbol is the ayame flower. The brewery aims to produce sake that has "the taste of rice and a sense of comfort". With the desire to be a truly local brewery, Tomono Shuzo continues to brew sake that brings comfort.

Sawanohana Nakadori Junmai Daiginjo Hitotsubueri

"Brewing every single drop with care"

This Junmai Daiginjoshu has an elegant aroma and a fresh taste. Drink it at a slightly chilled temperature.

Junmai Daiginjoshu	720ml
Alc.	16%
Rice type	Miyamanishiki
Sake meter value	-1.0
Acidity	1.5

Sawanohana Junmai Ginjo Himari

"As warm as a sunny day in winter"

Junmai Ginjoshu with a refreshing aroma and gentle flavor. It can be served slightly chilled, at room temperature or lukewarm.

Junmai Ginjoshu	1,800ml/720ml
Alc.	16%
Rice type	Hitogokochi
Sake meter value	+3.0
Acidity	1.6

Nanawarai Shuzo

Water source	Subsoil water from the Shinshu Kiso mountains range
Water quality	Soft

5135 Fukushima, Kiso-machi, Kiso-gun, Nagano 397-0001　TEL.0264-22-2073
E-mail: nk7781@titan.ocn.ne.jp　https://www.nanawarai.com/

Sake with umami that makes you smile

Drink this delicious sake, laugh seven times (nanawarai) and blow away the fatigue of traveling on a mountainous road.

Nanawarai Shuzo was founded in 1892. The brewery is located in Kisoji, Nagano, a region that gets so cold that a poet once said Kiso is "cold even in summer". The sake brewed with water from the Kiso mountain range is characterized by its soft taste, and has long been enjoyed by people living in the cold mountains. For the people of Kiso, Nanawarai's sake has become a part of their lives. Nanawarai, also the brand name of the brewery's sake, reflects sayings like "Good fortune comes to those who laugh", "laugh seven (nana) times and Shichifuku (Seven Deities of Good Fortune) will come". Nanawarai's sake is made with honesty without trying too hard to be unique. The sake, has umami and is smooth and refreshing. When served with simple dishes like wild vegetables, this sake can easily be the main part of the meal. When served with splendid dishes, it can step aside and bring out the best of the food. It is favored by many as versatile sake for meals.

普通酒
Nanawarai Kobai

"Aromatic honjozo that is smooth and refreshing"

This standard lineup aims for umami instead of labeling it sweet or dry. It is soft on the palate with a crisp finish that complements food.

Futsushu	1,800ml/720ml
Alc.	15%
Rice type	Nagano rice
Sake meter value	+2.0
Acidity	1.2

純米吟醸酒
Nanawarai Junmai Ginjo

"When served cold, you can sense the original umami of Miyamanishiki rice"

It has a subtle aroma, a gentle texture, and a fine, smooth taste. The original flavor of the rice and the softness of the water are in harmony

Junmai Ginjoshu	1,800ml/720ml
Alc.	15%
Rice type	Miyamanishiki
Sake meter value	±0
Acidity	1.5

Water source	Saikawa River subsoil water
Water quality	Medium-hard

Nishiida Shuzoten

1726 Komatsubara, Shinonoi, Nagano-shi, Nagano 381-2235 TEL.026-292-2047
E-mail: nishiida@mx2.avis.ne.jp http://w2.avis.ne.jp/~nishiida/

The only brewery in Nagano Prefecture using flower yeast

Brewing a one-of-a-kind sake by combining flower yeast and sake rice.

Nishiida Shuzo brews all of its products using flower yeast, which is extracted and isolated from flowers in the natural world. It is the only brewery in Nagano Prefecture that uses this type of yeast. Brands like "Shinanohikari" and "Sekizen" are made by combining various flower yeasts and sake rice. The brewery aims to brew sake with one-of-a-kind quality. In addition to sake, the brewery also produces fruit wine. The lineup includes "Mellows", white wine made from Niagara grapes grown on the company's own farm, Cidre, sparkling apple wine created in collaboration with a local orchard, and other fruit wines made from peaches, plums, and blueberries grown in Nagano Prefecture.

本醸造酒
Shinanohikari Honjozo

"A flavor you'll never get tired of. Enjoy it either cold or hot"
This sake is made with flower yeast from the gentian, the prefecture's flower. It has a smooth and refreshing taste with umami.

Honjozoshu	1,800ml
Alc.	15%
Rice type	Nagano rice
Sake meter value	+2.0
Acidity	1.6

純米酒
Sekizen Junmaishu

"Fresh acidity like a fruit"
Apple flower yeast This Junmaishu is brewed with apple flower yeast and has a refreshing fruit-like acidity. Best served cold.

Junmaishu	1,800ml
Alc.	15%
Rice type	Hitogokochi
Sake meter value	-2.0
Acidity	1.8

Miyasaka Jozo

Water source	Subsoil water from the Mt. Nyukasa
Water quality	Soft

1-16 Motomachi, Suwa-shi, Nagano 392-8686 TEL.0266-52-6161
https://www.masumi.co.jp

Connecting people, nature, and time

Sake brewed in the rich climate of Suwa is evolving into a global brand.

Local sake reflects the climate, traditions, and culture of the region. Each bottle contains a story about the origins of the sake brewery, the thoughts of the people who lived in each era, and the natural environment in which the rice and water used as ingredients come from. Miyasaka Jozo was established in 1662. "Masumi", the brand launched in the late Edo period, has the theme of "connecting people, nature, and time". The brewery is committed to activities related to local development, nature conservation, and cultural inheritance. As part of such activities, the brewery hold events in collaboration with the five sake breweries located along the Koshu road, and publishes the tabloid magazine Brew, which introduces the climate of the region and the lives of its people. The brewery hopes to contribute to a better future for the region because the local community always supports its sake production.

純米大吟醸酒
Junmai Daiginjo Yumedono

"Sake brewing that the Toji takes all responsibility for"

"This is one of the best Junmaishu brewed by Masumi. It has an elegant aroma and a deep flavor. Use a wine glass to enjoy the aroma.

Junmai Daiginjoshu	1,800ml/720ml
Alc.	15%
Rice type	Yamadanishiki
Sake meter value	-3.0
Acidity	1.5

純米吟醸酒
Junmai Ginjo Shikkoku KURO

"The theme is the glossy texture of Japanese lacquer work"

This sake has both translucence and fullness as well as a harmonious taste. It can be enjoyed cold, at room temperature, or lukewarm.

Junmai Ginjoshu	1,800ml/720ml
Alc.	15%
Rice type	Yamadanishiki/Miyamanishiki
Sake meter value	+5.3
Acidity	1.4

| Water source | Kiso River source subsoil water |
| Water quality | Medium-soft |

Yukawa Shuzoten

1003-1 Yabuhara, Oaza, Kiso-mura, Kiso-gun, Nagano 399-6201 TEL.0264-36-2030
E-mail: info@sake-kisoji.com https://yukawabrewery.com/en/

Making the most of the climate of the region at an elevation of 1,000 meters

Continued to brew delicious local sake for 370 years deep in the heart of Shinshu's Kiso Kaido.

A brewery that has continued to produce delicious local sake deep in the heart of Shinshu's Kiso Kaido for 370 years, Yukawa Shuzoten continues to follow traditional sake brewing methods while taking on the challenge of modifying techniques and flavors to meet the needs of the times. It stands quietly in a corner of Yabuharajuku, deep in the Shinshu-Kiso road, which famous poet Basho Matsuo once wrote about and renowned Ukiyoe artist Hiroshige Utagawa once painted. The brewery has a long history – it was established in 1650. There is an anecdote from the time of the 13th-generation head of the brewery. Poets of the Araragi school gathered at the Chinryukan on the brewery's premise to drink its famous sake. Despite its star-studded history, the brewery has not been complacent in its efforts to rejuvenate its brewers and has built a system that covers everything from brewing to sales channels. Kisoji has continued to brew richly flavored sake with a deep umami that only a local brewery can produce, using quality water from Kiso and rice from Shinshu. The people at the brewery are grateful to its predecessors for sticking with their occupation through to the 16th generation, and pledges to pass it on to future generations.

特別純米酒
Kisoji Tokubetsu Junmaishu

"Sake to drink every evening"
This is tokubetsu junmaishu with a mellow mouthfeel and a full-bodied flavor. The taste is full of rice flavor with sweetness and acidity.

Special Junmaishu	… 1,800ml/720ml
Alc.	15%
Rice type	Miyamanishiki by Shinshu

純米吟醸酒
Jyurokudai Kurouemon Junmai Ginjo Miyamanishiki

"The standard lineup of Jyurokudai Kurouemon"
The bitterness and astringency of Miyamanishiki are balanced by a refreshing acidity and sweetness. Great served cold, at room temperature, or warm.

Junmai Ginjoshu	…… 1,800ml/720ml
Alc.	17%
Rice type	…Miyamanishiki by Shinshu

Yonezawa Shuzo

Water source	Subsoil water from the Southern Al...
Water quality	Soft

4182-1 Okusa, Nakagawa-mura, Kamiina-gun, Nagano 399-3801 TEL.0265-88-3012
E-mail: jizake@imanisiki.co.jp https://imanisiki.co.jp/

Local Sake Nurtured in Minami Shinshu

Sake brewing begins with the care of terraced rice paddies to preserve the original landscape of Japan.

Yonezawa Shuzo is a small sake brewery located in Nakagawa Village in the southern part of Nagano Prefecture, founded in 1907. The name of the main brand, Imanishiki, comes from the fact that the first-generation head of the family, Yasotaro Yonezawa, used Imanishiki as a ring name in amateur sumo wrestling matches. Yonezawa Shuzo has been brewing sake that is loved by the local community, but in recent years, like many other sake breweries, its business has faced difficulty surviving. In 2014, it made a fresh start by becoming a group company of Ina Shokuhin Kogyo, known for its "Kantenpapa" brand. The brewery advocates for cultivation of rice utilizing terraced rice paddies and the protection of the beautiful scenery of Nakagawa Village. All of its sake, from regular sake to Junmai Daiginjo, is pressed in a traditional sake tank with great effort. The sake's soft, gentle taste has been highly evaluated in various competitions in Japan and abroad in recent years.

純米大吟醸酒
Imanishiki Junmai Daiginjo

"Serve it slightly chilled to enjoy the aroma"

This sake is made from mash that has been fermented at low temperature for a long time and then pressed in a sake tank with great care. It features a harmony of soft and refreshing acidity and rice flavor.

Junmai Daiginjoshu	... 1,800ml/720ml
Alc.	16%
Rice type	Miyamanishiki
Sake meter value	-1.0
Acidity	1.3

特別純米酒
Imanishiki Nakagawamura no Tamako Tokubetsu Junmai

"Slightly dry, sharp mouthfeel"

This is one of the Special Junmai lineups made from Miyamanishiki rice grown in Nakagawa village, pressed in a traditional sake tank. You can enjoy seasonal flavors four times a year depending on which season it was bottled.

Special Junmaishu	... 1,800ml/720ml
Alc.	16%
Rice type	Miyamanishiki
Sake meter value	+1.5
Acidity	1.8

Water source	Ground water on the premises of the company
Water quality	Soft

Asahi Shuzo

880-1 Asahi, Nagaoka-shi, Niigata 949-5494   TEL.0258-92-3181
https://www.asahi-shuzo.co.jp

Following the tradition of Echigo Toji and making new challenges for the future

The brewery in Niigata that produces the "Kubota" brand. Quality-driven in every way.

Asahi Shuzo was founded in 1830 under the name "Kubotaya". The brewery continues to preserve traditions of Echigo Toji, but it always looks beyond the current era and continues to pursue higher quality sake. The Kubota brand, which began in 1985, opened up a new market by offering a "light and dry taste" to meet the changing tastes of customers. The brewery is also working to protect the natural environment of Koshiji, which is essential for making delicious sake. The brewery is commited to creating a natural environment where fireflies can grow, an indicator of clean air and water. As our food culture becomes more and more diverse, Asahi Shuzo listens to its customers and proposes new tastes while maintaining its quality-oriented sake brewing. A single sip will fill your body and soul with joy. In order to make such sake, Asahi Shuzo will continue on the right path of sake brewing.

吟醸酒
Kubota Senju

"Light and refreshing, Kubota's standard"

This sake makes ordinary dishes extraordinary. You can taste the original flavor and acidity of the rice, with a subtle aftertaste and sweetness.

Ginjoshu	1,800ml/720ml/300ml
Alc.	15%
Rice type	Gohyakumangoku
Sake meter value	+5.0
Acidity	1.1

純米大吟醸酒
Kubota Junmai Daiginjo

"Fine, splendid aroma"
This Junmai Ginjoshu was created in pursuit of a new taste – it has a modern and sharp flavor. The aroma, sweetness, and sharpness are all in harmony.

Junmai Daiginjoshu	1,800ml/720ml/300ml
Alc.	15%
Rice type	Gohyakumangoku
Sake meter value	±0
Acidity	1.3

Ikeura Shuzo

Water source	Well water
Water quality	Soft

1538 Ryotaka, Nagaoka-shi, Niigata 949-4524 TEL.0258-74-3141
E-mail: sake@ikeura-shuzo.com https://ikeura-shuzo.com/

An elegant philosophy that sees sake as a culture

The brewery aims to handcraft sake using quality-oriented methods to make sure you will never get tired of drinking it.

Ikeura Shuzo was established in 1830, at the end of the Edo period (1603-1868), and is now in its seventh generation. It is said that the first-generation head, who came from the family of a village headman, started brewing sake when he learned that good water was gushing out of the ground in this area. In the early Showa period (1926-1989), Takaemon, the fifth- generation head of the brewery, who was a youth into egalitarianism, respected philosopher Gosonou Nomoto of Nagaoka, with whom he had a close relationship. The name of the brewery's flagship brand, "Warakugoson", means that we will be in harmony and joyful if we all respect each other. The name was chosen based on the advice of Masahiro Yasuoka, who was a scholar of the Chinese classics. It reflects the noble idea of leading the world to peace through sake brewing. Perhaps under the influence of Takaemon, his son the sixth-generation head, also wanted to spread the spirit and virtue of the monk Ryokan, whose final resting place was Nagaoka in Niigata Prefecture, to the world. He came up with the brands "Shingetsurin" and "Tenjotaifu".

特別純米酒

**Warakugoson
Tokubetsu Junmaishu**

"Sake rice Takanenishiki's umami"

This is easy to drink Junmaishu with a great taste that brings out the true flavor of the rice. Serve cold or lukewarm.

Special Junmaishu	1,800ml
Alc.	15.3%
Rice type	Takanenishiki
Sake meter value	+4.0
Acidity	1.4

純米大吟醸酒

Tenjotaifu Junmai Daiginjo

"Sake from the village of Ryokan"

An elegant sake with a rich aroma and flavor, carefully brewed using "Koshitanrei", a local rice suitable for sake brewing.

Junmai Daiginjoshu	720ml
Alc.	16.3%
Rice type	Chotanrei
Sake meter value	-1.0
Acidity	1.4

| Water source | Subsoil water from Mt. Hakuba |
| Water quality | Medium-hard |

Ikedaya Shuzo

1-3-4 Shintetsu, Itoigawa-shi, Niigata 941-0065 TEL.025-552-0011

Honored sake in the village of Bidan

The brewer of "Kenshin", a famous sake of Echigo born in the village of "Shiookuri Bidan".

Ikedaya Shuzo was established in 1812 in Itoigawa, the westernmost city of Niigata Prefecture, near the base of the "Salt Road" or Sengoku Kaido road, where the famous general Uesugi Kenshin sent salt to his enemy Takeda Shingen during the Battle of Kawanakajima. Itoigawa is designated as a "World UNESCO Geopark" and is known as the only place in Japan where jade, the national stone, is produced. The brewery uses carefully-selected rice suitable for sake brewing, such as Yamadanishiki, Gohyakumangoku, and Koshitanrei. It brews using subterranean water from the Himekawa River, which is fed by Mt. Hakuba in the Japanese Alps. Instead of producing sake in a large amount, it focuses on sake with a balanced sweetness and acidity that goes well with the local food culture. It also aims to create a new type of Niigata sake by incorporating the skills of veteran Echigo Toji brewers and the sensibilities of young sake brewers, while maintaining the handmade feel of a small brewery.

特別本醸造酒
Tokubetsu Honjozo Kenshin

"An authentic dry sake that goes well with Japanese food"

It has a clean, soft mouthfeel, and is a Niigata dry sake with a refreshing yet delicious taste suitable for evening drinking.

Special Honjozoshu	1,800ml
Alc.	15-16%
Rice type	Yukinosei
Sake meter value	+3.0
Acidity	1.3

大吟醸酒
Daiginjo Kenshin

"The sake that won the gold award for seven consecutive years"

A mild Ginjo aroma with a full mouthfeel, crisp acidity, and a refreshing finish.

Daiginjoshu	1,800ml
Alc.	16-17%
Rice type	Yamadanishiki
Sake meter value	+4.0
Acidity	1.3

Ishimoto Shuzo

Water source	Agano River system
Water quality	Soft

847-1 Kitayama, Konan-ku, Niigata-shi, Niigata 950-0116 TEL.025-276-2028
https://koshinokanbai.co.jp

The brewer of the super popular brand "Koshino Kanbai"

A dry sake with the condensed original flavor of rice. "Koshino Kanbai" is not a "mythical sake".

Ishimoto Shuzo, the brewer of the very famous brand "Koshino Kanbai", was founded in 1907. It is a sake that every sake lover would like to try at least once. The brewer says, "Koshino Kanbai is originally sake for people to enjoy as a casual evening drink". This brand is not mass-produced because the brewery wants to keep the priority on quality. For this reason, it is called "mythic sake" due to its unavailability. It is usually traded at a premium price, but it is not the brewer's intention to sell it at a high price. "Koshino Kanbai" is dry sake made from high-quality rice, with the original flavor of the rice concentrated in a clean and refreshing taste. It does not spoil the taste of food and can be enjoyed with meals. There is still a lack of balance between supply and demand, but the pride of the brewery will never change: "To be stubbornly dedicated to perfection, and to be Koshino Kanbai".

Koshino Kanbai Super-select Daiginjo

"The ultimate luxury"

The rice is polished to 30%. The Ginjo aroma is moderate and goes well with food. It has a clean and refreshing taste. You will never get tired of drinking this sake.

Daiginjoshu	720ml
Alc.	16.6%
Rice type	Yamadanishiki
Sake meter value	+6.0
Acidity	1.1

Koshino Kanbai Kinmuku Junmai Daiginjo

"A matured feel on the palate"

The rice polishing ratio is 35%. Thanks to low temperature fermentation and aging, you can feel the soft, matured quality of the sake in your mouth.

Junmai Daiginjoshu	720ml
Alc.	16.3%
Rice type	Yamadanishiki
Sake meter value	+3.0
Acidity	1.3

Water source	Subsoil water from the Gozu mountain system
Water quality	Soft

Echigozakura Shuzo

1-7-13 Yamaguchi-cho, Agano-shi, Niigata 959-2005 TEL.0250-62-2033
https://www.echigozakura.co.jp

A place where swans fly in Niigata's delicious local sake

Quality that exceeds customers' expectations is achieved with traditional techniques and the latest technology.

Located in Agano City, Niigata Prefecture, not far from Hyoko Lake designated as a Ramsar site for swans, Echigozakura Shuzo was established in 1890. The brewery is dedicated to making sake that exceeds customers' expectations. It aims to make customers feel that the pricing is great for the quality of sake they make. Due to the Niigata earthquake, the historic building of the brewery was completely renovated in 2009. It now has all the latest equipment, and can make new types of sake, while at the same time maintaining its traditional brewing techniques. The new brewery building was named "Hakuchogura" (hakucho means a swan) after the image of a swan flying over the surface of the lake. It has become a local tourist spot, too. The brewery's store offers tastings of unpasteurized sake and amazake (sweet sake), soy sauce made by a company in Agano City, as well as miscellaneous goods from the dyeing factory. People in the local community as well as tourists from outside the prefecture frequently visit the store.

大吟醸酒
Daiginjo Echigozakura

"Dry and crisp, brewed at low temperatures"

Brewed at a low temperature to maximize the flavor of the rice, with a splendid aroma and a clean finish that leaves a lasting impression.

Daiginjoshu	720ml
Alc.	15%
Rice type	Domestic rice
Sake meter value	+0.5
Acidity	1.3

大吟醸酒
Echigozakura 38 Daiginjo

"The premium umami of Yamadanishiki"

This is a premium version of Daiginjo, made with the finest ingredients and care. It has a splendid, elegant aroma and a soft mouthfeel.

Daiginjoshu	720ml
Alc.	16%
Rice type	Yamadanishiki
Sake meter value	-1.0
Acidity	1.5

Koshi Meijo

Water source	Subsoil water from the Sumon mountain range
Water quality	Soft

2-8 Tochiomachi, Nagaoka-shi, Niigata 940-0217 TEL.0258-52-3667
E-mail: koshinotsuru@extra.ocn.ne.jp https://koshimeijo.jp/

Protecting the local terraced rice paddies to produce sake rice

Pursuing a beautiful taste based on the tradition of handcrafted sake.

The motto of Koshi Meijo is, "Make sake that is delicious, not sake that sells". The brewery has been loved by the people of Tochio for a long time, and has been working on sake brewing while keeping close ties with local liquor stores and farmers. The brewery was established in 1934 after two businesses -- Yamakeya, founded in the Kyoho era of the Edo period, and Yamashiroya, founded in 1845 -- merged to form the current Koshi Meijo. The Tochio area, where Kenshin Uesugi spent his childhood, has a rich history. In 2017, the company began cultivating "Gohyakumangoku", a variety of rice suitable for sake brewing, in the abandoned terraced paddy fields of Tochio. In order to protect the local environment and continue brewing sake with local rice, the company continues to grow its own rice, aiming to brew sake that only Koshi Meijo, with its roots in the local community, can produce.

純米大吟醸酒
**Koshinotsuru
Junmai Daiginjo Domaine***

"The first step to revamp Koshinotsuru brand"

Splendid flavor. This sake is characterized by a soft mouthfeel and a moderate spiciness and sharpness that comes later..It is best served cold.

Junmai Daiginjoshu	1,800ml
Alc.	16%
Rice type	Gohyakumangoku
Sake meter value	±0-+2.0
Acidity	1.5-1.7

*The word "Domaine" is a French word indicating a plot or territory. In sake brewing, it means that everything from cultivation, brewing, maturing, and bottling is done by one producer.

純米大吟醸酒
**Yamashiroya Standard Class
Junmai Daiginjo**

"A gem with a smart, transparent flavor"

Characterized by a refreshing, gentle, and light acidity. The dryness accentuates the flavor, making it a good match for Japanese food as well as salty dishes.

Junmai Daiginjoshu	1,800ml
Alc.	14%
Rice type	Koshiibuki/Gohyakumangoku
Sake meter value	+7.0-+9.0
Acidity	1.5-1.7

Water source	Hime River subterranean water
Water quality	Medium-hard

Kaganoi Shuzo

2-3-5 Omachi, Itoigawa-shi, Niigata 941-0061   TEL.025-552-0047
E-mail: kaganoi@cocoa.ocn.ne.jp   http://www.kaganoi.co.jp

Kaga Clan's favorite brewery in Itoigawa

New brewery completed, sake production resumes Creating a "new" taste of Kaganoi.

Kaganoi Shuzo was founded in 1650. In the Edo period (1603-1868), Itoigawa was a post town for the Kaga Hyakumangoku daimyo to stay during their traveling as part of mandatory residence in Edo. Business came to a halt after a disastrous fire in Itoigawa in December 2016. In the spring of 2018, the new brewery was completed and brewing resumed. The new building allows its brewers to better assess the condition of the raw rice compared to before the fire. With the motto of "brewing sake that is true to the basics" in mind, the brewery is working to create a "new" Kaganoi flavor by utilizing the features of the new brewery building while considering the compatibility of the brewing water, which is medium-hard water.

純米吟醸酒
Kaganoi Junmai Ginjo

"A perfect match for seafood"
Junmai Ginjoshu made with Gohyakumangoku sake rice and Kyokai 901 yeast from Niigata Prefecture.

Junmai Ginjoshu	1,800ml/720ml/300ml
Alc.	15%
Rice type	Gohyakumangoku
Sake meter value	+3.0
Acidity	1.6

純米吟醸酒
Kaganoi Junmai Ginjo Takanenishiki

"Made with Takanenishiki, rice suitable for sake brewing"
Only a few breweries use "Takanenishiki" as their sake rice. This Junmai Ginjoshu has a mild aroma and rich flavor with a mellow umami taste.

Junmai Ginjoshu	1,800ml/720ml
Alc.	15%
Rice type	Takanenishiki
Sake meter value	+4.0
Acidity	1.7

Kikusui Shuzo

750 Shimagata, Shibata-shi, Niigata 957-0011 TEL.0254-24-5111
E-mail: customer@kikusui-sake.com https://www.kikusui-sake.com

Water source	Shibata City water
Water quality	Soft

More than just taste
Committed to passing on not only the taste but the culture of sake

Sake brewing based on tradition and new ideas more than just the pursuit of taste.

Kikusui Shuzo was founded in 1881 in the city of Shibata, Niigata Prefecture. In addition to the high-quality rice harvested in the northern Echigo Plain, there are abundant underground water veins around the Kaji River. Combined with pure underground water from the melting snow of the Iide mountain range, the river provides the perfect environment for sake brewing. Kikusui Shuzo is known for a lineup of unique sake, including "Funaguchi Kikusui Ichiban Shibori", Japan's first undiluted sake in an aluminum can, and "Kikusui no Karakuchi (dry)", a classic Niigata local sake. The brewery also develops new products in order to accommodate to customers' preferences that change with the times. Currently, the brewery exports its products to twenty countries. The golden cans of Funaguchi Kikusui Ichiban Shibori, which look like the color of the tassels of rice in autumn, eloquently explain the place and climate in which the sake is made, and convey the local terroir of Shibata to the world.

本醸造酒
Kikusui no Karakuchi

"A sharp dryness with umami"

Dry sake with a light flavor that is easy to drink. Its sharp dryness is well balanced with the umami of this Honjozoshu.

Honjozoshu	…1,800ml/720ml/500ml/300m/180ml
Alc.	15%
Rice type	Niigata rice
Sake meter value	+7.0
Acidity	1.3

本醸造生原酒
Funaguchi Kikusui Ichiban Shibori

"An aroma and flavor like fruit"

This is Japan's first unpasteurized sake in an aluminum can. It is rich in the fresh aroma and umami that only freshly squeezed sake can provide.

Honjozoshu	1,500ml/500ml/200ml
Alc.	19%
Rice type	Niigata rice
Sake meter value	-3.0
Acidity	1.8

Water source	Subsoil water from the Yashiro River
Water quality	Soft

Kiminoi Shuzo

3-11 Shimocho, Myoko-shi, Niigata 944-0048 TEL.0255-72-3136
E-mail: mail@kiminoi.co.jp   http://www.kiminoi.com/

In the past and in the future
It's always been, and it always will be, "elegant Yamahai"

Sake brewed with lactic acid bacteria of the brewing house pursuing one-of-a-kind umami.

Kiminoi Shuzo is located along the old Hokkoku Kaido road at Araijuku. Because Arai has long been prone to fires, it is unclear when exactly the brewery was established. But the oldest document mentioning the brewery says it was founded in 1842. One of the main features of Kiminoi's sake brewing is the time-consuming and labor-intensive process of the Yamahai method. It is a process that takes time to ferment using natural lactic acid bacteria. The bacteria is naturally found in the brewing house, producing "life" in a mysterious way. This is called "Kuratsuki lactobacillus Jikomi", and the brewery has inherited the "elegant Yamahai" brew with a unique flavor. With this specialty of Yamahai brewing, the brewery continues to produce sake for meals that is well-received by fans both in Japan and abroad.

純米吟醸酒

**Kiminoi
Yamahai Junmai Ginjo**

"Aromatic umami made possible by the Yamanai method"

This sake is brewed with lactic acid bacteria that is naturally found in the brewing house. It has a rich umami and cleanliness that will keep you coming back for more.

Junmai Ginjoshu	1,800ml/720ml
Alc.	15.5%
Rice type	Gohyakumangoku
Sake meter value	+4.0
Acidity	1.4

純米酒

Kiminoi Yamahai Junmai

"Junmaishu with a mild flavor"

Yamahai sake brewed with lactic acid bacteria that is naturally found in the brewing house. It was awarded the Platinum Prize in the Junmai Sake category of KURA MASTER 2020.

Junmaishu	1,800ml/720ml
Alc.	15.5%
Rice type	Koshitanrei/Niigata rice
Sake meter value	+2.0
Acidity	1.7

Kusumi Shuzo

Water source	Spring water
Water quality	Soft

1537-2 Ojiyama, Nagaoka-shi, Niigata, 949-4511 TEL.0258-74-3101

An important brewery in snowy Niigata that boasts artisanal skills of Echigo Toji

Revived high-quality rice using the remaining seeds. The brewery associated with the manga "Natsuko no sake" (Natsuko's Sake).

Founded in 1833 by Sakunosuke Kusumi, the first generation of the family, the brewery has been producing sake for more than 180 years in a small village blessed with nature in snowy Niigata, maintaining the tradition of handmade sake. Using natural water from the mountain behind the brewery (designated as Niigata Prefecture's high-quality water) as brewing water, the representative brand "Kiyoizumi" is brewed with the masterful skills of Echigo Toji. Daiginjoshu "Kame no O" was launched in 1980 by the sixth-generation chief using rice grown by the family over a period of three years from just 1,500 seeds of the famous "Kame no O" rice, the production of which had been stopped before World War II. The story of this rice and sake brewing became the motif of the manga "Natsuko's Sake" and was made into a TV drama series. It has been awarded the top spot among sake breweries in Japan in the Parker Point rating by Robert Parker, a wine critic.

純米吟醸酒
Kiyoizumi Nanadaime

"Young brewers' dream coming true"
The seventh-generation president of the brewery and its young brewers worked together to create sake with a flavor of "a flower blooming in the field". The taste is similar to high-quality white wine.

Junmai Ginjoshu	1,800ml/720ml
Alc.	15%
Rice type	Yamadanishiki

純米吟醸酒
Kopirinko Kopirinko

"Kopirinko of men, Kopirinko of women"
The term "kopirinko" was coined by fermentologist Takeo Koizumi to describe the way people savor Japanese sake.

Junmai Ginjoshu	300ml
Alc.	15%

Takeda Shuzoten

Water source	Company's own well water
Water quality	Somewhat soft

171 Kamikofunatsuhama, Oogata-ku, Joetsu-shi, Niigata 949-3114 TEL.025-534-2320
https://www.katafune.jp/

The renowned sake brand "Katafune" is "umakuchi" sweet sake with plenty of body

The sake is made with rich, high-quality brewing water that brings out the characteristics of the rice.

In 1866, at the end of the Tokugawa Shogunate era, Seizaemon Takeda started a sake brewing business in Kamikofunatsu known as a harbor town. The name of the brewery's main brand, "Katafune," originated from the "lagoons" (kata) and "boats" (fune) that dotted the sand dunes. The special feature of "Katafune" is its rich and delicious taste. Compared to the so-called Tanrei dry sake that has become so popular these days, Katafune's flavor is fuller and rounder, with a gentle sweetness that is in harmony with its dryness. In order to bring out the characteristics of the rice, the brewery asks contract farmers to produce no more than eight bales per hectare. The high-quality, abundant brewing water is filtered through sand dunes. With this rice, water, and the philosophy of "never cut corners at work," the brewery continues to make sake that will please its customers.

特別本醸造酒
Katafune Tokubetsu Honjozo

"The winner of the highest award at the IWC"

A deep flavor that goes down well. A fragrant sake that can be served warm or cold, and is appreciated by all.

Special Honjozoshu	1,800ml/720ml
Alc.	15.6%
Rice type	Koshikagura/Koshiibuki
Sake meter value	-2.0
Acidity	1.3

純米酒
Katafune Junmai

"You can taste the flavor of the rice"

This sake is made only from rice and rice malt. It won the top prize in the Junmaishu category of the Kanto-Shinetsu Regional Taxation Bureau's sake competition.

Junmaishu	1,800ml/720ml
Alc.	15.7%
Rice type	Koshikagura/Koshiibuki
Sake meter value	-3.0
Acidity	1.7

Tochikura Shuzo

274-3, Otsu 1chome, Oozumi-machi, Nagaoka-shi, Niigata 940-2146 TEL.0258-46-2205
E-mail: t-tochi@niks.or.jp

Water source	Natural water
Water quality	Soft

The treasure of a brewery is its relationships with people

A brewery that continues to strictly adhere to traditional sake brewing techniques in Nagaoka, where the spirit of "100 bales of rice" originated.

Since its establishment in 1904, Tochikura Shuzo has been producing sake in Ozumi in the western part of Nagaoka City. The quality of sake the brewery aims for has a strong flavor and can be enjoyed every evening without growing tired of it. To achieve this, the brewery is committed to meticulous hand-crafting in the important processes of producing koji, mash, and malt. Particularly in the process of making koji, the Toji himself takes responsibility, while teaching the know-how to younger brewers. Brewing water is drawn fresh from a side well in the mountain behind the brewery, and the sake rice is locally grown. The name of the main brand, "Komehyappyo" (meaning 100 bales of rice), is derived from a legend that the Nagaoka clan sold the 100 bales of rice they received after the Boshin War to establish a school, instead of feeding people. The importance of fostering human resources is the same for sake brewing. For the marketing of sake, the brewery has established a "direct sales system," whereby it distributes sake directly to trusted retailers in and outside of the prefecture.

純米吟醸酒
Kuranotakara Rokuroji Junmai Ginjo Namazake

"The best of brewing techniques bearing the name of a master brewer"

This sake is named after Rokuroji Goh, a master brewer who was selected as one of the artisans of Niigata. It has a moderate aroma and a good flavor.

Junmai Ginjoshu	1,800ml/720ml
Alc.	16%
Rice type	Domestic rice
Sake meter value	-1.0

普通酒
Komehyappyo Dento no Aji

"Futsushu with no cutting corners whatsoever"

A sharp, refreshing flavor but also has natural sweetness and umami. It is a reasonably priced sake for evening drinking, but the rice polishing ratio is less than 60%, the same as Ginjoshu.

Futsushu	1,800ml
Alc.	15-16%
Rice type	Gohyakumangoku
Sake meter value	+5

| Water source | Subsoil water from Hakkai mountain system |
| Water quality | Soft |

Hakkai Jozo

1051 Nagamori, Minamiuonuma-shi, Niigata, 949-7112 TEL.0800-800-3865
https://www.hakkaisan.co.jp/

A region blessed by the gods for brewing sake

From Niigata in the midst of deep snow, the brewery aims to raise the standard of Japanese sake.

Hakkai Jozo was established in 1922 and is the brewer of Hakkaisan, one of Niigata's most famous brands. Minamiuonuma City, located in the Chuetsu region of Niigata Prefecture, has heavy snowfall in winter. The brewing water is called "Raiden-sama no Kiyomizu" (meaning Raiden's sacred water), or subterranean water from the Hakkai mountain system. The stable, low-temperature water brought about by the snow that falls in Uonuma, Niigata, enables the production of clean, light, and refreshing sake without any peculiarities. The brewers say, "This is the place gods created for brewing sake". Hakkai Jozo hopes to raise the standard of Japanese sake by improving the quality of all types of sake. Its goal is to raise the quality of futsushu (ordinary sake with added alcohol) to that of ginjo-shu, and ginjo-shu to that of daiginjo. The brewery is determined to make good sake available to as many people as possible. With this commitment passion, the number of fans of the famous Hakkaisan sake continues to increase.

大吟醸酒
Daiginjo Hakkaisan

"The brewer's skill shines through in this sake for meals"
High-quality Yamadanishiki and Gohyakumangoku selected for this sake are polished to 45%. This Daiginjo is brewed with handmade koji and spring water from Mt. Hakkai.

Daiginjoshu	1,800ml/720ml/300ml/180ml
Alc.	15.5%
Rice type	Yamadanishiki/Gohyakumangoku etc
Sake meter value	+5.0
Acidity	1.2

スパークリング日本酒
Binnai Nijihakkoshu Shirokoji Awa Hakkaisan

"Brewed with white Koji used for making shochu"
This fizzy sake is characterized by the delicate bubbles produced by in-bottle fermentation and the refreshing acidity created by the fermentation process.

Sparkling sake	720ml/300ml
Alc.	12%
Rice type	Gohyakumangoku
Sake meter value	+1.0
Acidity	6.5

Masukagami

Water source	Subsoil water from Mt. Awagatake
Water quality	Soft

1-1-32 Wakamiya-cho, Kamo-shi, Niigata 959-1355 TEL: 0256-52-0041
E-mail: info@masukagami.co.jp https://www.masukagami.co.jp/

One-of-a-kind sake brewing with a playful spirit

Aiming to be a One-of-a-kind local sake brewery with quality and unique product lineups.

Kamo City, Niigata Prefecture, is located on the south side of the Niitsu Hills, in a scenic area with views of Mt. Awagatake and Mt. Yahiko. The city has a thriving woodworking industry where paulownia cabinets are made, and is known as the "Little Kyoto of Hokuetsu" for its charming townscape. Masukagami is located on the banks of the Kamo River, which runs through the center of Kamo City, and has been in the sake brewing business since its establishment in 1892. The brand name "Masukagami" is derived from a waka poem in the Manyoshu collection of Japanese poems. There are more than eighty sake breweries in Niigata Prefecture, which produces many famous brands. The sake brewers' association in Niigata has set "tanrei" (light flavor with a clean finish) as a guideline for the quality of Niigata sake. Masukagami also advocates sake that is tanrei and has umami. The brewery is known for its unique product lineups such as "Alphabet line", including the "F40" brand, and "Kamenozoki".

普通酒
Masukagami F40

"Futsushu (F) with rice polished to 40%"

This is an unusual Futsushu (ordinary sake with additives) that uses rice polished to as much as 40%. It has a slightly sweet and soft taste with a crisp finish.

Futsushu	1,800ml/720ml
Alc.	15%
Rice type	Rice suitable for brewing
Sake meter value	-2.0
Acidity	1.1

本醸造酒
Kamenozoki

"Enlightening and revolutionary sake in a jar"

This is luxury sake in a jar. Use a ladle to scoop it out. Popular during the year-end and New Year holidays.

Honjozoshu	1,800ml
Alc.	17%
Rice type	Koshifubuki
Sake meter value	+4.0
Acidity	1.3

Water source	Asahi mountain range subsoil water
Water quality	Soft

Miyao Shuzo

5-15 Kamikatamachi, Murakami-shi, Niigata 958-0873 TEL.0254-52-5181
E-mail: miyao.sake@shimeharitsuru.com https://www.shimeharitsuru.co.jp

Light and delicious
The brewery known for "Shimeharitsuru"

The brewery has inherited an honest way of sake brewing over generations.

Miyao Shuzo was founded in 1819 in Murakami City, the northernmost castle town in Niigata Prefecture. At the end of the Edo period, the company also served as a shipping wholesaler for Kitamaebune (cargo ships that sailed the Japan Sea during the Edo period). At the brewery, there are valuable documents from that time, such as old navigation charts with detailed descriptions of Hokkaido's ports and other old documents. There is also a book on the secret of sake brewing called Shuzo Denju Himitsu no Maki written by Matayoshi Miyao, the second generation of the brewery, which shows that the brewery was committed to making high-quality sake from that time. For the brewing water, pure subsoil water from the Asahi mountain range scooped from a well on the brewery's premises is used. Locally-grown rice is used as the main ingredient. The rice produced in Murakami City and Iwafune County is known for its high quality, but Gohyakumangoku, a rice suitable for sake brewing, is also of equally high quality and contributes to the deliciousness of the Shimeharitsuru sake.

純米吟醸酒
Shimeharitsuru Jun

"Junmai Ginjoshu with a lovely flavor"
This Junmai Ginjoshu has an elegant aroma, a mellow taste unique to Junmai, and a clean aftertaste. local sake that goes well with carp dishes, a specialty of Murakami City.

Junmai Ginjoshu	1,800ml/720ml/300ml
Alc.	15%
Rice type	Gohyakumangoku
Sake meter value	+2.0
Acidity	1.5

大吟醸酒
Shimeharitsuru Gold Label

"A mellow, light, and dry Daiginjoshu"
This seasonal sake is produced in limited quantity. Brewed with Yamadanishiki rice. It has a splendid and elegant aroma and a full flavor, but is light and refreshing.

Daiginjoshu	1,800ml/720ml
Alc.	16%
Rice type	Yamadanishiki
Sake meter value	+5.0
Acidity	1.2

Murayu Shuzo

| Water quality | Soft |

1-1-1, Funato, Akiha-ku, Niigata-shi, Niigata 956-0116   TEL.0250-38-2028

We want our customers to use their senses to taste our sake
We dare not disclose the specifications of our sake

A small brewery with a production capacity of 20,000 bottles producing fine, clear sake.

Founded in 1948 in Kosudo, Niigata City, Murayu Shuzo is a small brewery with a production capacity of 200 koku(1 koku = 100 bottles). The brewery does not increase its production volume so that it can pay attention to all of the production processes. The limited-edition brand "Murayu" was launched in 2002 by Kensuke Murayama, and is characterized by a fine, transparent, elegant sweetness that overturns the image of Niigata sake. After graduating from the Junior College of Tokyo University of Agriculture, Murayama worked as an apprentice at other breweries before joining his family's Murayu Shuzo, which had been temporarily closed for two years at the time. With no Toji (master brewer) at the brewery, he struggled to develop his own style of sake making. The specs of the sake are not disclosed to the public, because the brewers think that it is up to the drinker to decide whether or not it is good.

大吟醸酒
Hanakoshiji Daiginjo

"Sake that brings out the best of the flavor of rice"

This is a gem produced by Kensuke Murayama with his heart and soul. It has a splendid, fruitlike Ginjo aroma with an elegant sweetness.

Daiginjoshu	1,800ml/720ml
Alc.	15-16%
Rice type	private
Sake meter value	private
Acidity	private

純米大吟醸酒
Murayu Tokiwa Label Junmai Daiginjo

"A clean aftertaste"

A balance of fine, elegant sweetness and moderate acidity. This Junmai Daiginjoshu that evokes the flavor of good quality white wine.

Junmai Daiginjoshu	1,800ml/720ml
Alc.	15%
Rice type	private
Sake meter value	private
Acidity	private

Water source	Mt. Sumondake subsoil water
Water quality	Soft

Yukitsubaki Shuzo

3-14 Naka-machi, Kamo-shi, Niigata 959-1351 TEL.0256-53-2700
https://www.yukitsubaki.co.jp

A small brewery's handcrafted sake

The brewery aims to make tasty sake you can drink everyday.

Yukitsubaki Shuzo, located in Kamo City, Niigata Prefecture, was established in 1806. The brewery values "accessible sake brewing" above all else, and insists on making sake without mechanization in the traditional way. The city of Kamo, also known as the little Kyoto of Hokuetsu, is famous for its native snow camellia trees, which have been designated as prefectural trees, and the brewery hopes that its approach to sake brewing will be like the snow camellia trees that endure the harsh winds and snow, and bloom when the snow melts. The brewery's main brand, "Koshino Yukitsubaki," was named after this philosophy. In 2011, the brewery became a "Junmai" brewery, producing only Junmaishu, in order to emphasize the brewery's uniqueness and characteristics, and continues to brew Junmai sake that is full of the brewer's passion for flavor and clarity. In recent years, Yukitsubaki has also taken on the challenge of creating products that are rooted in the local community, such as yeast from the flowers of the Camellia japonica, used for brewing.

Junmai Daiginjo Genshu Tsukino Tamayura

"Unprocesed sake brewed with time and effort"

This sake is made entirely from Yamadanishiki rice. Its fruity aroma and full flavor are in harmony. Best served chilled.

Junmai Daiginjoshu	720ml
Alc.	17%
Rice type	Yamadanishiki
Sake meter value	-2.0
Acidity	1.3

Koshino Yuki Tsubaki Junmai Ginjo Yukitsubaki Kobo Jikomi

"Junmai brewed with original yeast"

This sake is brewed with yeast collected from the snow camellia in Kamoyama Park. The aroma is fresh and floral, and the taste is refreshing.

Junmai Ginjoshu	720ml
Alc.	16%
Rice type	Niigata rice
Sake meter value	±0
Acidity	1.6

Watanabe Shuzoten

Water source	Shiroyama water system
Water quality	Soft

1197-1 Negoya, Itoigawa-shi, Niigata 949-0536 TEL.025-558-2006
E-mail: houjyougura@nechiotokoyama.jp https://nechiotokoyama.jp/

Domaine style sake reflecting the climate of Nechi Valley

The brewery started growing its own sake rice, "Gokyakumangoku" and "Koshitanrei," in 2003.

Heijuro Watanabe started Watanabe Shuzo in 1868, and Yoshiki Watanabe, the current president, is the sixth generation. The brewery is located in Itoigawa City, Niigata Prefecture, one of Japan's leading rice producing areas. Along the Nechi River, with its source on Mt. Amakazari-- one of the 100 most famous mountains in Japan, there is a rural landscape called Nechi Valley. In 2003, the brewery started cultivating its own sake rice, "Gohyakumangoku" and "Koshitanrei", which are unique to Niigata, and producing domaine- style sake that reflects the climate of Nechi valley. The goal is to establish the value of region, variety, quality, and vintage in the world of sake. The rice fields opened by the family's ancestors belong to the whole valley. The brewery believes that it is also responsible for preserving the beautiful landscape for the next generation.

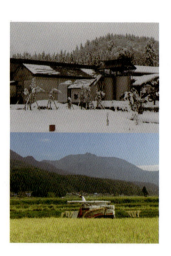

Nechi Otokoyama Junmai Ginjo

"A taste you will never get tired of drinking"

This Junmai Ginjoshu features a soft and fragrant rice flavor. The sake rice used is Gohyakumangoku and Koshitanrei, both locally grown in Nechi.

Junmai Ginjoshu	720ml
Alc.	15%
Rice type	Gohyakumangoku/Koshitanrei
Sake meter value	private
Acidity	private

Nechi Nechidani Gokyakumangoku

"Each bottle has a different taste according to the year of production"

This sake is made from Gohyakumangoku and features a vintage flavor that expresses the differences in the quality of the rice in each harvest year.

Junmai Ginjoshu	720ml
Alc.	16%
Rice type	Gohyakumangoku
Sake meter value	private
Acidity	private

Guide to Sake Breweries and Famous Sake
Hokuriku Region
Toyama, Ishikawa, Fukui

Food culture of the Hokuriku region

Toyama: Surrounded by mountains and Toyama Bay, which borders the Sea of Japan, Toyama is famous for its heavy snowfall. In Toyama Bay, fishing for cold yellowtail and firefly squid thrives, and "Kaburazushi" (yellowtail sushi), "buri-daikon" (yellowtail and daikon radish), and "Okizuke" (firefly squid pickled in tamari soy sauce) are well known. White shrimp is also enjoyed as sashimi, kombu-jime, or tempura. River fish is also eaten, as is "Masuzushi" (trout sushi) made with Sakuramasu from the Jinzu River, and "Iwadofu" (rock tofu) made with soybeans and clear water.

Ishikawa: Agriculture thrives, and dishes using lotus root include "hasumushi" (steamed lotus) and "Surinagashijiru" (soup made from ground lotus root). There is also "Kaburazushi" made with salted mackerel, "Fugunokasuzuke" made with pickled blowfish egg nests, and colorful fu (wheat gluten), which adds a colorful touch to stewed dishes. A type of seasoning made by fermenting and maturing salted fish and shellfish for more than a year is known as "Ishiru," one of the three major fish sauces in Japan. The different flavors of the ingredients used in each region are also appealing.

Fukui: Rice production flourishes, but Wakasa Bay has a variety of dishes using abundant seafood, such as crab and mackerel. Crab is a specialty of Fukui Prefecture known as "Echizen crab". Mackerel seasoned in salted rice-bran paste is called "Heshiko" and is also a preserved food. Shojin ryori, or Buddhist vegetarian cuisine, is also popular in Fukui. The prefecture is home to Eiheiji, a large Buddist temple, which has influenced local cuisine with vegetarian elements.

Sansyouraku Shuzo

Water source	Well water
Water quality	Soft

678 Kaminashi, Nanto-shi, Toyama 939-1914　TEL.0763-66-2010
E-mail: info-san@sansyouraku.jp　http://www.sansyouraku.jp

Drinking sake should be fun

Gokayama, a region hidden in the depths of winter, nurtures sake that is fun to drink and delicious.

Sansyouraku Shuzo was founded in 1880. Gokayama, where the brewery is located in the southwestern part of Toyama Prefecture, is a land of steep mountains and deep valleys. In winter, the area receives heavy snowfall of more than two meters, and the traditional Gasshozukuri villages scattered around Gokayama were registered as a UNESCO World Heritage Site in 1995. The harsh climate of Gokayama is the essence of Sansyouraku's flavor. The water used to brew this sake gushes out from the virgin beech forest that protects the village from avalanches. The severe cold of winter and the large amount of snow that falls Japan nurture Sansyouraku's sake. The company name "Sansyouraku" comes from an ancient Chinese story which goes that a monk and two of his old friends were having so much fun drinking sake that they crossed the "Tiger Valley," which they had no intention of crossing, and they laughed (San means three, and Syouraku means to have fun and laugh). The philosophy of Sansyouraku is "sake that is meant to be enjoyed with pleasure".

普通酒
Sansyouraku Jyousen

"Sake that is loved the most in Gokayama"

A blend of Yamahai honjozo (Japanese brewed sake), this sake has a wide range of flavors that go well with wild game meat dish and other mountain delicacies. It can be enjoyed both cold and warm.

Futsushu	1,800ml/720ml
Alc.	15%
Rice type	Gohyakumangoku/Prefectural rice
Sake meter value	+4.0
Acidity	1.6

純米酒
Sansyouraku Yamahai Junmai

"Clean umami and acidity"

This sake is characterized by the goodness of Yamahai and its acidity. The soft acidity complements the food, and the fruity aroma keeps you drinking. It is best served lukewarm.

Junmaishu	720ml
Alc.	16%
Rice type	Yamadanishiki
Sake meter value	+3.0
Acidity	1.8

| Water source | Shogawa river system subsoil water |
| Water quality | Soft |

Kiyoto Shuzojo

12-12, Kyo-machi, Takaoka-shi, Toyama 933-0917 TEL.0766-22-0557

Our sake brewing philosophy is "do not tolerate falsehoods"

The brewing house with a good old-fashioned atmosphere has been registered as a national tangible cultural property.

The brewery was founded in 1906. Keisuke Seito, the founder of the brewery, was a member of the cavalry during the Russo-Japanese War. He named the brewery "Kachikoma" (winning horse) in commemoration of Japan's victory, and started brewing sake at the current location. Later, in the 1960s, when the brewery's sake was almost forgotten due to the trend of mass production and mass sales during the period of rapid economic growth, local volunteers organized an event to rediscover the hometown through local products. As they enjoyed local sake and local cuisine, they were inspired by people's desire to drink Kachikoma's Ginjo sake, and released a Daiginjo product. The logo on the label was created by the late Masuo Ikeda, an artist with whom the brewery had a close relationship. Since then, the company has continued to refine its Ginjo, aiming to produce authentic sake that is rooted in people's daily lives, without being overly fanciful.

大吟醸酒
Kachikoma Daiginjo

"An elegant, fruity aroma and refreshing mouthfeel"

This sake has a Ginjo aroma that fills the mouth with a plump, clear taste. It is also refreshing to drink.

Daiginjoshu	1,800ml/720ml
Alc.	17%
Rice type	Yamadanishiki
Sake meter value	+4.0
Acidity	1.4

純米吟醸酒
Kachikoma Junmai Ginjo

"Gentle aroma and deep flavor"

With a gentle, fruity Ginjo aroma and a deep flavor that only Junmaishu can provide, this sake has an exquisite, mild aftertaste.

Junmai Ginjoshu	1,800ml/720ml
Alc.	16%
Rice type	Yamadanishiki
Sake meter value	+2.0
Acidity	1.4

Takasawa Shuzojo

Water source	Pure water from Goishigamine
Water quality	Medium-soft

18-7 Kitaomachi, Himi-shi, Toyama 935-0004 TEL.0766-72-0006
E-mail: akebono@p1.cnh.ne.jp https://ariiso-akebono.jp/

A brewery that views the dawn over the Ariiso Sea

A sake brewery that preserves the old-fashioned handcrafting method of "Funeshibori".

The brewery was founded in 1872 by Riemon Takasawa, who came to Himi in the late Edo period and started brewing sake. This area is ideal for sake brewing for many reasons; Toyama Bay in front of the brewery, with one of Toyama's 100 peaks, Mt. Goishigamine rising behind it. Also the abundance of well water and the sea breeze "Ainokaze" blowing in from Toyama Bay. The brewery uses local, high-quality sake rice, and cools the steamed rice using "Ainokaze", then the Funeshibori method is applied, brewing and pressing the sake entirely in the tank, in order not to stress or damage the sake. The brewery's sake goes perfectly with the abundant seasonal seafood caught in Toyama Bay, a natural fish tank that stretches out in front of the brewery. The main brand name is "Ariiso Akebono". The word "Ariiso" means "rough sea", which was used in old Japanese poetry to refer to Toyama Bay.

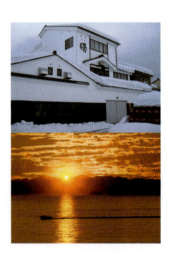

Ariiso Akebono
Junmai Daiginjo

"Calm Ginjo aroma, unique to Hokuriku"

This sake is made with 100% Yamadanishiki produced in Nanto, Toyama. It has a calm Ginjo aroma unique to Hokuriku, a refined rice flavor, and a full-bodied elegance.

Junmai Daiginjoshu	1,800ml/720ml
Alc.	16%
Rice type	Yamadanishiki
Sake meter value	+2.0
Acidity	1.5

Ariiso Akebono
Hatsuarashi Junmai Ginjo

"The Sake Rice "Tomi no Kaori" 100% produced in Himi"

This sake is made from 100% "Tominokaori", a sake rice produced in Himi. It is characterized by the splendid ginjo aroma of Kanazawa yeast, elegant flavor and sharpness.

Junmai Ginjoshu	1,800ml/720ml
Alc.	16%
Rice type	Tominokaori"
Sake meter value	+4.0
Acidity	1.5

| Water source | Subsoil water from the Shogawa River |
| Water quality | Medium-hard |

Tateyama Shuzo

217 Nakano, Tonami-shi, Toyama 939-1322 TEL.0763-33-3330
E-mail: tateyama@tateyama-brewing.co.jp https://www.sake-tateyama.com

The largest brewery in Toyama, a rice producing region

Use only carefully selected rice suitable for sake and brew carefully by hand.

Tateyama Shuzo, one of the largest sake breweries in the Hokuriku region, was founded in 1830, and is located in the Tonami Plain in the beautiful rural landscape of Shikii Village in western Toyama. The brand was named after "Tateyama", one of the three most famous mountains in Japan. With a belief in pursuing quality, the brewery uses subsoil water from the Shogawa River in the Hakusan mountain range and high-quality raw rice selected by the brewery's own color sorter. Sake brewed from these ingredients goes well with fresh Hokuriku seafood, such as white shrimp and grilled white fish, and is supported by the residents of Ishikawa and Toyama. In recent years, the company has expanded its business nationwide, mainly in the Tokyo metropolitan area, and won various awards in the last twenty years, including twenty gold medals (fifteen top honors) at the Regional Taxation Bureau's Appraisal Conference, nine gold medals at the National New Sake Appraisal Conference, and gold medals at the IWC's Sake Competition in 2017, 2018, and 2019.

大吟醸酒
Daiginjo Tateyama

"Rich aroma and graceful texture"
A smooth and elegant taste with a full-bodied aroma and a graceful scent on the palate,. The flavor of the rice has been pursued to the limit. Best served chilled.

Daiginjoshu	720ml
Alc.	15.8%
Rice type	Yamadanishiki
Sake meter value	+3.0
Acidity	1.1

純米大吟醸酒
Junmai Daiginjo Tateyama

"A refreshing and soothing Ginjo aroma"
This sake has a refreshing and soothing Ginjo aroma, and a rich sweetness that harmonizes with its mellow flavor. It can be enjoyed in a wide range of ways, both cold and warmed.

Junmai Daiginjoshu	1,800ml/720ml
Alc.	15.8%
Rice type	Yamadanishiki
Sake meter value	+1.0
Acidity	1.3

Fukutsuru Shuzo

Water source	Subsoil water from the North Alps
Water quality	Ultra-soft

2352 Nishimachi, Yatsuo-machi, Toyama-shi, Toyama 939-2355 TEL.076-455-2727
E-mail: seisyu@kazenobon.co.jp https://www.kazenobon.co.jp

Brewed with organically grown rice Trustworthy sake

Made from organically grown Koshihikari rice.

Founded in 1848 in the town of Etchu-Yatsuo in Toyama, the current brewer is the eighth generation. The brewery is known as "Fukuku" in the area, and supports the Hikiyama festival in the spring and the Owarakazenobon festival in the fall as a long-established company living and working with the community. Since the 1950's, the town has promoted organic farming, livestock production and liquor processing. Fukutsuru makes the only organic sake in the prefecture, and its quality has won the trust of many people. In recent years, the "Hachiou" series of low-milling Junmai Genshu has been well received and has grown to include a second major brand after "Kazenobon". The company has been working with the local JA (Japan Agricultural Cooperatives) to produce sake rice using the traditional Yamahai and Kimoto brewing methods. While inheriting the tradition, Fukutsuru also takes on the challenge of a new style. The brewery strives to contribute to the local recycling through local production for local consumption, and they will continue to promote the appeal of sake.

純米酒
Yuuki Junmaishu Kazenobon

"Mellow, deep and rich"
This organic sake is made from 100% JAS-certified organic "Koshihikari" rice.

Junmaishu	720ml
Alc.	15%
Rice type	Koshihikari
Sake meter value	±0
Acidity	private

純米酒
Junmai Genshu Hachiou

"Delicate sake with low rice polishing"
Junmai Genshu made with 78% polished "Tentakaku", rice grown in Toyama and brewed under exquisite temperature control, from start to finish.

Junmaishu	720ml
Alc.	18%
Rice type	Tentakaku
Sake meter value	+1.0
Acidity	private

Water source	Subsoil water from the Hakusan Tedorigawa river system
Water quality	Medium soft

Kobori Shuzoten

1-Wa47, Tsurugi-honmachi, Hakusan-shi, Ishikawa 920-2121 TEL.076-273-1171
E-mail: h.jougen@manzairaku.co.jp   https://www.manzairaku.co.jp/

Kaga, Tsurugi
Famous sake from the foot of Mt. Hakusan

"Manzairaku" is a brand name that expresses the desire to bring happiness to customers.

The brewery has been making sake for about 300 years, since the Kyoho era (1716-1734) of the Edo period. In the deep snow of the Hokuriku region, they use "Yamadanishiki", "Gohyakumangoku" and other varieties of rice, and have worked with local farmers to revive "Hokuriku No.12". It is a rice suitable for sake brewing, but had disappeared completely. In 2001, the company completed construction of the Ginjo Brewery "Hakusan". Their desire was to establish a handcrafting brewery for the 21st century that continues to make genuine sake, and also to make good sake in a quiet place blessed with water and nature. Kobori Shuzoten is equipped with the latest facilities and uses cedar trees cut down during construction as building materials, which won it the Good Design Award in 2002. Their latest facilities make it possible to brew a high quality sake, made with the heart and hands of a craftsman.

純米酒
Manzairaku Tsurugi Yamahai Junmai

"Refreshing and simple local sake"
The sake is made from "Gohyakumangoku" which is produced at the foot of the Hakusan. It has a strong, consistent taste with a good amount of umami and acidity. It also has a nice crisp aftertaste.

Junmaishu	1,800ml/720ml
Alc.	16%
Rice type	Gohyakumangoku
Sake meter value	+80
Acidity	2.2

純米酒
Manzairaku Jin Junmaishu

"Local sake certified by GI Hakusan"
The sake is made from "Hokuriku No. 12", a rare variety of rice grown independently. This Junmaishu has the aroma of sake rice, a mild acidity, and a soft richness.

Junmaishu	1,800ml/720ml
Alc.	16%
Rice type	Hokuriku No. 12
Sake meter value	+4.0
Acidity	1.8

Kano Shuzo

Water source	Hakusan water system
Water quality	Soft

I-6 Yokaichimachi, Kaga-shi, Ishikawa 922-0336 TEL.0761-74-1551
E-mail: h.kano@jokigen.co.jp   http://www.jokigen.co.jp

A brewery that has been watching over Ishikawa's history and sake culture

Sake brewed with tradition and historical romance that matches local ingredients.

The brewery was founded in 1819 in Kaga, Ishikawa, overlooking the sacred mountain Hakusan. The company's history as a landowner watching over the area for generations goes hand in hand with its sake brewing. The starting point of Kano Shuzo is its commitment to people, rice, and water. The brewery uses "Yamadanishiki" rice harvested from its own rice paddies, and the "Hakusui Well" which springs up in the area, as its brewing water. This is the reason why Kano Shuzo is sometimes called a rural sake brewery. It is said that many locals go out of their way to draw water from the "Hakusui Well" for tea and coffee. The name of the brewery's flagship brand "Jyokigen" comes from a poem written by the fourth-generation head of the brewery at a party celebrating a bumper harvest with the villagers one year: "Yaegiku (double chrysanthemums) and sake are as good as Jyokigen (eternal good mood)."

純米大吟醸酒
Jyokigen Junmai Daiginjo

"Junmai Daiginjoshu with the ultimate in depth and umami"

This Junmai Daiginjoshu is made by fermenting sake at low temperature for a long time and then maturing it further. The rich aroma and full-bodied taste are excellent.

Junmai Daiginjoshu	1,800ml/720ml
Alc.	16%
Rice type	Yamadanishiki
Sake meter value	+2.0
Acidity	1.4

純米酒
Jyokigen Yamahai Junmai

"Richness and sharpness unique to Yamahai"

The yeast is cultivated using lactic acid bacteria from the natural world, an ancient method of production. It has a rich, full-bodied taste with a sharp finish.

Junmaihu	1,800ml/720ml
Alc.	16%
Rice type	Gohyakumangoku
Sake meter value	+3.0
Acidity	1.8

| Water source | Hakusan water system |
| Water quality | Medium-hard |

Kikuhime

Ta-8 Tsurugi-shinmachi, Hakusan-shi, Ishikawa 920-2126 TEL.076-272-1234
E-mail: webmaster@kikuhime.co.jp https://www.kikuhime.co.jp

Since the Azuchi-Momoyama period
The taste of tradition

A sake brewery uncompromising in its quest to make delicious sake and take on new challenges.

The brewery was founded during the Tensho period (1573-1592) in the Azuchi-Momoyama era, which makes it one of the oldest sake breweries in Japan. The name of the company was "Koyanagiya" at the time of its establishment, but it was reorganized into "Limited Partnership Yanagi Shuzo-ten" in 1902, and then into the current "Kikuhime Limited Partnership" in 1928. In 1968. The company improved its entire brewing process, including raw materials, to competition level and released "Daiginjo". In 1978(Showa 53), the company was the first in Japan to release "Yamahai Jikomi Junmaishu" a unique Junmaishu with a lot of rice flavor, using "Yamahai sake mash". This brewing method takes time and effort to grow sake mash, and is not suitable for mainstream "Tanrei" type sake, so it almost completely disappeared. However, "Kikuhime" revived it and now has many fans.

普通酒
Kikuhime Futsushu Kiku

"The old-fashioned "Yamahai" brewing method"

This ordinary sake is brewed using the traditional Yamahai brewing method, and has a rich color, taste, and aroma, making it very drinkable.

Futsushu	1,800ml
Alc.	15-16%
Rice type	Yamadanishiki
Sake meter value	-2.0
Acidity	1.5

大吟醸酒
Kikuhime Daiginjoshu

"Uses the highest quality of Yamadanishiki"

This is a pioneering sake that has changed the world of Ginjo sake. It has a unique aroma of maturity, a mellow flavor, and an enhanced texture.

Daiginjoshu	1,800ml
Alc.	17-18%
Rice type	Yamadanishiki
Sake meter value	+5.0
Acidity	1.2

Kuze Shuzoten

Water source	Own groundwater / Shozu spring water
Water quality	Hard / Soft

I-122 Shimizu, Tsubata-machi, Kahoku-gun, Ishikawa 929-0326 TEL.076-289-2028
E-mail: info@choseimai.co.jp   https://www.choseimai.co.jp

Rice cultivated by the company "Chosei-mai"

The only sake brewery in Japan that has integrated production, from rice cultivation to sake brewing.

This sake brewery has been growing its own original sake rice "Choseimai" in its own rice fields and brewing sake with that rice since it was established in 1786. "Choseimai" has larger grains than regular rice such as "Koshihikari", and has more shinpaku (opaque part made of starch), which is suitable for sake brewing. Kuze Shuzo uses two different types of water to brew sake; hard water drawn from their own well and soft water from a local spring called "Shozu". It is said that brewing with hard water produces a firm and robust masculine sake, while brewing with soft water produces a soft, fluffy feminine sake. At Kuze Shuzo-ten, you can enjoy these two different flavors. It is the only sake brewery in Japan that has been producing sake from rice cultivation through to brewing since its establishment. It has won five awards at the National New Sake Competition, including two gold medals.

特別純米酒
Chosemai
Tokubetsu Junmaishu

"Special Junmaishu made without filtration"

This sake is brewed using hard and soft water, and has a firm, yet full flavor from the rice.

Special Junmaishu	720ml
Alc.	15%
Rice type	Chosemai (In-house cultivated rice)
Sake meter value	+4.0
Acidity	1.5

大吟醸酒
Notoji Daiginjo

"This Daiginjo sake has a masculine flavor"

"Yamadanishiki" rice is polished to 40% and brewed with the brewery's own groundwater (hard water 7.62 degrees). The aroma is delicate and the taste is smooth.

Daiginjoshu	720ml
Alc.	17%
Rice type	Yamadanishiki
Sake meter value	+2.0
Acidity	1.2

Water source	Well water
Water quality	Somewhat soft

Sakurada Shuzo

So-93, Takojima-machi, Suzu-shi, Ishikawa 927-1204 TEL.0768-82-0508
E-mail: info@sakurada.biz https://www.sakurada.co.jp

A sake brewery in a fisherman's town

Aiming to make sake that will be loved from generation to generation.

Sakurada Shuzo is a small sake brewery run by a family of four, located in Takojima, a fishing town in Suzu City at the tip of the Noto Peninsula. It was established in 1915, and has continued to brew sake in connection with the local fishermen, who drink a lot of it. About 90% of the sake produced at the brewery is consumed in Suzu City, and 60% of that is consumed in the Takojima district. At Sakurada Shuzo, the owner is the Toji (master brewer) himself, and the brewing is done by hand in the traditional way, using the so-called "brewer's yeast" that lives in the brewery. The part of the brewery facing the main street is used as a store, and local people come to buy sake for celebrations, festivals, and everyday enjoyment. Sakurada Shuzo will continue to brew sake together with the locals so that their children and grandchildren will be able to enjoy it in the future.

本醸造酒

Noto Josen Honjozo Hatsuzakura

"Sake that you can enjoy on a daily basis"

Delicious Honjozo, either warm or cold. Locally, it is valued as a sake for weddings, funerals, and celebrations.

Honjozoshu	1,800ml/720ml
Alc.	16%
Rice type	Gohyakumangoku/Ishikawamon/Sticky rice
Sake meter value	±0
Acidity	1.2

特別純米酒

Tokubetsu Junmaishu Taikei

"Sake to celebrate the joy of a big catch"

The Toji followed a process of trial and error and finally reached the quality needed to replicate the brand at the time of its founding.

Special Junmaishu	1,800ml/720ml
Alc.	16%
Rice type	Yamadanishiki/gohyakumangoku
Sake meter value	±0
Acidity	2.0

Hakutou Shuzoten

Water source	Mountain behind the brewery
Water quality	Soft

24 Kamimachi, Fugeshi-machi, Wajima-shi, Ishikawa 928-0077 TEL.0768-22-2115
E-mail: info@hakutousyuzou.jp http://www.hakutousyuzou.jp

White chrysanthemums bloom in Wajima, Okunoto

A small-scale production of 230 bottles per year.

This sake brewery was founded in the early 18th century as a shipping wholesaler and started brewing sake in the late Edo period. The name of the brewery was changed to "Shiragiku". It was taken from "Shirakabeya", the name of the company at the time, and "Chrysanthemum Wine" which is drunk during the "Chouyounosekku" Festival. Later, in order to avoid confusion with similar sake names, "Okunotonoshiragiku" became the official trademark. In 1996, the eighth-generation head of the brewery began to serve as Toji (master brewer), and now the ninth-generation head, a husband and wife team, are working on their ideal sake production. Although the brewery was severely damaged by the Noto Peninsula Earthquake in 2007, they retooled the facilities and rebuilt with the aim of producing sake that is loved by the local community and that complements the cuisine. In recent years, the brewery has asked local farmers in Wajima City to grow rice suitable for sake brewing, and is still committed to working with the local Noto community.

Okunotonoshiragiku Junmai Ginjo

"Gentle sweetness and rich sake"
This sake is characterized by its elegant, gentle aroma and sweetness. Best served cold or at body temperature. It has won good results in overseas sake competitions.

Junmai Ginjoshu	1,800ml/720ml
Alc.	16-17%
Rice type	Yamadanishiki/Gohyakumangoku
Sake meter value	-4
Acidity	1.5

Okunoto no Shiragiku Tokubetsu Junmai

"Good taste and aroma of Yamadanishiki"
A refreshing bananalike aroma with a well-balanced flavor and acidity. This sake can be enjoyed cold or hot.

Special Junmaishu	1,800ml/720ml
Alc.	15-16%
Rice type	Yamadanishiki/Gohyakumangoku
Sake meter value	+3.5
Acidity	1.5

Water source	subsoil water from the Dainichizan
Water quality	Soft

Hashimoto Shuzo

I-184 Iburihashimachi, Kaga-shi, Ishikawa 922-0331 TEL.0761-74-0602
E-mail: webmaster@judaime.com https://judaime.com/

Kaga's local sake is a result of the climate of Hyakumangoku

Kaga's sake is brewed by Noto Toji, a master sake brewer who is extremely particular about water and rice.

Hashimoto Shuzo is located in Kaga, home to famous hot spring resorts, such as Yamashiro Onsen and Yamanaka Onsen. Since its founding in 1760, the brewery has been making the most of its refined sake brewing techniques, and is proud to continue brewing sake as a part of traditional Japanese culture. Looking into the history of the brewery, the ancestors of the Hashimoto family can be traced back to the samurai of the Heike clan. When the current 10th generation took over as the representative of the brewery, he produced a Junmai Daiginjoshu called "Jyudaime", which is a compilation of the history of the Hashimoto family, probably because he never forgot the spirit of the Japanese samurai in his heart. The sake is brewed by the Toji, who is extremely particular about water, rice, and the brewing method, such as wooden tank pressing and long-term aging at low temperatures. His mindset does not allow compromise, and the techniques are the reason for the truly genuine sake. The spirit of the first Toji still lives on in Hashimoto Shuzo's local sake.

Jyudaime Umajyuku Honjozo

"Taste and aging that exceed the price"

A soft Ginjo aroma and an umami taste. A gentle aftertaste that will tickle your taste buds. It can be enjoyed in a wide range of styles, from cold to warm.

Honjozoshu	1,800ml/720ml
Alc.	17%
Rice type	Domestic rice
Sake meter value	+3.0
Acidity	1.6

Higashi Shuzo

Water source	Subsoil water from the Mt.Hakusan system
Water quality	Soft

35 Nodamachitei, Komatsu-shi, Ishikawa 923-0033 TEL.0120-47-2302
E-mail : info@sake-sinsen.co.jp http://www.sake-sinsen.co.jp

Chrysanthemum Sake of Kaga, "Sinsen"

Brewing good quality sake in small quantities by focusing on local production.

Higashi Shuzo was established in 1860. This sake brewery is located in Komatsu City, Ishikawa Prefecture and is known for its famous sake, "Sinsen". In the Taisho era (1912-1926), there were about 300 breweries in Ishikawa, and Higashi Shuzo was one of the ten largest ones. However, Higashi downsized and become one of the smaller breweries among the thirty-five in operation today. Their primary focus has shifted to production of high quality sake, mainly Daiginjo and Junmai, and they are dedicated to making sake that is unique to the region by using water, rice, and people all from Ishikawa. This is the motto of Higashi Shuzo today. The subsoil water from Mt. Hakusan is used as brewing water, and the rice and yeast are locally produced in Ishikawa. In 2009, the sake brewery and twelve of its buildings were designated as registered tangible cultural properties. Today it works together with other industrial tourist sites to make Komatsu more lively.

Sinsen Junmai Ginjo Blue label

"Fruity apple aroma"

This is a mellow, dry Junmai Ginjoshu with a fruity apple aroma. You can enjoy the taste and acidity of Kanazawa yeast and the refreshing aftertaste. It was awarded the Kura Master 2020 Platinum Prize.

Junmai Ginjoshu	1,800ml/720ml
Alc.	15%
Rice type	Yamadanishiki

Sinsen
Junmai Ginjo Umakuchi

"Sake to drink in a wine glass"

It has a clear sweet taste with umami spreading in the mouth. But the aftertaste is crisp and dry. It is a good sweet sake to drink cold.

Junmai Ginjoshu	1,800ml/720ml
Alc.	14.5%
Rice type	Gohyakumangoku/Yamadanishiki
Sake meter value	-12.0

| Water source | Subsoil water from Mt. Hakusan |
| Water quality | Medium soft |

Fukumitsuya

2-8-3 Ishibiki, Kanazawa-shi, Ishikawa 920-8638 TEL.076-223-1161
E-mail: press@fukumitsuya.co.jp https://www.fukumitsuya.co.jp/

The first sake brewery in the industry to introduce a back label

In 2001, the brewery declared itself a Junmai brewery.

The brewery was founded in 1625 and later purchased by Tasuke Shioya, a pawnbroker in Kanazawa during the Anei era. In 1803, the seventh generation changed the name of the brewery from "Shioya" to "Fukumitsuya", after the place where the previous generation came from. In order to make delicious and light sake, they are particular about rice. Sake rice is carefully chosen according to the desired taste; "Yamadanishiki" is mainly used for Ginjo-style sake, "Kinmonnishiki" for maturing sake, and "Fukunohana" for koji rice to bring out the flavor of the sake. Fukumitsuya can do this thanks to the continuous research with farmers on cultivation beginning with soil preparation. The brewery adopted the "village rice system", and has purchased all their rice directly from farmers for more than 60 years. In 2001, the brewery declared itself a Junmai brewery which only makes Junmai products. They will continue to brew natural sake with a focus on high-quality rice.

Kagatobi Yamahai Junmai Chokarakuchi

"Rich taste and sharpness"

Yamahai is a traditional brewing method. This is a super dry sake with exquisite acidity and deep richness.

Junmaishu	1,800ml/700ml/300ml
Alc.	16%
Rice type	Contract cultivated rice/ Shuzo suitable rice
Sake meter value	+12.0
Acidity	2.0

Kuroobi Yu-Yu Tokubetsu Junmai

"A "relaxed" taste"

This is a dry sake with a crisp, mellow flavor, made with both Ginjo and Junmai brewing techniques, and aged slowly in the brewery.

Special Junmaishu	1,800ml/700ml/300ml/180ml
Alc.	15%
Rice type	Yamadanishiki/Kinmonnishiki
Sake meter value	+6.0
Acidity	1.6

Yoshida Shuzoten

Water source	Subsoil water from the Mt. Hakusan System
Water quality	Medium-hard

41 Yasuyoshi-machi, Hakusan-shi, Ishikawa 924-0843 TEL.076-276-3311
E-mail: info@tedorigawa.com   https://tedorigawa.com/

History and Innovation of a Sake Brewing Village

Returning to the origins of local sake brewing and striving to make sake for the future.

Yoshida Shuzoten was founded in 1870. There used to be more than a dozen sake breweries in the area, but the Great Depression of the late Taisho Era (1912-1926) led to the closure of many of them, and by the beginning of the Showa Era (1926-1989), only the Yoshida Shuzo remained. The brewery itself suffered many crises due to wars and fires, but each time they rallied. In 2020, they celebrated their 150th anniversary. In recent years, with the goal of returning to the basics of local sake brewing, Yoshida Shuzoten has contracted with thirty farmers around the brewery for their harvest of sake rice, promising a price higher than usual. This is to prevent the farmers from leaving their farms, and also to protect the rice fields and the water. Particularly with regard to water, which is the lifeblood of sake, Yoshida donates a portion of their sales to the "Hakusan Tedorigawa Geopark" in order to protect Mt.Hakusan, the source of the water. They are also making efforts in making the brewing process more sustainable, using wastewater treatment facilities and shifting power to 100% renewable energy.

Tedorigawa Yamahai Junmai

"Yamahai but fresh"
The specialty of Noto Toji (chief brewer) is the Yamahai style. While expressing a freshness unique to "Tedorigawa", it also has an elegant richness and flavor.

Junmaishu	1,800ml/720ml
Alc.	15%
Rice type	Yamadanishiki/Gohyakumangoku
Sake meter value	private
Acidity	private

Yoshidagura "u" Ishikawamon

"A new style of Yamahai"
Natural and gentle sake brewed with "Modern Yamahai". It has the gentle sweetness and refreshing taste unique to Ishikawamon.

Junmaishu	1,800ml/720ml
Alc.	13%
Rice type	Ishikawamon
Sake meter value	private
Acidity	private

| Water source | Subsoil water from the Kuzuryu River |
| Water quality | Soft |

Kokuryu Shuzo

1-38 Matsuoka-Kasuga, Eiheiji-cho, Yoshida-gun, Fukui 910-1133   TEL.0776-61-6110
E-mail: info@kokuryu.co.jp   http://www.kokuryu.co.jp/

Local sake loved for over 200 years in Fukui

Once designated by the feudal government as an encouraged industry for its good water. It still produces a gentle taste.

Kokuryu Shuzo was founded by Ishidaya Nizaemon in 1804 in Edo. The ruling clan had already designated sake brewing as an encouraged industry because of the good water. The source is the sacred Hakusan mountain range, one of the three most famous mountains in Japan. Then it becomes the subsoil water in the Kuzuryu River, the largest river in Fukui Prefecture. The softness of the water gives a light and supple feel, which is ideal for making beautiful, full-bodied Ginjoshu. In 1975, the seventh-generation head of the brewery applied the techniques he learned in France and released "Kuroryu Daiginjo Ryu". It attracted much attention because it was the first Daiginjo sake to be sold commercially. The delicious taste of "Kuroryu" and "Kuzuryu" brands, which have been loved in Fukui for over 200 years, are spreading throughout the world.

大吟醸酒
Kuroryu Daiginjo Ryu

"A long seller for 45 years"
The passion of the brewers and their meticulous care and efforts have borne fruit in this gem. This Daiginjoshu has been loved for nearly 50 years since its release in 1975.

Daiginjoshu	720ml
Alc.	16%
Rice type	Yamadanishiki
Sake meter value	+4.0
Acidity	1.0

大吟醸酒
Kuzuryu Daiginjo

"Delicious Daiginjo when warmed"
This Daiginjo has a deep, refined flavor that is perfect for heating up. It is best served lukewarm to hot.

Daiginjoshu	720ml
Alc.	15%
Rice type	Gohyakumangoku grown in Fukui
Sake meter value	+4.0
Acidity	1.0

Tajima Shuzo

Water source	Spring water "Oshouzu" from Mt.Asuwa
Water quality	Soft

1-3-10, Momozono, Fukui-shi, Fukui 918-8051 TEL.0776-36-3385
E-mail: info@fukuchitose.com https://www.fukuchitose.com/

Tradition and Innovation as a Yamahai Brewery

Seeking the true flavor of sake and sticking to the traditional "Yamahai" brewing method.

"Fukuchitose" began in the late Edo period (1840s). Initially, the brewery was located in Shimizu-cho (formerly Omori, Shizu Village) but due to flooding problems, it relocated and resumed business in Momozonocho in 1953. The current brand name "Fukuchitose" is derived from the good memories of Chitose Town (now Asuwa 2-chome, Fukui City), where the brewery first moved. The sake brewed by "Fukuchitose" is "traditional and innovative". The traditional brewing method, "Yamahai jikomi" which uses lactic acid bacteria from the natural world, is still used today. Although it requires several times more effort and time than modern methods, "Fukuchitose" believes that the best sake can only be brewed using the "Yamahai" method. Since 2013, "Fukuchitose" has made with rice produced entirely in Fukui, and has also taken up the challenge of using awine-making method and storing sake in barrels, aiming to expand the new possibilities of sake and become a one-of-a-kind brewery.

純米大吟醸酒
Fukuchitose Fuku

"Some of the best sake made by Yamahai brewing method"

It is characterized by a mellow and splendid aroma and the clean acidity unique to Yamahai brewing. Best served chilled in a wine glass.

Junmai Daiginjoshu … 1,800ml/720ml
Alc. 15%
Rice type .... Koshinoshizuku
Sake meter value .... +3.0

純米酒
PURE RICE WINE

"A miraculous collaboration of rice and wine yeast"

This junmai wine is also called "Koshihikari for drinking". When chilled and poured into a glass, it has a golden color and tastes just like white wine.

Junmaishu 1,800ml/720ml
Alc. 12%
Rice type .... Koshihikari
Sake meter value .... -25.0
Acidity 5.0

Water source	Mt. Hakusan system
Water quality	Soft

Nambu Shuzojo

6-10 Motomachi, Ono-shi, Fukui 912-0081 TEL.0779-65-8900
https://www.hanagaki.co.jp/

The best sake brewed with Echizen's famous water and rice

The philosophy of the brewery is to deliver higher quality sake to the world through handcrafting.

The brewery is located in Echizen-Ono, a village known for its famous "Oshouzu" one of Japan's 100 best waters. The village is also one of Japan's leading producers of the sake rice "Gohyakumangoku". The brewery was founded in 1733 when it was a large hardware store on Shichiken-street in Ono. They started brewing sake in 1901, and chose the brand name "Hanagaki" from a verse in the noh song "Hanagatami". The current brewery's philosophy is "to send out higher quality sake made by hand", and they carefully brew only what they can see. While paying close attention to the ingredients and respecting the traditional techniques, they also strive to expand the varieties of sake. In recent years, the brewery has been focusing on researching aged sake, and has been using a rare sake rice "Kamenoo". Although the brewery is small, they brew sake with a great deal of attention to detail.

大吟醸酒

Daiginjo for a competitive show, Kyuukyoku no Hanagaki

"The best limited edition of the year"

This is the ultimate Daiginjo with a splendid aroma, a dignified sweetness, a moderate taste range, and a deep aftertaste. This sake is the crystallization of the brewery's technology.

Diginjoshu	1,800ml/720ml
Alc.	17%
Rice type	Yamadanishiki

純米酒

Junmai Nigori

"Okuechizen's Nigorishu goes well with winter food"

The silky smooth mash brings out the subtle sweetness and the deliciousness of the rice, while the soft acidity gives the taste a tightness.

Junmaishu	1,800ml/720ml
Alc.	14%
Rice type	Gohyakumangoku/Hanaechizen
Sake meter value	-20.0
Acidity	1.8

Manaturu Shuzo

Water source	Hakusan water system
Water quality	Mild soft

11-3 Meirin-cho, Ono-shi, Fukui 912-0083 TEL.0779-66-2909
E-mail: info@manaturu.com http://www.manaturu.com

Never resting on tradition New challenge

The prestige of the brewery and the pride of the master brewer.

Manaturu Shuzo is a long-established brewery that has been located in Echizen-Ono, a small city in the Hokuriku region, since the Horeki period (1751-1764) in the middle of the Edo era. Echizen-Ono is located in the upper reaches of the Kuzuryu River, a fan-shaped basin with abundant water and greenery surrounded by 1,000-meter high mountains on all sides. A specialty of Echizen-Ono is Gohyakumango-ku, a rice suitable for sake brewing. The clear water of "Oshozu", one of the 100 best waters selected by the Ministry of the Environment, combined with the deep snow and cold climate of the Hokuriku region, makes it a wonderful place to brew sake, which is said to be "a place created by God for sake brewing". It is a small brewery that does not rely on machines and produces all of its sake by hand. While preserving traditional methods, Keisuke Izumi, the brewery's chief brewer, has adopted a policy of making all of his products Ginjo-standard in order to ensure high quality.

大吟醸酒
Daiginjo LOUI

"Exquisite Daiginjo with a cute label"

A fresh aroma like apples, a slightly elegant sweetness, and a cool taste. It has a clean finish and a beautiful aftertaste.

Daiginjoshu	1,800ml/720ml
Alc.	17%
Rice type	Yamadanishiki
Sake meter value	+4.0
Acidity	1.5

純米大吟醸酒
SOW

"A refreshing sensation like the sound of rain"

The clean, elegant sweetness and refreshing citrus acidity are in perfect balance. It has a refreshing and innovative taste.

Junmai Daiginjoshu	1,800ml/720ml
Alc.	13%
Rice type	Gohyakumangoku
Sake meter value	-15.0
Acidity	3.4

Water source	Own spring water
Water quality	Medium-hard

Miyakehikoemon Shuzo

21-7 Hayase, Mihama-cho, Mikata-gun, Fukui 919-1124 TEL.0770-32-0303
E-mail: hayaseura@ever.ocn.ne.jp https://www.fukuisake.jp/

Crisp and refreshing aftertaste
The best "men's sake" in the Hokuriku region

Delicious ingredients and delicious sake from Wakasa Bay. The spirit of supporting a port town.

Miyakehikoemon Shuzo was founded in 1718, and the current successor, Norihiko, is the 12th- generation head. Mihama in Fukui Prefecture prospered as a port town for fishing and commerce, and the fishermen who were deeply religious, wanted to have a local sake as their sacred wine. The sake produced by Miyakehikoemon Shuzo can be called a "men's sake" that supports such fishermen. In Wakasa, there are many natural and delicious foods, and sake is essential to make food taste even better. This background and the climate are the pride of the brewery, and the reason for the characteristics of their sake. Norihiko says that he was disappointed when his former colleagues at the Tokyo University of Agriculture brought their own Ginjoshu, but what he saw was just ordinary sake. By pursuing "the kind of sake that only this brewery can make", Miyakehikoemon has grown to attract attention from all over the country.

Hayaseura Junmaiorizake urazoko

"A Junmai made of natural lees"
The supernatant is smooth and dry, but when the residue is mixed in, a mellowness spreads in the mouth. This sake looks like the Japan Sea in winter as seen from the brewery.

Junmai Orizake	1,800ml/720ml
Alc.	18%
Rice type	Suitable rice for making sake grown in Fukui
Sake meter value	+10.0
Acidity	1.8

Hayaseura Daiginjo

"Packed with mountain and sea flavors"
This is a men's sake with the rich aroma and umami of a Daiginjo and a crisp finish. It also has a hint of the sea of Hayase.

Daiginjoshu	1,800ml/720ml
Alc.	16%
Rice type	Yamadanishiki
Sake meter value	-0.5
Acidity	1.3

Yasumoto Shuzo

Water source	Subsoil water from the Hakusan mountain range
Water quality	Medium-hard

7-4 Yasuhara-cho, Fukui-shi, Fukui 910-2167   TEL.0776-41-0011
http://www.yasumoto-shuzo.jp

Quality First
A brewery that produces sake in small quantities

We do not pursue production volume at all, but spare no effort to make sake that gives satisfaction.

Yasumoto Shuzo was founded in 1853 at the end of the Edo period (1603-1868), It is located in the Yasuhara-cho area of Fukui, the home of Yoshikage Asakura who built up prosperity during the Warring States period. Since this era, the Yasumoto family has been engaged in various businesses, such as money exchange, farming, and forestry. The concept of the brewing process is to make only Jumaishu using 100% rice grown in Fukui, with good water and traditional techniques. The brewery uses subsoil water drawn from their well (about 200 meters underground) which comes from the Hakusan mountain range. They continue to brew quality-first sake that is suitable for meals, using the traditional "Funeshibori" technique in which the mash is placed in individual sake bags and squeezed. In 1889, ten earthen storehouses and brewing warehouses were registered as tangible cultural properties of Japan. The brewery inherits this cultural tradition and continues to strive for regional revitalization.

Hakugakusen Junmai Daiginjo NUREGARASU

"A delicacy like jet-black hair"
"Nuregarasu" is the name of the traditional Japanese color that represents the most beautiful black hair. It has a gentle aroma and a transparent, glossy, soft flavor.

Junmai Daiginjoshu	720ml
Alc.	15%
Rice type	Ginnosato
Sake meter value	+5.0
Acidity	1.65

Hakugakusen Junmai Daiginjo ROKOU

"Beautiful flavor without any miscellaneous taste"
This is an excellent product made by assembling different rice polish ratios to create an aroma, flavor, and aftertaste.

Junmai Daiginjoshu	720ml
Alc.	15%
Rice type	Ginnosato
Sake meter value	+4.0
Acidity	1.65

Guide to Sake Breweries and Famous Sake

Tokai Region

Gifu, Shizuoka, Aichi, Mie

Food Culture in the Tokai Region

Gifu: Depending on the region, beef cattle, known as Hida Beef, are raised in the mountains of the northern part of the prefecture. In the southern part of the region, where the plains are wider, the specialties are ayu (sweetfish) from the large rivers.

Shizuoka: With its warm climate, Shizuoka produces many vegetables, including tea and tangerines. Tuna and skipjack tuna are landed at Yaizu Port, and the skipjack tuna is salted and dried and eaten as a preserved food. Sakura-ebi (cherry shrimp) from Suruga Bay are often eaten as dried shrimp, but they are also boiled or eaten raw as kakiage (fried shrimp). In the lakes and estuaries, eels and suppon (softshell turtle) are actively cultivated, and they are popular as "kabayaki", "shirayaki", and hot pot dishes.

Aichi: Agriculture is thriving, with cabbage and other vegetables being produced in abundance, while shellfish, seaweed and sea eel are cultivated in the area. Various dishes using soybean paste such as "Hitsu-mabushi" and "Hatcho-miso" are enjoyed. "Nagoya kochin" and "Kishimen noodles" are also sometimes served with miso-flavored dipping sauce.

Mie: Along the coast of the Shima Peninsula, the fishing industry is thriving, and there are many dishes using seafood, such as Ise shrimp, abalone, turban shells, saury, bonito, tuna, and clams. "Ise udon" is a thick udon noodle dish with a black sauce made of tamari soy sauce and broth. As for livestock, Matsuzaka Beef is widely known.

Adachi Shuzojo

Water source	Subsoil water from the Nagara River system
Water quality	Soft

3-21-10 Kotozuka, Gifu-shi, Gifu 500-8222 TEL.058-245-3658
E-mail: hinode@kinkazan.com   http://www.kinkazan.com

The entire process of sake brewing is done by hand

Most of the brewing process is done by hand with uncompromising brewing techniques.

The brewery was established in 1861. They continue to brew bold sake in a region blessed with natural environment. Abundant subsoil water from the "Three Kiso Rivers", which includes the Nagara River, flows through the Nobi Plain, and the Ibuki Oroshi wind blows in winter, and there is high-quality sake rice being produced. In the past, sake was brewed in the traditional way by the Toji (master brewer) of Echigo until mechanical equipment was utilized to keep up with the recent evolution of various types of sake. However, the brewery decided to get rid of it. Most of the processes now are done by hand, and all the brewing is done in small batches. It is rare that everything from washing the rice to cooling it down is done by hand these days, but this is the reason why the brewing process at Adachi is so uncompromising. As a small brewery, they strive to promote sake, welcoming visitors who wish to tour it (there is currently a limit on the number of visitors). Adachi shows every corner of the brewery so visitors can experience as much as possible with their five senses.

純米吟醸酒

Kinkazan Soukyuu Muroka Nama Genshu

"Perfect sake to drink with a meal, made with the brewer's soul"

This is a thick and rich Junmai Genshu that allows you to taste the wild flavor of Hidahomare. It goes well with Japanese food, and you'll never get tired of drinking it.

Junmai Ginjoshu	1,800ml/720ml
Alc.	17%
Rice type	Hidahomare
Sake meter value	private
Acidity	private

純米酒

Kinkazan Nakagumi Junmai Nama Genshu

"This is an unpasteurized sake that tastes good in a wine glass"

The balance of umami and acidity is exquisite. It has a mouthfeel and taste like white wine, and goes well with meat dishes and fried foods.

Junmaishu	1,800ml/720ml
Alc.	18%
Rice type	Hidahomare
Sake meter value	+7.0
Acidity	1.8

| Water source | Subsoil water from the Hida mountain range |
| Water quality | Soft |

Kaba Shuzojo

6-6 Ichinomachi, Furukawa-cho, Hida-shi, Gifu 509-4234 TEL.0577-73-3333
E-mail: kaba@yancha.com https://www.yancha.com

Enjoy the long winter Local Sake of Hida

A sake that will accompany you when you are happy, sad, or lonely.

Hida's oldest sake brewery, Kaba Shuzojo, was founded in 1704. The current brewer is in the 13th generation. As Hida is a snowy region, the culture of sake is deeply rooted in daily life. Many people in Hida Furukawa, where the brewery is located, have a strong attachment to their local brand, and restaurants in the area rarely sell sake from other regions. There is also a strong connection between festivals and sake, and the view of all the sake offerings lined up at the festival hall is astonishing. This is how much the local people love their sake. The Kaba Shuzo-jo is also close to the local community and has been brewing local sake for them, using the rice and water from Hida. The flagship brand, "Shiramayumi" is made with the aim of producing a sake that will always be there for you when you are happy, sad, or lonely.

純米大吟醸酒

Shiramayuimi
Junmai Daiginju Homare

"The finest taste that has been carefully finished"

This is a full-bodied sake with a gorgeous yet gentle and elegant aroma that fully expresses the attractive characteristics of "Hidahomare". It has been highly evaluated worldwide.

Junmai Daiginjoshu	1,800ml/720ml
Alc.	16%
Rice type	Hidahomare
Sake meter value	-1.0
Acidity	1.7

本醸造酒

Hida no Yanchasake

"The sake that sake lovers will love"

Good match for the winter in Hida. At room temperature, it is sharp, light, and slightly dry, but when warmed, the rice flavor increases and a gentle sweetness can be enjoyed.

Honjozoshu	1,800ml/720ml
Alc.	15%
Rice type	Hidahomare
Sake meter value	+2.0
Acidity	1.6

Kawashiri Shuzojo

Water source	Subsoil water from the Kuraiyama Mountains system
Water quality	Soft

68 Kaminino-machi, Takayama-shi, Gifu 506-0845 TEL.0577-32-0143
E-mail: mail@hidamasamune.com http://www.hidamasamune.com

Handmade in small quantities
Aged sake directly from the brewer

Strive for a mellow taste by using locally grown rice.

The brewery was founded in 1839, at the end of the Edo period (1603-1868). In the 1970s, they abolished the use of sugar and liquid seasoning, as well as vat sales (subcontracting to major companies) in order to put quality first in the old townscape where production is limited, and then shifted their focus on direct sales. The brewery believes that local sake must be brewed with locally grown rice. They use only "Hidahomare", a rice grown in Hida that is suitable for sake brewing. Originally the brewery's sake had a full-bodied flavor, but when it was new, it tasted rough so it needed to be kept for a long time before it mellowed out. This is why the seventh-generation head and the previous generation established a style of sake maturing, and specialize in maturing sake using only locally grown rice, storing it for several years.

純米酒
Jukusei Koshu Junmai Yamahida

"A slight bitterness in the aftertaste increases umami"

Cold sake with a refreshing and firm taste. When it is warmed, it has a sweet, rich taste. The difference in taste for different temperatures is remarkable for this Junmai Koshu.

Junmaishu	1,800ml/720ml
Alc.	15.3%
Rice type	Hidahomare

本醸造酒
Jukusei Koshu Honjozo Tenon

"Good balance of rich taste and aroma"

It has an aroma reminiscent of steamed chestnuts, bananas and pancakes. The acidity is moderate, and this aged sake is full-bodied but easy to drink.

Honjozoshu	1,800ml/720ml
Alc.	15.3%
Rice type	Hidahomare

| Water source | Well water |
| Water quality | Soft |

Kuramoto Yamada

3888-2 Yaotsu, Yaotsu-cho, Kamo-gun, Gifu 505-0301 TEL.0574-43-0015
E-mail: info@kura-yamada.com https://www.kura-yamada.com/

A land of mountains, hills and clear water, and the Toji's passion to make sake

Toji Masaki Uno continues to take on new challenges, aiming for a sake that can be enjoyed for a long time.

Founded in 1868, Kuramoto Yamada is located in Gifu Prefecture, surrounded by forests, mountains and water. The Toji (master brewer) is Masaki Uno, who was born in 1968. He was influenced by Toji Takakura, whom he met after joining the company, and changed his mind about sake brewing. He says, "I was naive and made too many compromises. Takakura's sake brewing is meticulous, careful, and faithful to the basics. I was awakened by that". Uno became a Toji in 2003, and his masterpiece, "Junmai Daiginjo Tamakashiwa", won the Gifu Prefecture Governor's Award and the National Gold Award. However, he does not indulge himself, and continues to take on new challenges to further improve his skills. His goal is to make sake that is easy on the throat and can be enjoyed for a long time without getting tired of drinking it. As a small brewery, the careful brewing of sake is bearing fruit.

純米大吟醸酒
Junmai Daiginjo Tamakashiwa

"A masterpiece of Toji Masaki Uno"

This sake is brewed with 35% polished "Yamadanishiki" rice and carefully fermented at low temperature. It won the gold medal at the National New Sake Competition in 2019, and was selected as one of the top five in its category by Kura-Master (France) in 2019.

Junmai Daiginjoshu	720ml
Alc.	17%
Rice type	Yamadanishiki
Sake meter value	+2.0
Acidity	1.3

純米吟醸酒
Junmai Ginjo Tamakashiwa

"The best aroma and flavor"

The rice is Yamadanishiki polished to 45%. This is a masterpiece of Kuramoto Yamada, with the aroma and flavor unique to a Junmai Ginjo.

Junmai Ginjoshu	1,800ml
Alc.	17%
Rice type	Yamadanishiki
Sake meter value	+2.0
Acidity	1.3

Komachi Shuzo

Water source	Subsoil water from the Nagara River
Water quality	Soft

2-15 Soharalbuki-cho, Kakamigahara-shi, Gifu 504-0851 TEL.058-382-0077
E-mail: info@nagaragawa.co.jp   https://www.nagaragawa.co.jp/

The sake brewery associated with Komachi
Brew sake with natural music

A brewery in Gifu named after the legend of "the sacred water that healed Komachi".

Founded in 1894. Located in the northern part of the Nobi Plain, the name of the brewery, Komachi Shuzo, is derived from the local legend of "the sacred water that healed Ono-no-Komachi". The brewery's main brand "Nagaragawa" is named after one of the clearest streams in Gifu. The subsoil water from the river flows underneath the brewery and it is used for brewing along with Gifu's sake rice, "Hidahomare". Gifu is an "akamiso (red miso)" culture area, and the brewing process is designed to bring out the "umami" in the sake to match this food culture. The brewery is also known as one that "makes sake with natural music". In order to make the brewing environment as close to the natural state as possible, they play environmental music that elicits alpha waves in the brewery. They started to export overseas from an early stage, and has been delivering "Gifu local sake" to the world for over 20 years.

純米酒
Nagaragawa Junmaishu

"A mellow sake that you won't get tired of drinking"

This Junmaishu has a mellow taste with a focus on the flavor of the rice. It is slightly dry with a refreshing aftertaste, so you will never get tired of drinking it.

Junmaishu	1,800ml
Alc.	14-15%
Rice type	Gifu rice
Sake meter value	+5.0
Acidity	1.7

吟醸酒
Nagaragawa Sparkling Nigori

"Semi-sweet and lightly effervescent"

Although this is n unfiltered sake, it is fermented in the bottle and the taste is not too sweet. It is a refreshing, slightly fizzy drink.

Junmai Ginjoshu	720ml/300ml
Alc.	17-18%
Rice type	Hidahomare grown in Gifu
Sake meter value	-4.0
Acidity	1.7

Water source	Subsoil water of the Ibi River
Water quality	Soft

Sugihara Shuzo

1 Shimoiso, Ono-cho, Ibi-gun, Gifu 501-0532 TEL.0585-35-2508
E-mail: sugihara@feel.ocn.ne.jp https://www.sugiharasake.jp/

Persuade the predecessor to rebuild the brewery

A sincere desire of those who left Japan to preserve the culture of sake.

The company was founded in 189. It is known as "the smallest sake brewery in Japan", with an annual production of about 60 koku (equivalent to only 6,000 bottles). Yoshiki Sugihara, a former Japan Overseas Cooperation Volunteers (JOCV) member and the current fifth-generation sake brewer, was driven by the desire to preserve sake as a part of Japanese culture. In 2003, he returned to his parents' brewery to persuade his predecessor to rebuild the brewery, who was thinking of closing. Since he had no training experience in other breweries, Sugihara used a process of trial and error. After meeting Hiroki Takahashi, a staff member of the Gifu Agricultural Experiment Station (also called a "rice geek"), he succeeded in cultivating a new breed of sake rice, "Ibinohomare", with the cooperation of local farmers. He completed the brand "Ibi" in 2009. The brewery's motto is "Making true local sake by doing everything locally, from the production of raw materials, to the brewing and the enjoyment of sake".

特別純米酒
Tokubetsu Junmaishu Ibi Funaba Muroka Nama Genshu

"Easy to drink and sharp sweetness"

It has an aroma of ripe apples, a round sweetness, and a light amber color, like a fine dessert wine.

Special Junmaishu … 1,800ml/720ml	
Alc.	16%
Rice type	Ibinohomare
Sake meter value	-3.0
Acidity	1.8

特別純米酒
WHITE Ibi

"This sake is made with a portion of white rice malt. Easy-to-drink sourness"

This is the most balanced of the "Ibi" series. The citric acid from the white malt gives it a sourness that does not overpower the sweetness, making it rich and refreshing.

Special Junmaishu … 1,800ml/720ml	
Alc.	16%
Rice type	Ibinohomare
Sake meter value	-3.0
Acidity	.2.5

Chigonoiwa Shuzo

Water source	Underground water
Water quality	Ultra-soft

2177-1 Dachi-cho, Toki-shi, Gifu 509-5401 TEL.0572-59-8014
E-mail: info2@chigonoiwa.com https://chigonoiwa.jp

Prosperity in a big way With a wish

"Chigonoiwa" brewed with local rice certified as one of the best 100 terraced rice fields in Japan.

The name of the local sake "Chigonoiwa" comes from the giant rock "Chigoiwa", which is designated as a natural monument by Toki City. This mysteriously shaped rock, estimated to weigh 13,125 tons, is said to have bestowed a baby on a childless couple. When the company was established in 1909, the founder, Shigezo Nakashima, named the sake brewed in this area "Chigonoiwa" in honor of this rock and also with a wish for a thousand years of happiness and prosperity that would grow like a child. The sake is brewed with the traditional techniques of the Echigo Toji (master brewer). The brewing water is ultra-soft water drawn from 45 meters below the brewery. The rice used for the sake, such as "Sakaoritanadamai" and "Hidahomare", is locally grown. The Sakaoritanada rice field is recognized as one of the 100 best terraced rice fields in Japan.

純米吟醸酒・原酒
Chigonoiwa Junmai Ginjo Genshu Sakaoritanada Tanadamai Jikomi

"Rich and dry with the power of raw sake"

It is made from rice grown in the Sakaoritanada rice field, one of Japan's 100 best terraced rice fields, and uses the minimum of agricultural chemicals. Unfiltered sake in winter, filtered and heated in summer.

Junmai Ginjoshu	1,800ml/720ml
Alc.	17.5%
Rice type	Sakaoritanadamai"
Sake meter value	+9.0
Acidity	1.9

大吟醸酒
Chigonoiwa Daiginjo

"The secret of Chigonoiwa's sake brewing"

A clean, dry and sharp taste, with a fruity aroma and a refreshing throat feel.

Daiginjoshu	1,800ml/720ml
Alc.	16.8%
Rice type	Hidahomare
Sake meter value	+6.0
Acidity	1.4

| Water source | Groundwater from the Hida Mountains |
| Water quality | Soft |

Tenryou Shuzo

1289-1 Hagiwara, Hagiwara-cho, Gero-shi, Gifu 509-2517 TEL.0576-52-1515
E-mail: info@tenryou.com https://www.tenryou.co.jp/

In Hida, there is a sake that speaks of Hida

The spirit of the first-generation brewer who loved Hida is still alive in this famous sake, brewed in Gifu.

Tenryou Shuzo was founded in 1680, at the beginning of the Edo period (1603-1868). At first, the founder was a peddler under the name of Hinoya Sahee, but later set up store in Hida and started making sake while selling goods. In 1955, the business was incorporated into the present-day "Tenryou Shuzo". The brewery uses "Hidahomare" which is the best rice for sake brewing grown in the natural environment of Hida. Then they draw the groundwater flowing from the Hida mountain range from thirty meters below ground. These are the key to brew their main brand, a special Junmaishu "Tobikiri" and a Daiginjo "Hidahomare Tenryou". The company continues to brew sake with the spirit of the founder who loved the nature of Hida. Tenryou Shuzo is number one in Gifu Prefecture in terms of the percentage of rice used, the percentage of rice polished in-house, and the percentage of natural water used. Tenryo Shuzo continues to maintain this dedication.

Tokubetsu Junmaishu Tobikiri

"Best served cold with sashimi, and warm with hot pot dish"

This is a dry Junmaishu with a rich aroma and body that will envelop you in the richness of modern tastes.

Special Junmaishu	... 1,800ml/720ml
Alc.	15-16%
Rice type	Hidahomare
Sake meter value	+4.0
Acidity	1.4-1.6

Junmai Ginjo Hodahomare Tenryou

"A splendid aroma that goes cleanly down the throat"

This sake is made with Hida's special rice "Hidahomare", which is suitable for brewing. The taste becomes richer after the summer.

Junmai Ginjoshu	...... 1,800ml/720ml
Alc.	15-16%
Rice type	Hidahomare
Sake meter value	+3.0-+5.0
Acidity	1.3-1.5

Hakusen Shuzo

Water source	Hida River
Water quality	Soft

28 Nakakawabe, Kawabe-cho, Kamo-gun, Gifu 509-0304 TEL.0574-53-2508
https://www.hakusenshuzou.jp/

"Care" is the life of raw materials.

A delicious sake with a complex flavor, brewed with handmade koji.

The brewery is located in Kawabecho, on the banks of the Hida River, which flows through the mountains of Hida. Hakusen brewing is not influenced by trends. As the name of the town (Kawabe=near the river) suggests, it is blessed with delicious water. The brewery was founded in the late Edo period, but the exact date of its establishment is unknown. According to the brewery, "It's because our ancestors were indifferent about leaving their mark". During the Edo period, they mainly made mirin (sweet cooking rice wine), and began brewing sake in 1902. Hakusen Shuzo believes that the basis of brewing is the hands (their attitude of "care"). They spend time on every brewing process, from the selection of ingredients, "koji" making, brewing, aging, all the way to the finish. All of this makes each drop of sake a complex and delicious drink. It is the philosophy of Hakusen Shuzo to carry on the history cultivated by their predecessors and continue to produce safe, reliable, and delicious sake for the next generation.

純米大吟醸酒
Kuromatsu Hakusen Junmai Daiginjo Fuku

"Crystal of the best materials and the best techniques"

This is a true Junmai Daiginjoshu that the floral and clean aroma will spread in the mouth.

Junmai Daiginjoshu	1,800ml/720ml
Alc.	16%
Rice type	Yamadanishiki
Sake meter value	-4.0

純米吟醸酒
Kuromatsu Hakusen Junmai Ginjo Hana

"Dry and handmade in the cold winter"

A mild and elegant aroma with a moderate richness and a refreshing, smooth taste. The harmony of acidity and sweetness creates a mellow taste.

Junmai Ginjoshu	1,800ml/720ml
Alc.	16%
Rice type	Gohyakumangoku

Water source	Subsoil water from the Nagara River
Water quality	Soft

Hayashi Honten

2239 Nakashinkano-cho, Kakamigahara-shi, Gifu 504-0958 TEL.01-58-382-1238
E-mail: hayashihonten@eiichi.co.jp   http://www.eiichi.co.jp/

Japanese Sake Culture to the World

Aiming to expand the enjoyment of premium sake to the world.

Founded in 1920, this sake brewery has been brewing sake in Gifu, a country with clear streams and one of the richest forests in Japan. To brew sake, they use the famous Japan Alps subsoil water, which is ultra-soft and produces a gentle taste with a lustrous flavor. The high-quality sake rice harvested in the World Agricultural Heritage area is cultivated with water from the Nagara River, one of the three clearest rivers in Japan, where ayu fish are caught and presented to the Emperor. Hayashi Honten's goal is to spread the traditional culture of Japanese sake to people around the world with premium sake brewed in this climate and soil. The main brand "Hyakujuro" is named after Ichikawa Hyakujuro, the home-grown Kabuki actor, who donated 1,200 sakura trees ninety years ago, with the hope of producing sake that people can enjoy drinking under the "Hyakujyuro Sakura" trees in full bloom.

Hyakujyuro Junmai Daiginjo New Moon

"Addltive-free sake brewed with the power of lactic acid bacteria"

A refreshing aroma like bergamot. It has a beautiful acidity and a faint sweetness derived from lactic acid bacteria, with a clean and light aftertaste.

Junmai Daiginjoshu	720ml
Alc.	15%
Rice type	Gifuhatsushimo

Hyakujyuro Junmai Ginjo Yamahai Jidai

"Sweet and bewitching richness and umami"

The aroma is buttery and full. The rich and complex flavor of this sake is a blend of richness, umami and acidity.

Junmai Ginjoshu	1,800ml/720ml
Alc.	15%
Rice type	Gohyakumangoku

Nunoya Hara Shuzojo

Water source	Sacred Mt. Haku
Water quality	Soft

991 Shirotori, Shirotori-cho, Gujo-shi, Gifu 501-5121 TEL.0575-82-2021
E-mail: nunoya-sake@bridge.ocn.ne.jp http://genbun.sakura.ne.jp

A Gift from Flowers
Sake made with flower yeast

The head of the brewery, a Toji (master brewer), is particular about the richness and aroma of natural flower yeast.

Nunoya Hara Shuzojo was founded in 1740. The brand name "Genbun" refers to the name of era at the time of its founding. Mt. Haku is one of the three sacred mountains in Japan along with Mt. Fuji and Mt. Tate, and has been worshiped as a god of water since ancient times. The Nagara River, one of the clearest rivers in Japan, originates at the foot of this sacred mountain. Nunoya Hara Shuzojo is the northernmost sake brewery along the Nagara River, located in Shirotoricho, Gujo City, Gifu Prefecture, where the worship of Mt. Haku is deeply rooted. The brewery uses local rice from Gifu, and subterranean water from the Hakusan mountain range drawn from a well on the premises, which has been used since the brewery's founding, Also used is Tokyo University of Agriculture's flower yeast, which is isolated from flowers in nature. Currently, all the sake produced at the brewery is brewed with natural flower yeast.

普通酒
Genbun

"A classic sake of Nunoya, made with flower yeast"

Nunoya's standard sake brewed with flower yeast has a moderate aroma and a refreshing taste. It goes well with "Keichan", a local dish flavored with miso, soy sauce, and garlic.

Futsushu	1,800ml/720ml
Alc.	15.5%
Rice type	Akitakomachi grown in Gifu
Sake meter value	+4.0
Acidity	1.3

大吟醸酒
Hanakoubo Kiku Daiginjo

"The flower word for chrysanthemum is "noble"; it suits this sake"

This Daiginjoshu has a splendid sweet aroma reminiscent of flowers and a sharp, clean aftertaste. It goes well with grilled ayu fish.

Daiginjoshu	720ml
Alc.	15.5%
Rice type	Akitakomachi by Gifu
Sake meter value	+4.0
Acidity	1.4

| Water source | Northern Alps |
| Water quality | Soft |

Harada Shuzojo

10 Kamisanomachi, Takayama-shi, Gifu 506-0846 TEL.0577-32-0120
E-mail: info@sansya.co.jp http://www.sansya.co.jp

Kyoto Culture and Edo Culture are fused in Hida

Sake brewed with abundant pure underground water from the Northern Alps and high-quality Hida rice.

Since 1855, the end of the Edo period, the brewery has been making sake under the name "Chougoro Utsueya" in the old castle town of Hida Takayama, which was under the direct control of the Tokugawa Shogunate. Originally, Chougoro was the headman of the neighboring town of "Utsuenosho", but he quickly took notice of sake production in Nada and Fushimi, which are close to Kyoto. He dedicated himself to sake production in order to change his business from a headman to a sake brewer. The brewery's sake, " Sansha is another way to read the characters for dashi floats in the Takayama Matsuri, one of the three most beautiful festivals in Japan. "Sansha Karakuchi", in particular has earned a good reputation from Japanese-style restaurants and pubs in Hida Takayama. The brewery is also a member of the Tokyo University of Agriculture's Flower Yeast Research Group, and continues to take on the challenge of brewing sake with flower yeast.

普通酒
Sansha Karakuchi

"Okuden: Hida-style cold brewed in winter"

This sake is perfect for meals, bringing out both the refreshing transparency and the taste of rice. Monde Selection Gold Medal.

Futsushu	1,800ml
Alc.	15-16%
Rice type	Hidahomare/Akitakomachi
Sake meter value	+2.0
Acidity	1.4

純米吟醸酒
Sansha Junmai Ginjo Hanakoubo Zukuri

"Sweetness and aroma brought out by flower yeast"

The local rice "Hidahomare" is brewed with Abelia flower yeast to produce a sake with a refreshing sweetness and fruity aroma. Best served cold.

Junmai Ginjoshu	720ml
Alc.	15-16%
Rice type	Hidahomare
Sake meter value	+3.0
Acidity	1.4

Hirase Shuzoten

82 Kamiichinomachi, Takayama-shi, Gifu 506-0844 TEL.0577-34-0010
E-mail: info@kusudama.co.jp http://www.kusudama.co.jp/

A famous sake nurtured by the harsh climate of Hida

A sake brewery named after "Yakudama" (a ball of medicine) that is said to dispel evil spirits.

It is not known when Hirase Shuzoten was founded, but the name of the founder was written in 1623 in an obituary at Bodai temple. After 390 odd years, the brewery is in its 15th generation in Hida Takayama today. Hida has long been a place where "Hida artisans" went to Kyoto in lieu of taxation, and much of the Kyoto culture was introduced that way. According to the shakekabu (an official certificate of sake brewing) from 300 years ago, there were sixty-four breweries in Takayama, and they were actively making sake. In the spring festival, people pray for a good harvest, and in the autumn festival, people rejoice in the harvest. "Kusudama" is loved by the people of Hida who live in a harsh climate. Hirase Shuzo's technique of carefully brewing sake using local rice matches the tastes of the local people.

純米酒
Kusudama Tezukuri Junmai

"Good chemistry with Hida beef and Hoba miso"
There is a slight bitterness and astringency from the rice, but it combines with a high acidity to create a deep flavor. Best served cold or lukewarm.

Junmaishu	720ml
Alc.	15.5%
Rice type	Hidahomare
Sake meter value	+5.0
Acidity	1.6

大吟醸酒
Kusudama Daiginjo

"Soft sake brewed in the harsh winter of Hida Takayama"
Made with subterranean water from the North Alps, this sake has a smooth, elegant taste with a fine texture. Best served cold.

Daiginjoshu	720ml
Alc.	16%
Rice type	Yamadanishiki
Sake meter value	+2.0
Acidity	1.2

| Water source | Groundwater from the Mikuni mountain range |
| Water quality | Soft |

Michisakari

2919 Kasahara-cho, Tajimi-shi, Gifu 507-0901 TEL.0572-43-3181
E-mail: info@michisakari.com http://www.michisakari.com/

Easy to drink Mizuguchi Sake

Dry sake not influenced by current trends and maintains its traditional taste.

During the Anei era (1772-1781), Tetsuji Mizuno from Owari Province, opened a brewery in present-day Tajimi City in Gifu Prefecture, which led to the birth of a famous sake, "Michisakari". This dry sake was popularly known as "Kin Maruo", "Gin maruo", and "Sumi Maruo" until the middle of the Meiji era (1868-1912). The name was changed to "Kogane" in the first year of the Showa era. After that, the brewery started calling only high-grade sake "Michisakari". However, in the 1950's and 60s, sweet sake became the mainstream of the times and "Michisakari" faced a difficult time. Despite the adversity, Kokichi Mizuno, the owner at the time, sought the ideal dry sake and released "Michisakari Tokkyushu(Special Grade Sake)" with a sake strength of +10, which was very rare at the time. This caught the attention of the writer Tatsuo Nagai, and the name "Michisakari" spread to dry sake lovers all over Japan.

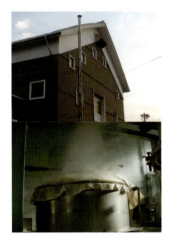

大吟醸酒
Michisakari Chotoku

"A clear, translucent flavor"
This Daiginjo has a clean, dry, delicate and deep taste and aroma. There is no extra taste, and the flavor of the rice is strong.

Daiginjoshu	1,800ml/720ml
Alc.	15-16%
Rice type	Miyamanishikii/Akitakomachi
Sake meter value	+16.0-+17.0%
Acidity	1.0

純米大吟醸酒
Michisakari Junmai Daiginjo

"Junmai Daiginjoshu with a characteristic sour taste"
This Junmai Daiginjoshu is full of "Michisakari"'s character with a mild aroma and refreshing acidity that enhances the taste.

Junmai Daiginjoshu	1,800ml/720ml
Alc.	15-16%
Rice type	Miyamanishikii/Akitakomachi
Sake meter value	+12.0-+13.0%
Acidity	1.3

Watanabe Shuzoten

Water source	Miyagawa River system / Arashiro River system
Water quality	Medium-hard

7-7 Ichinomachi, Furukawa-cho, Hida-shi, Gifu 509-4234 TEL.0577-73-2347
E-mail: info@kusudama.co.jp https://www.sake-hourai.co.jp

Hida severe cold-weather brewing since the brewery's founding

Brew sake with humanity and honesty to make the most of the life of the rice.

Watanabe Shuzoten is located in the Furukawa Basin of Gifu Prefecture, surrounded by the mountains of the Northern Alps, which are over 3,000 meters above sea level. The fifth generation head, Kyuemonsho, started brewing sake in 1870. He was originally a raw silk merchant, but started brewing sake because he couldn't forget the taste of the liquor that he had during a trip to Kyoto. The sake he brewed was well received and referred to as "droplets from jewels". It was named "Horai", taken from a line in the chant "Tsurukame". What "Horai" pursues is "authentic sake brewing, where skilled brewers honestly brew sake that is good to drink". They emphasize tradition and handcrafting, and continue to be very particular about raw materials and brewing techniques. The brewery is now in its 152nd year, and has developed the "Morohaku Ginjo style" that has been handed down since its founding.

Horai Junmai Ginjo Kaden Tezukuri

"Feel the taste of rice as it is"
Deep richness in its softness, with the five flavors (sweet, sour, salt, bitter and pungent taste) in harmony. A Junmai Ginjoshu with a graceful and honest taste that is not showy.

Junmai Ginjoshu	720ml
Alc.	15%
Rice type	Hidahomare
Sake meter value	+3.0
Acidity	1.4

Horai Rice Junmai Daiginjo Irootoko

"The "Romanée Conti" of the sake world"
This sake was created after a certain No. 1 host (cocktail waiter) begged for the best sake to intoxicate women. It has a sweet taste with a Fruity aroma.

Junmai Daiginjoshu	720ml
Alc.	15%
Rice type	Yamadanishiki
Sake meter value	+3.0
Acidity	1.3

| Water source | Subsoil water from the Oigawa River |
| Water quality | Soft |

Isojiman Shuzo

307 Iwashigashima, Yaizu-shi, Shizuoka 425-0032 TEL.054-628-2204
http://www.isojiman-sake.jp/

World-class Artistic Sake

Innovative sake brewing that goes well with French and Italian cuisine.

Isojiman is located in Yaizu, Shizuoka Prefecture, overlooking Suruga Bay, with Mt. Fuji to the east and the Southern Alps to the northwest. It is no exaggeration to say that Isojiman's sake brewing process begins and ends with washing rice using the subsoil water of the clear Oigawa River. The natural aroma of white peaches, musk melons, ripe bananas, and passion fruit gently blend into the sake brewed in the natural environment of Shizuoka Prefecture's mild climate and abundant water. Therefore, it goes well with French and Italian cuisine, cheese, tomatoes, and even desserts. The Ginjoshu in particular is best served in a wine glass, not too cold. Isojiman, born from the harmony of carefully selected rice, famous water, and the skill and passion of the brewers, is a world-class sake.

純米大吟醸酒
Isojiman Emerald

"The aroma changes with temperature"

You can enjoy the fresh impression when the package is first opened, and the change in aroma and roundness on the tongue as time passes in the glass.

Junmai Daiginjoshu	1,800ml/720ml
Alc.	16.2%
Rice type	Tokujyo Yamadanishiki 50%
Sake meter value	+3.0
Acidity	1.2

大吟醸酒
Isojiman Daiginjo

"Brings out the scent of nature as it is"

This Daiginjo sake is a marriage of seafood and a variety of other dishes. During a meal, it enhances the flavor of food.

Daiginjoshu	1,800ml
Alc.	16.2%
Rice type	Tokujyo Yamadanishiki 45%
Sake meter value	+6.0
Acidity	1.2

Takashima Shuzo

Water source	Subsoil water from Mt. Fuji
Water quality	Soft

354-1 Hara, Numazu-shi, Shizuoka 410-0312 TEL.055-966-0018
E-mail: info@hakuinmasamune.com http://www.hakuinmasamune.com

Contributes to the local community Local pure rice brewery

Local sake that can only be made in this region is the best communication tool.

The Takashima Shuzo was founded in 1804. The main brand sake "Hakuin Masamune" was named by Teshu Yamaoka, who visited Numazu as an imperial envoy in 1884 when Zen master Hakuin of the Rinzai sect was given the posthumous title of Masamune Kokushi. The brewery is committed to contributing to the local community and uses "Homarefuji" sake rice, which is grown only in Shizuoka Prefecture. In addition, the brewery has eliminated the use of alcohol-added sake since 2012 in order to contribute to local farmers by using as much rice as possible, and to become a brewery which only brews Junmaishu. Takashima is always conscious of the fact that "local sake is part of the food culture" and the basis of its brewing is to make sake that can be drunk without being too assertive. The brewery continues to carefully brew sake that can only be made locally with the hope that the sake will become the best communication tool.

Hakuin Masamune Homarefuji Junmaishu

"Warm sake is delicious with meals"

A refreshing sake with a mild aroma and a sharp, clean taste that emphasizes the flavor of the rice. It is made with 100% "Homarefuji" rice from Shizuoka Prefecture.

Junmaihu	720ml
Alc.	15%
Rice type	Homarefuji
Sake meter value	private
Acidity	private

Hakuin Masamune Junmaishu Kimoto Homarefuji

"A gentle sake that you will never get tired of drinking"

This sake has a soft, light taste. It has a soft and light mouthfeel, and is good both cold and warmed.

Junmaishu	720ml
Alc.	15%
Rice type	private
Sake meter value	private
Acidity	private

Water source	Head waters of the Oigawa River
Water quality	Medium-hard

Sugii Shuzo

4-6-4 Koishigawa-cho, Fujieda-shi, Shizuoka 426-0033 TEL.054-641-0606
E-mail: wbs47338@mail.wbs.ne.jp   http://suginishiki.com

Traditional techniques
The Challenge of Yamahai Brewing

Exploring the flavor of sake by utilizing the power of nature with kimoto and yamahai moto.

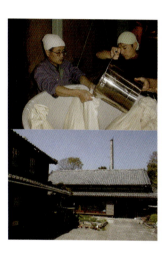

Sugii Shuzo was founded in 1842 by Saisuke Sugii, a branch of the Sugii family, who started brewing sake in Takasu Village (now Koishigawacho, Fujieda City). The name of the sake was "Kamegawa" until the middle of the Meiji era, "Sugimasamune" until the Taisho era, and "Suginishiki" in the early Showa era. They still practice careful sake brewing by steaming rice using the traditional "koshiki" steamer and making koji by hand. While based on Shizuoka-style ginjo brewing, they have a wide variety of sake, from dry to rich types. Since 2000, the chief of "Suginishiki" has taken the lead without inviting a Toji (master brewer), brewing sake without compromise, together with his other employees. In 2003, the brewery began to take on the challenge of brewing Yamahai Junmaishu, and at the 2012 U.S. National Sake Appraisal, "Suginishiki Kimoto Tokubetsu Junmai Nakadori Genshu" won the gold medal.

Suginishiki Tamasakae Yamahai Junmai

"Ancient Japanese technique, Yamahai Junmai"

The taste is deep, and the moderate acidity adds to the flavor. The rice used for the sake is "Tamasakae", which is sour and full-bodied.

Junmaishu	1,800ml
Alc.	15.6%
Rice type	Tamasakae
Sake meter value	+6.0
Acidity	1.7

Suginishiki Yamadanishiki Kimoto Tokubetsu Junmaishu

"A pale, fruity aroma"

"Yamadanishiki" rice is used to make this sake. It is characterized by a full flavor with a good harmony of acidity and sweetness.

Special Junmaishu	1,800ml
Alc.	15.8%
Rice type	Yamadanishiki
Sake meter value	+2.0
Acidity	1.7

Doi Shuzojo

Water source	Spring water of the Takatenjin
Water quality	Soft

633 Onuki, Kakegawa-shi, Shizuoka 437-1407 TEL.0537-74-2006
E-mail: doisake@plum.ocn.ne.jp   https://kaiunsake.com/

Shizuoka Yeast Home of HD-1

Lucky sake with a rake-shaped label A brewery loved by the locals for the celebratory sake, "Kaiun".

Doi Shuzojo, founded in 1872, still retains its original building and is often referred to as a historical building in Shizuoka Prefecture. Since 2003, Doi Kiyoaki, 4th generation kuramoto, and Hase Shokichi, famous Toji (master brewer), have won the gold medal at the Zenkoku Shinshu Kanpyokai (National New Sake Competition) for seven consecutive years. However, Hase's sudden death stopped the winning streak for a year. With the great efforts of a young master brewer, Shimba Minori who inherited brewing skills from him, the brewery won the gold medal again the following year. The brewery is also known as the birthplace of Shizuoka yeast HD-1, and uses it in all of its sake production, bringing out the flavors and aroma that can only be brewed in Shizuoka Prefecture. The main brand name is "Kaiun Iwaizake" with its lucky rake label. The price is reasonable, but the quality of all the products, from Honjozo to Daiginjo, is very high.

Kaiun Daiginjo
Den Hase Shokichi

"A sake named after Hase Shokichi, famous Toji"

This Daiginjo bears the name of Hase Shokichi, a master brewer who the brewery thinks of as one of Noto's Four Heavenly Kings. The delicate flavor stands out when cooled.

Daiginjoshu	1,800ml/720ml
Alc.	17-18%
Rice type	Yamadanishiki
Sake meter value	private
Acidity	private

Kaiun Iwaizake
Tokubetsu Honjozo

"Celebration sake from the time of its founding"

This is an excellent product carefully brewed with a rice polishing ratio of 60%, the same as that of Ginjoshu. It is highly evaluated by fans and restaurants all over Japan.

Honjozoshu	1,800ml/720ml
Alc.	15-16%
Rice type	Yamadanishiki
Sake meter value	+4.0
Acidity	1.3

Water source	Subsoil water from Mt. Fuji
Water quality	Soft

Fujinishiki Shuzo

532 Kamiyuzuno, Fujinomiya-shi, Shizuoka 419-0301 TEL.0544-66-0005
E-mail: info@fujinishiki.com https://www.fujinishiki.com

Shining in the evening light Like Mt. Fuji

A taste created using traditional techniques, the rich nature of Mt. Fuji and the latest technology.

Fujinishiki Shuzo was founded in the Edo Genroku era (1688-1704) and has been in business for eighteen generations. The brewery is located at the foot of Mt. Fuji and brews sake using the clear underground water of Mt. Fuji. The brewery is equipped with a huge refrigerator and a thorough quality control system. Over the past ten years, it has won first prize in the Shizuoka Seishu Kanpyokai (Shizuoka Sake Competition) five times, and has also won the highest gold medal in the Monde Selection of the World Food Competition for seven consecutive years for its Daiginjo, as well as gold medals for its Ginjo and Junmaishu. The company's motto is to make the best use of the good water, and put quality first. They are also making efforts to develop products to promote the local community. It is said that more than 10,000 people of all ages from all over the country gather at the annual sake brewing festival held in March. The taste of new sake (provided free of charge) must be special if you have a view of Mt. Fuji under the blue sky in the spring.

純米酒
Fujinishiki Junmaishu

"The historic taste of Fujinishiki"

Shizuoka yeast, developed in Shizuoka Prefecture, is used. It has a light taste unique to the Fuji spring water used for brewing.

Junmaishu	1,800ml/720ml
Alc.	15.5%
Rice type	Domestic rice
Sake meter value	+3.0
Acidity	1.4

特別純米酒
Fujinishiki Tokubetsu Junmaishu Homarefuji

"New type of sake rice with hidden potential"

This sake is made with the newly developed sake rice "Homarefuji". It has a perfect balance of robust taste and aroma.

Special Junmaishu	1,800ml/720ml
Alc.	16.5%
Rice type	Homarefuji
Sake meter value	+2.0
Acidity	1.5

Yamanaka Shuzo

Water source	Subsoil water from the Akaishiyama Mountain
Water quality	Soft

61 Yokosuka, Kakegawa-shi, Shizuoka 437-1301 TEL.0537-48-2012
http://www5a.biglobe.ne.jp/~yamanaka/

Famous Sake of Tootoumi "Aoi Tenka"

Brewed to become the world's best sake, in honor of the Tokugawa family, with their hollyhock crest.

Yamanaka Shuzo started from Yamanaka Shokichi Shoten, which was established by Omi merchant Shokichi Yamanaka in the Bunsei Era (1818-1831) in Omiya, Fuji-gun, Suruga Province (present-day Fujinomiya City). He started brewing sake in 1831 and later established a brewery in Yokosuka, Kakegawa City. In 1929, the brewery in Yokosuka became an independent branch of the current Yamanaka Shuzo. In the Edo era, the area had abundant good-quality water, and flourished as a castle town of the Enshu Yokosuka clan. Since Yokosuka Castle was the front line of the battle in which Tokugawa Ieyasu toppled Takeda to conquer the whole country, the brewery's main brand was named "Aoi Tenka" in the hope that the young Toji (master brewer) of the time would be able to take the top position. In fact, the following year, the young Toji won the gold medal at the Zenkoku Shinshu Kanpyokai (National New Sake Competition) for three consecutive years, as well as other prestigious awards at various other competitions.

大吟醸酒
Aoi Tenka Daiginjo

"Highly acclaimed at various competitions"

This is a sake with a splendid aroma that brings out the full flavor of the rice. It is brewed with techniques inherited from the previous generation chief, Takashi Yamanaka, who graduated from Tokyo University of Agriculture.

Daiginjoshu	1,800ml/720ml
Alc.	15-16%
Rice type	Yamadanishiki
Sake meter value	+3.0
Acidity	1.1

純米吟醸酒
Aoi Tenka Homarefuji Junmai Ginjo

"Junmai Ginjoshu is brewed entirely with fune-shibori"

The use of flower yeast gives it a gentle, soft floral aroma. It has a rich and refreshing taste.

Junmai Ginjoshu	1,800ml/720ml
Alc.	15-16%
Rice type	Homarefuji
Sake meter value	+1.0
Acidity	1.3

| Water source | Subsoil water from the Chita Peninsula hillsides |
| Water quality | Soft |

Sawada Shuzo

4-10 Koba-cho, Tokoname-shi, Aichi 479-0818 TEL.0569-35-4003
E-mail: sawadasyuzou@hakurou.com   http://www.hakurou.com

The passionate heart of the craftsman
Tokoname's Sake

A traditional sake brewery in Tokoname, one of the six oldest in Japan.

Sake brewing in Chita began in 1688. It is said to have started when Kinoshita Niemon, a merchant, made a medicinal sake called "Houmeishu" and presented it in a jar. Around 1697, shipments to Edo (Tokyo) began, and Chita developed into a major production city. On the other hand, Sawada Shuzo was founded in 1848 at the end of the Edo era. The first-generation head, Giheiji Sawada, established a sake brewery in this area blessed with good water, and expanded its sales channels to Nagoya, Shizuoka, and Tokyo by ship. In the late Meiji era (1868-1912), Giheiji, the second generation, succeeded in developing a new method of brewing sake by adding lactic acid, which became the basis for the revolutionary "sokujo moto" method of preventing sake spoilage. Since then, the company has been focusing on quality first by brewing sake that emphasizes the flavor of the rice and is faithful to the basics in order to make sake that complements food.

純米酒
Junmai Hakurou

"Sake with a strong taste of rice"
This sake is brewed using "Wakamizu" rice cultivated with reduced pesticides in the brewery's own rice field, with traditional tools and methods, such as steaming with a wooden steamer and making koji with the tray koji method (koji-buta).

Junmaishu	1,800ml/720ml/300ml
Alc.	15%
Rice type	Wakamizu
Sake meter value	+4.0
Acidity	1.6

本醸造酒(生原酒)
Kurouto dakeshika nomenu Sake

"Seasonally produced Nama Genshu"
This is a historic product that was the first raw sake to be released in Japan. This unfiltered, unpasteurized sake has a rich body and taste.

Honjozoshu	700ml
Alc.	18%
Rice type	Gohyakumangoku
Sake meter value	-1.0~+2.0
Acidity	1.6

Sekiya Jozo

Water source	Spring water in Okumikawa
Water quality	Strongly soft

22 Taguchi machiura, Shitara-cho, Kitashitara-gun, Aichi 441-2301 TEL.0536-62-0505
E-mail: info@houraisen.co.jp https://www.houraisen.co.jp/

Brewed with tradition and logic Aichi's famous sake

All rice is polished in-house. Making efforts to create a comfortable environment.

Sekiya Jozo was established in 1864, and is famous in Aichi Prefecture for its local sake "Houraisen". The brewery's approach to sake brewing is "Wajoryoshu" (which means harmony brews good sake). Good sake is produced in a good environment where the brewers can work comfortably. The harsh processes of sake brewing which requires a lot of physical strength are mechanized, and the human labor saved by the mechanization is used to collect and analyze data for reproducible sake brewing. As a result, the brewery is able to maintain a stable supply while maintaining high quality, and has become the most popular and technologically advanced brewery in Aichi. Sekiya Jozo is particular about the rice, and polishes all of it in-house. Other popular services include tours of the brewery, sales by weight, and "made-to-order" sake where you can choose the flavor, aging period, and label to create your own original sake.

Horaisen
Junmai Daiginjo Kuu

"Splendid aroma and elegant sweetness"

"Yamadanishiki" rice polished in-house is fermented at low temperatures. It has a mellow aroma reminiscent of fruit and an elegant taste that brings out the flavor and sweetness of the rice.

Junmai Daiginjoshu	1,800ml/720ml
Alc.	15.5%
Rice type	Yamadanishiki
Sake meter value	+2.0
Acidity	1.4

Houraisen Junmai Daiginjo
Maka

"Junmai Daiginjoshu nurtured by the climate of Okumikawa"

High quality sake made by polishing "Yumesansui" to 30% and carefully drawing out the flavor of the rice.

Junmai Daiginjoshu	1,800ml/720ml
Alc.	16.5%
Rice type	Yumesannsui (in-house cultivation)
Sake meter value	+1.0
Acidity	1.3

| Water source | Kiso River system |
| Water quality | Soft |

Naito Jozo

52-1 Kabutoshinden Takasuka, Sobue-cho, Inazawa-shi, Aichi 495-0022 TEL.0587-97-1171
E-mail: naitojouzou@h5.dion.ne.jp   https://www.naitojouzou.com

Nambu Toji's tradition of handcrafting

Local sake brewing that uses prefectural rice, led by home-grown employees.

The brewery has been making sake for more than 190 years, since it was founded in 1826. Originally, the Toji (master brewers) from Niigata and Iwate brewed ordinary sake, however, they have changed to brew only specially designated sake since 2011, based on the idea of brewing sake rooted in the local community; The brewery mainly uses rice produced in Aichi Prefecture, and the sake is brewed by home-grown employees. Naito aims to optimize its brewing process by installing the latest air conditioning and refrigeration systems, while handing down old techniques, such as the box-koji method, which is becoming rare even among breweries in Japan. The concept of the flagship brand "Kiso Mikawa", is to create a flavor that goes beyond the image of a sake that can be enjoyed with food, so that both the sake and the food accompany each other. For several years now, the company has been focusing on exporting sake overseas, and as of 2021, they export mainly specially designated sake to about 10 countries.

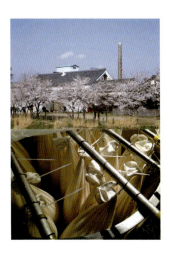

大吟醸酒
Kiso Sansen Daiginjo

"The best sake brewed in the cold season"
A clean and elegant taste with a subtle fruity aroma that lingers in the mouth. It is best served cold or slightly chilled.

Daiginjoshu	720ml
Alc.	15%
Rice type	Yamadanishiki
Sake meter value	±0
Acidity	1.2

純米吟醸酒
Junmai Ginjo Kiso Sansen Niji no Shirabe

"Only highly polished prefectural rice is used"
It is characterized by a rich, fruity aroma with a slight sweetness. It is best served cold on the rocks with Japanese food.

Junmai Ginjoshu	720ml
Alc.	15%
Rice type	Yumesansui
Sake meter value	-4.0
Acidity	1.6

Maruishi Jozo

Water source	Otogawa River system
Water quality	Medium-soft

6-3-3 Naka-machi, Okazaki-shi, Aichi 444-0015 TEL.0564-23-3333
E-mail: maruishi@014.co.jp http://www.014.co.jp

In the land associated with Lord Ieyasu
Chasing two rabbits

Sake brewing in search of the best balance of conflicting "taste and aroma" and "acidity and umami".

Maruishi Jozo was founded in 1690 in Okazaki, Aichi Prefecture, the birthplace of Ieyasu Tokugawa. The brewery has been producing sake in this city, full of history and nature surrounded by mountains and rivers for 332 years. During the Pacific War, most of the other breweries in the area were destroyed in an air raid on Okazaki, but the remaining miso storehouses were rebuilt and sake production was resumed. Maruishi Jozo's sake brewing is conscious of "sweetness, aroma, and acidity". "Sweetness" refers to the natural sweetness of the rice, "Aroma" to the fragrant aroma of sake, and "Acid" to the strength and clarity of the aftertaste. The concept of the new brand "Nito" is, "Only those who chase two rabbits will get two rabbits". True to this concept, the sake is made to have the best balance of two conflicting elements such as "taste and aroma", "acidity and umami" and "sweet and dry".

純米大吟醸酒
Junmai Daiginjo Nito Omachi 48

"A sake with an elegant taste"

It has a lily-like aroma, a rich fruitiness and a good acidity with a good balance between the umami and sweetness, which is unique to Omachi.

Junmai Daiginjoshu	1,800ml/720ml
Alc.	16%
Rice type	Omachi

純米酒
Junmai Nito Banzai 70

"Use of re-released rare rice "Banzai""

The aroma of banana chips, sweetness, a hint of bitterness, and a pleasant acidity are well balanced. Locally grown rice is used to produce this Okazaki terroir.

Junmaishu	1,800ml/720ml
Alc.	16%
Rice type	Banzai

| Water quality | Soft |

Banjo Jozo

41 Nishikadota, Oodaka-cho, Midori-ku, Nagoya-shi, Aichi 459-8001 TEL.052-621-2185
E-mail: nakagawa@kuheiji.co.jp https://kuheiji.co.jp/

Dramatic Sake Brewing: Innovation in a nutshell

Revolution by the 15th-generation chief, Kuheiji Kuno. The birth of "Kamoshibito Kuheiji".

Banjo Jozo was founded in 1789. Successive generations of the family have called themselves "Kuheiji" and sake brewing began from the ninth generation, passed on to the present-day fifteenth generation. The main brand "Kamoshibito Kuheiji", launched by the fifteenth generation is very popular both in Japan and abroad. It is served at a three-star restaurant in Paris. In a word, Banjo Jozo is innovative. If they want to brew sake with their own rice, they start rice farming and cultivate it themselves. If they want to change drinks enjoyed during a meal from sake to white wine or red wine, they start making wine in Burgundy. If they want to feel the characteristics of the rice, they use the same rice-polishing ratio and brewing method, but brew using a different rice variety such as "Yamadanishiki" or "Omachi". This is a sake brewery that Japan is proud to present to the world.

Kamoshibito Kuheiji Betsu Atsurae Junmai Daiginjo

"The taste is full of nobility"

This Junmai Daiginjoshu is brewed with "Yamadanishiki" from Hyogo's special A district, polished to 35%. It has a juicy rice flavor.

Junmai Daiginjoshu	720ml
Alc.	16%
Rice type	Yamadanishiki
Sake meter value	private
Acidity	private

Kamoshibito Kuheiji Hi no Kichi Junmai Daiginjo

"Rich, elegant flavor"

This is the flagship Junmai Daiginjoshu in the "Kamoshibito Kuheiji" series. The true value of this flavor is best expressed in a wine glass.

Junmai Daiginjoshu	720ml
Alc.	16%
Rice type	Yamadanishiki
Sake meter value	private
Acidity	private

Yamazaki

Water source	Subsoil water from the foot of the Sangane mountains
Water quality	Soft

57 Kakida, Nishihatazu-cho, Nishio-shi, Aichi 444-0703   TEL.0563-62-2005
E-mail: info@sonnoh.co.jp   https://www.sonnoh.co.jp

Rooted in Hazu
A long-established sake brewery

A long journey to finding an authentic taste... Uncompromising craftsmanship is our pride.

Founded in 1903, the brewery makes sake using rice grown in Aichi Prefecture. The company name is Yamazaki Limited Partnership, but the trade name is "Sonnou Kuramoto". In Aichi Prefecture, there are four designated rice-growing regions; "Yumesansui" "Wakamizu" "Yumeginga" and "Yamadanishiki", and the Sonno Kuramoto uses more than 98% Aichi-grown rice, mainly from these regions. In particular, the brewery works closely with the local rice farmers in the Hazu district of Nishio City, and hosts a sake rice study session with a technical advisor from Aichi Prefecture in July and September. After planting in July, the brewery asks for guidance on pest control and when to use fertilizer. In September, before harvest, the brewery checks the growth status by analyzing rainfall and temperature. Through these efforts, the rice farmers deepen their understanding of the need to supply good sake rice, which leads them to work together to produce better sake.

Seishu YumesansuiJuwari Oku Nama (Seasonal products)

"Splendid aroma and rich, dense umami"

With an alcohol content of over 18 degrees, this is the most fragrant Junmai Ginjoshu with a "thickness" that you have probably never experienced before. Best served cold.

Junmai Ginjoshu	1,800ml/720ml
Alc.	18.5%
Rice type	Yumesansui
Sake meter value	+2.0
Acidity	1.8

| Water source | Kiso River |
| Water quality | Medium-soft |

Watanabe Shuzo

83 Michishita, Kusahira-cho, Aisai-shi, Aichi 496-8015 TEL.0567-28-4361
E-mail: w_shuzo@clovernet.ne.jp

Handmade sake with 100% love

Sake brewed in an environment where the cold wind "Ibuki Oroshi" blows on the Nobi Plain.

Watanabe Shuzo is located in Aisai City, the most western part of Aichi Prefecture, in the middle of the Nobi Plain. The area has long been known as a water town and is blessed with abundant water resources from the clear waters of the Kiso River. During the best season of brewing, the winter wind called "Ibuki Oroshi" blows from the northwestern direction bringing cold fresh air, and a high-quality rice is grown in the fertile ground. Along with these essential elements, sake is brewed by hand. In the land of three natural blessings, "water, climate, and rice", sake is brewed entirely by family members who pour their love into the sake. They touch steamed rice, stir the koji, nurture the yeast starter, and brew the mash—all with their hands. The sake is made with the hope that it will grow slowly and healthily. It is delivered to the drinker as if it were a child the brewers raised themselves.

特別純米酒
Kaho no Sake

"The taste of local Aichi rice"
The rice is washed by hand and the koji is handmade. This Junmaishu can be enjoyed cold or warmed all year round.

Special Junmaishu	720ml
Alc.	15.8%
Rice type	Wakamizu
Sake meter value	±0
Acidity	1.6

大醸造酒
Hiraisami Masamune
Kuromatsu Genshu

"Taste the umami on the rocks"
This sake is made by hand from steaming, to koji making, to nurturing yeast starter, to brewing the mash. It is a taste for connoisseurs that goes well with a strong-flavored dish like miso cutlets.

Honjozoshu	1,800ml
Alc.	19.6%
Rice type	Hyogoyumenishiki
Sake meter value	+4.0
Acidity	1.9

Ota Shuzo

Water source	Subsoil water from the Nunobiki mountain range
Water quality	Medium-soft

1365-1 Kaminosho, Iga-shi, Mie 518-0121   TEL.0595-21-4709
E-mail: ota@hanzo-sake.com   https://www.hanzo-sake.com

Sake nurtured by the climate of Iga

The brewery of "Hanzo", a local sake brewed in the climate and nature of Iga.

Ota Shuzo was founded in 1892. The brewery is located in the middle of the Iga Basin, surrounded by lush rice fields with egrets flying around. Only during the cold season do they produce a small amount of sake rooted in the local community and using as much manual labor as possible. Surrounded by mountains on all sides, the land is said to have been at the bottom of Lake Biwa about 4 million years ago. The clayey, fertile land and the basin's unique climate produce high-quality sake rice, and the high-quality subsoil water brought down from the mountains is used for brewing. The main ingredients are "Kamino-Ho" sake rice from Mie Prefecture, "Yamadanishiki" rice from Iga, and "Ukonnishiki" rice grown by a local contract farmer. The "Hanzo" series are brewed in Igaueno, a place associated with Hattori Hanzo, using a variety of yeasts including "MK-3" and "MK-1" which are developed in Mie Prefecture.

HANZO Junmai Daiginjo Kaminoho

"Elegant sake enjoyed in a wine glass"

An elegant Ginjo aroma and a gentle sweetness that stands out when chilled. You can enjoy clean mouthfeel and a plump taste.

Junmai Daiginjoshu ... 1,800ml/720ml/300ml	
Alc.	15%
Rice type	Kaminoho
Sake meter value	+2.0
Acidity	1.6

HANZO Tokubetsu Junmaishu Igaukon Nishiki

"Specially brewed Junmai rice wine "Ukonnishiki" from Iga"

A robust umami of rice and a crisp finish. It is a gentle dry sake with a soft flavor. Best served cold or warmed at a wide range of temperatures.

Special Junmaishu ... 1,800ml/720ml/300ml	
Alc.	15%
Rice type	Ukonnishiki
Sake meter value	+5.0
Acidity	1.6

| Water source | Suzuka mountain range |
| Water quality | Slightly-hard |

Ogawa Honke

1425 Itsushiki, Kawage-cho, Tsu-shi, Mie 510-0306 TEL.059-245-0013
E-mail: info@ogawahonke.com http://www.ogawahonke.com

The secret, the tradition, the inheritance, and the legend of Ginjo-zukuri method

From washing the rice to bottling, everything is done by hand with the utmost care.

Founded in 1868, Ogawa Honke is located in Tsu City, Mie Prefecture. From washing the rice to bottling, every step of the sake brewing process is done by hand with the utmost care. The spirit that has been handed down from the time of its founding to the fourth-generation head, Masuto Ogawa. The main brands are "Daiginjo Yachiyo", "Junmai Ginjo Den" and "Junmai Nama-shu Toyotsugo Nama". All of them use rice from Mie Prefecture and subsoil water from the Suzuka mountain range to produce a rich taste with a hint of acidity. "Junmai Ginjo Den" is known for its secret, traditional, and legendary Ginjozukuri, which is known for its inherent flavor of rice. Narazuke, which is made from sake lees produced during the sake brewing process, is also popular. It takes at least three years for sake to be ready to drink.

純米吟醸酒
Junmai Ginjo Den

"Taste the true flavor of rice"
Junmaishu is made with secrets, tradition and the inheritance of Ginjo-zukuri method. A dry sake with well-balanced flavor and richness.

Junmai Ginjoshu	1,800ml
Alc.	16%
Rice type	Mie rice
Sake meter value	+3.0
Acidity	private

純米酒
Toyotsugo Nama Junmaishu

"The taste of freshly squeezed freshness"
Unpasteurized sake (Junmai Nama Genshu) that is matured and stored at low temperatures to retain the freshness of freshly pressed sake. It is light and easy to drink dry.

Junmaishu	720ml
Alc.	14%
Rice type	Mie rice
Sake meter value	+3.0
Acidity	private

Takijiman Shuzo

Water source	Akame 48 waterfalls
Water quality	Soft

141 Kashiwara, Akame-cho, Nabari-shi, Mie 518-0464 TEL.0595-63-0488
http://www.takijiman.jp/

The sake of Iga Loved in the land of Iga

Uncompromising sake brewing with good rice, good water, and the attitude of the brewery.

Founded in 1868, Takijiman Shuzo is located in the Iga Basin, in the mountains of Mie Prefecture, on the border with Nara Prefecture. The area is also famous for the Iga Ninja, and blessed with abundant nature. Near the brewery is the valley of the Akame 48 Waterfalls, a national park that has been selected as one of the top 100 waterfalls in Japan. Good sake is produced from good rice, good water, and with the attitude of the brewery. Takijiman is committed to uncompromising sake brewing, using mainly "Yamadanishiki" rice, which is grown under contract in the Iga Basin, and the subterranean water from the Akame 48 falls. Blessed with a climate that varies greatly in temperature and good quality water, this region is known for its good rice and sake. By looking at the large amount of local consumption, it is clear that the sake produced in Iga is loved by the locals and has been passed down for generations to be drunk every day.

純米酒
Karakuchi Junmai Hayase

"A taste you will never get tired of"

This sake has a clear mouthfeel like flowing water from the waterfall, and a pleasantly light crisp finish. This Junmaishu can be enjoyed during a meal.

Junmaishu	1,800ml/720ml
Alc.	15%
Rice type	Yamadanishiki/Gohyakumangoku
Sake meter value	+10.0
Acidity	1.4

純米大吟醸酒
Junmai Daiginjo

"The flavor of the rice is fully expressed"

This Junmai Daiginjoshu uses only "Yamadanishiki" from Iga. It is a gem with the potential of Iga rice and the clarity which is unique to Takijiman.

Junmai Daiginjoshu	1,800ml/720ml
Alc.	16%
Rice type	Yamadanishiki
Sake meter value	+4.0
Acidity	1.3

| Water source | Shakagatake in the Suzuka mountains range |
| Water quality | Soft |

Hayakawa Shuzo

468 Ojima, Komono-cho, Mie-gun, Mie 510-1323 TEL.059-396-2088
E-mail: info@hayakawa-syuzo.com https://hayakawa-syuzo.com

A sake that is gentle and mellow
A sake that enriches the soul

Local sake brewing with all Junmai and tank pressing (funeshibori).

Hayakawa Shuzo, established in 1915, is located in Komonocho, the north central part of the Mie Prefecture. The Suzuka mountain range spreads out gracefully in front of the brewery. The cold "Suzuka Oroshi" (a cold wind that comes down from the Suzuka mountains) and the high-quality underground water from the Shakagatake (a mountain range in the Suzuka mountains) play a major role in the brewing process of Hayakawa Shuzo. Komonocho is also known for its rice cultivation. In Tagomitsu district, "Kami-no-Ho" is grown under contract as the prefecture's rice suitable for brewing sake. The brewery makes two types of sake using this rice: Junmai Ginjo, and Tokubetsu Junmai. All the sake are brewed from pure rice and pressed in a tank, the way the brewery insists on. The main brand, "Tabika" comes from the Tabika River that runs beside the brewery. Hayakawa continues to make high-quality sake, aiming to be a sake brewery that is close to the local community and that local customers can recommend with confidence to the rest of the prefecture and the entire world.

Junmai Ginjo Muroka Nakadori Nama Tabika Omachi

"Rich flavor unique to Omachi"
This sake is characterized by a splendid and gentle aroma of apples, and a rich, deep flavor with gentle mouthfeel of Omachi. It also has a refreshing aftertaste.

Junmai Ginjoshu	1,800ml/720ml
Alc.	16%
Rice type	Omachi

Tokubetsu Junmaishu Tabika Omachi

"A filling umami"
This Junmaishu is characterized by a mild, gentle umami and a clean aroma. With the mild aroma, the clean taste can be enjoyed.

Special Junmaishu	1,800ml
Alc.	15%
Rice type	Omachi

Gensaka Shuzo

Water quality: Soft

346-2 Yanagihara, Odai-cho, Taki-gun, Mie 519-2422 TEL.0598-85-0001
E-mail: info@gensaka.com https://www.gensaka.com/

Made from rice grown in Mie Prefecture
Sake brewing that gives back to the local community

The brand "Sakaya Hachibei" is named after the founder, Gensaka Hachibei.

Gensaka Shuzo was established at its present location in 1805, at the end of the Edo era. It was around the time that Tadataka Ino began surveying the country. The business has remained in the family, and the current president, Arata Gensaka, represents the sixth generation to lead the brewery. Their flagship brand, "Sakaya Hachibei" was named after the founder "Gensaka Hachibei". Most of the rice used to make it is produced in Mie Prefecture. The brewery gives back to the local community by brewing sake. About 60% of the sake is delivered to customers within Mie Prefecture, and it is sold not only at restaurants but also at hotels and souvenir stores in tourist areas such as Iseshima. The current production volume is about 900 koku, while there are five people involved in the brewing process, including the president, Arata Gensaka. The total number of employees, including those in charge of delivery, is ten. Despite the small number of employees and limited facilities, the company strives to improve the quality of the sake every year.

Sakaya Hachibei Isenishiki Yamahai Junmaishu

"Sharp sweetness and three-dimensional bitterness"

This sake was brewed by reviving "Isenishiki", a sake rice native to Mie Prefecture. Served at the 2016 G7 Ise Shima Summit.

Junmaishu	1,800ml/720ml
Alc.	15-16%
Rice type	Isenishiki
Sake meter value	+4.0
Acidity	1.5

Sakaya Hachibei Isenishiki Junmai Ginjoshu

"Bitterness with umami"

This is a crystal color with a hint of yellow from six months of bottle aging. A rich and dry Junmai Ginjoshu.

Junmai Ginjoshu	1,800ml/720ml
Alc.	15-16%
Rice type	Isenishiki
Sake meter value	+5.0
Acidity	1.7

| Water source | Subsoil water of the Suzuka mountain range |
| Water quality | Medium-soft |

Moriki Shuzojo

41-2 Senzai, Iga-shi, Mie 518-0002 TEL.0595-23-3040
E-mail: ugk26398@nifty.com   https://morikishuzo.co.jp/

Sake brewing is making rice

A brewery that sticks to traditional brewing methods in Iga, a rice-growing and sake producing village.

Moriki Shuzojo, located in the northern part of Iga City, Mie Prefecture, was founded in 1897. The Iga region, generally known as the "home of the ninja", has long been blessed with abundant water and rice, and most of the "Yamadanishiki" in the prefecture is grown in Iga. Since 1998, Moriki Shuzojo has stopped adding alcohol to its brewing process, and produces only "Junmaishu" (pure rice sake), which is made entirely by hand with the tray koji method (kojibuta). The brewery's aim is to make sake that is less about fragrance and more about flavor: ideal to enjoy—sometimes cold, sometimes heated with food. The brewery's policy is to produce sake with as much care as possible, given its small production of 300 koku. For brands such as "Hanabusa", the brewery uses pesticide-free "Yamadanishiki" grown in its own fields by contract farmers in Iga.

Hanabusa Kimoto Junmai Daiginjo

"Ultimate sake brewed with traditional yeast mash"

You can enjoy the clean taste, the strength of kimoto , and the depth of aging. The labels are hand finished one by one by Yo Tanimoto, a master of Iga pottery.

Junmai Daiginjoshu ... 1,800ml/720ml
Alc. 16%
Rice type .... Yamadanishiki
Sake meter value .... +2.0
Acidity 1.7

Rumiko-no-Sake Tokubetsu Junmaishu 9-go Koubo

"Junmaishu with a clean, firm taste"

This is a sake that is suitable for a wide range of food and can be enjoyed on a daily basis. The label was designed by Akira Oze, the manga author of "Natsuko no sake".

Special Junmaishu ... 1,800ml/720ml
Alc. 15%
Rice type .... Yamadanishiki
Sake meter value .... +7.0
Acidity 1.6

Guide to Sake Breweries and Famous Sake
Kinki Region
Shiga, Kyoto, Osaka, Hyogo, Nara, Wakayama

Food Culture in the Kinki Region

Shiga: Funa (crucian carp), biwa trout, and shijimi clams are harvested from Lake Biwa, and funa-zushi or tsukudani is a popular food. Omi beef and duck are also famous local delicacies.

Kyoto: Kyoto was the capital of Japan in ancient times, and this can be seen in its food culture. Yuba (bean curd), fu (wheat gluten), and tofu, which are used in kaiseki and vegetarian dishes, are also used in everyday household cooking. Since hamo can be transported live for a long time, it is eaten in a variety of dishes in Kyoto, which is far from the Seto Inland Sea, using cooking techniques such as boning and parboiling.

Osaka: Known as "the kitchen of the nation" Osaka has been the birthplace of a variety of dishes such as "hamo ryori", "ehomaki", and "oshizushi", which are made from famous products from all over Japan.

Hyogo: The Seto Inland Sea is famous for its octopus, sea bream and other fish dishes. "Tai somen" served with somen noodles, and "tai meshi" are known as luxurious dishes that bring good luck by using a fish with a head. There are also dishes using various local ingredients, such as black beans from Tamba and Kobe beef.

Nara: Once being the capital of Japan, temples and aristocratic cuisine are inherited in the food culture. "Yoshino kuzu (arrowroot kudzu)", "Nara pickles", and a nabe dish which uses wild boar meat from the mountains, are well known.

Wakayama: Various marine products such as whale and moray eels have been used in daily dishes such as "tatsuta-age" and simmered dishes. Many large-head hairtails are landed, and the large-sized high-class fish, "kue", which is a member of the sea bass family, is used for sashimi and hot pot dishes. At Shirahige Shrine, a "Kue Festival" is held in hopes of bountiful fishing. The cultivation of fruits such as plums and oysters is also flourishing.

| Water source | Well water from the underground streams of Mt. Suzuka range |
| Water quality | Soft |

Emishiki Shuzo

1-7-8 Hon-machi, Minakuchi-cho, Koka-shi, Shiga 528-0031 TEL.0748-62-0007
E-mail: jizake@emishiki.com http://www.emishiki.com

A long-established brewery that produces new sake with a young sensibility

Looking to create a completely new "sweet taste supremacy" while inheriting tradition.

Founded in the early Meiji period Koka City, Shiga Prefecture, one of the old Tokaido Highway post stations, the Emishiki Shuzo has 120 years of tradition as a brewer of slightly sweet sake suited to the lightly flavored cuisine of the region. The name of the brewery, "Emishiki" was given by Sengoro Takeshima, the second-generation head of the brewery, with the hope that the brewery would "make people laugh every day. The current fifth-generation brewer is Atsunori Takeshima, who studied brewing at Tokyo University of Agriculture. With his young sensibility, he pursues "pop and fun sake" as a "new form of delicious sake" and adopted the concept of "sweetness supreme". There is a spirit of challenge in the sake industry. The company has developed a product lineup divided into three series, including the Monsoon Series, a uniquely designed kijo-shu that pursues the supreme sweetness and intense flavor of rice.

Emishiki Kijoshu Monsoon Tamasakae Nama Genshu

"Rich sweetness, transparent acidity"

A new type of sake made with the latest technology in Kijoshu brewing. No.30: Aroma of melon and herbs from yeast.

Special Junmaishu	750ml
Alc.	17%
Rice type	Tamsakae From Shiga

Emishiki Kijoshu Monsoon Yamadanishiki Namagenshu

"Rich but clean aftertaste"

A new quality of sake that makes full use of the latest technology in Kijoshu brewing. No.19: Tropical banana aroma from yeast.

Special Junmaishu	750ml
Alc.	17%
Rice type	Yamadanishiki from Shiga

Kitajima Shuzo

Water source	Subsoil water from Mt. Suzuka range
Water quality	Soft

756 Hari, Konan-shi, Shiga 520-3231   TEL.0748-72-0012
E-mail: info@kitajima-shuzo.jp   http://kitajima-shuzo.jp/

Not just preserving tradition, Evolution is the key

Mastering sake with two brands: the standard "Miyosakae" and the challenging "Kitajima".

Konan City in Shiga Prefecture is located southeast of Lake Biwa. The cold winters unique to a lake country, the Hiei oroshi (local wind), the soft, high-quality subsoil water from the Suzuka Mountains to the east, and the high-quality Omi rice from Shiga, the home of rice, create the perfect environment for sake brewing. Kitajima Shuzo was founded in 1805, the second year of the Bunka era, and for more than 200 years since its establishment, the brewery has been using traditional methods and reliable techniques. The lessons learned from "Kimoto-zukuri" are applied and practiced in all aspects of sake brewing. Kitajima Shuzo is a sake brewing company that pays attention to the details that are not written on the label, and does not allow any compromise. Kitajima Shuzo's two brands, the standard "Miyosakae" and the challenging "Kitajima," are all about evolving tradition.

大吟醸酒
Daiginjo Miyosakae

"Fruity, with a splendid aroma"
A limited-edition Daiginjoshu made with 35% Yamadanishiki polished to the highest possible level for submission to national competitions.

Daiginjoshu	1,800ml/720ml
Alc.	16-17%
Rice type	Yamadanishiki
Sake meter value	+5.0
Acidity	1.5

純米吟醸酒
Kitajima Karakuchi Kanzen Hakko

"Sharp, cool Junmai Ginjo"
This is a dry sake made by breeding a strong yeast and letting the saccharification and fermentation proceed smoothly until the sugar content is gone.

Junmai Ginjoshu	1,800ml/720ml
Alc.	15%
Rice type	Tamasakae
Sake meter value	+12.0
Acidity	1.7

Water source	Underground
Water quality	Slightly soft

Furukawa Shuzo

1-3-33 Yagura, Kusatsu-shi, Shiga 525-0053 TEL.077-562-2116
E-mail: rscrj393@ybb.ne.jp https://www.furukawashuzo.com/

A small brewery with local roots

A traditional sake brewery born and raised in Kusatsu-juku on the Tokaido.

In the Edo period (1603-1868), the Kusatsu River was not allowed to be bridged and had to be crossed by foot. When it rained or flooded, the Kawagoe Hitozoku would assist travelers by charging three yen per person for the bridge. The Kusatsu River is also called "Sand River" because a large amount of sediment is washed away along with the water when it rains heavily. The sediment accumulated on the riverbed and caused flooding, and as a result of endlessly raising the banks to prevent this, the riverbed became a ceiling river, higher than the surrounding plains. This ceiling river flows through Kusatsu-juku, which is located at the junction of Tokaido and Nakasendo roads. The Furukawa Shuzo, which grew up alongside the inn town, is the only sake brewery still in existence in Kusatsu-juku. For generations, the brewery has been dedicated to using the water, climate, and Omi rice of Kusatsu, and has painstakingly brewed its famous sake, "Tenjogawa" and "Munehana" by hand in the traditional manner.

本醸造原酒
Honjozo Tenjogawa

"A sake that can only be tasted locally"

Made from 100% organic rice grown locally in Kusatsu without the use of pesticides or chemical fertilizers. This is a safe sake that is good chilled or even better heated.

Honjozo Genshu	1,800ml/720ml
Alc.	19.9%
Rice type	Nipponbare
Sake meter value	-1.0
Acidity	1.6

特別純米酒
Tokubetsu Junmai Tenjogawa Genshu Mizukagami 100%

"Subtle sweetness and lingering taste"

The label reads, "The rice used is 100% environmentally friendly agricultural products". The rice "Mizukagami" used for the production made its debut in 2013.

Special Junmaishu	1,800ml/720ml
Alc.	17.9%
Rice type	Mizukagami
Sake meter value	-1.0
Acidity	1.7

Kizakura

Water source	Groundwater in Fushimi, Kyoto
Water quality	Medium-hard

223 Shioya-machi, Fushimi-ku, Kyoto-shi, Kyoto 612-8046 TEL.075-611-4101
https://kizakura.co.jp/

A unique latecomer Original ideas and actions

A brewery that produces "Kyoto sake with Kyoto rice" in Fushimi, one of Japan's leading sake brewing regions.

Founded in 1925, Kizakura is located in Fushimi, one of Japan's most famous sake brewing areas, with an abundance of groundwater. The company is a latecomer to Fushimi, where there are many long-established sake breweries. The name of the company is derived from "yellow cherry blossoms", whose petals scattered in the wind are considered to be elegant. The company is involved in the revival of Kyoto Prefecture's original rice, "Iwai", and the cultivation of "Kyo no Kagayaki", and is committed to local production for local consumption of " Kyoto sake with Kyoto rice". Sake made from high quality water and locally grown sake rice has an elegant aroma and is mild on the palate. The brewery continues to protect the culture of sake brewing by advancing traditional techniques that are supported by research.

Hanashofu Daiginjo Kizakura

"This Daiginjo uses Kizakura's original Ginjo yeast"

Characterized by its soft and splendid flower-like aroma. Smooth and rich flavor is mouth-filling. Serve at room temperature or chilled.

Daiginjoshu	1,800ml/720ml
Alc.	16%
Rice type	Yamadanishiki
Sake meter value	±0
Acidity	1.1

Kizakura Kyono Shizuku Junmai Ginjo Iwaimai

"The rice is made from "Iwai", a sake rice from Kyoto"

You can taste the smoothness, rice flavor, and richness unique to "Iwai" rice. This is a premium dry sake to drink on a special day.

Junmai Ginjoshu	720ml/300ml
Alc.	16%
Rice type	Iwai
Sake meter value	+2.0
Acidity	1.5

Water source	Spring water from Shiroyama
Water quality	Soft

Kinoshita Shuzo

1512 Kouyama, Kumihama-cho, Kyotango-shi, Kyoto 629-3442 TEL.0772-82-0071
https://www.sake-tamagawa.com/

Many of our products are shipped as aged sake, and our customers call this "secondary sake brewing"

A sake brewery near a clear stream brews sturdy sake that is suitable for aging.

Kinoshita Shuzo was founded in 1842 in Kumihama, Kyotango City in the northern Kyoto district. The brand name "Tamagawa" comes from the Kawakami Tanigawa River right next to the brewery, which was a clear stream lined with gravel. At that time, it was customary to regard rivers and lakes as sacred, and the name "Tamagawa" was derived from the fact that the river was like jade (very clean). Tamagawa has the characteristics of "delicious," "long-lasting," "interesting changes," and "nurturing," and has in common that it is a strong sake suitable for aging. Freshly pressed sake is considered to be the starting point (although there are some products that are shipped at the new sake stage), and most of them are shipped after two to three years of maturation in order for customers to experience the changes over time. We hope that our customers will develop their own favorite " Tamagawa" through "secondary sake brewing".

純米酒
Tamagawa Shizen Jikomi Junmaishu (Yamahai) Muroka Namagen-shu

"Producing limited editions using different rice and methods"
Rich in acid and amino acid, this Junmaishu is produced from a yeast-less yeast mash. Available in Gohyakumangoku, Omachi, and white label versions.

Junmaishu	1,800ml/720ml
Alc.	20%
Rice type	Hyogokitanishiki
Sake meter value	private
Acidity	private

純米酒
Tamagawa Time Machine 1712

"Applying the methods of the Edo period"
A sake version of a dessert wine, excellent with ice cream. It goes surprisingly well with "heshiko", a specialty of Tango.

Junmaishu	360ml
Alc.	14.0-14.9%
Rice type	Hyogokitanishiki
Sake meter value	private
Acidity	private

Gekkeikan

Water source	Underground water in southern Kyoto-shi
Water quality	Medium-hard

247 Minamihama-cho, Fushimi-ku, Kyoto-shi, Kyoto 612-8660 TEL.0120-623-561
https://www.gekkeikan.co.jp/

Brewing innovative sake using Fushimi water

Aiming for health and creating pleasure through the science of sake.

The company was founded in Fushimi, Kyoto in 1637 by Jiemon Okura, the first generation. The name of the company at that time was "Kasakiya," and the name of the sake was "Tama no izumi". Fushimi's geological formation of sand and gravel has a good amount of minerals dissolved in the water, which help to produce sake with a good balance of flavor and sweetness. Gekkeikan is characterized by its creativity. In 1910, at a time when sake was being bottled in barrels, Gekkeikan developed a "preservative-free bottled sake" and introduced Japan's first four-season brewing system, in which sake is brewed throughout the year. In 1902, it began exporting sake, and to date, it has done business with about 50 countries, expanding its sake culture to the world.

純米大吟醸酒
Cho-tokusen Horin Junmai Daiginjo

"Glamorus ginjo flavor and smooth taste"
This is a Daiginjoshu that enhances the taste of Kyoto cuisine, making the most of the ingredients, when enjoying kagai and ryotei restaurants in Kyoto.

Junmai Daiginjoshu	720ml
Alc.	16%
Rice type	Yamadanishiki/Gohyakumangoku

本醸造酒
Tokusen

"Elegant aroma and a refined, full flavor"
This honjozo-shu harmonizes with food without interfering with its taste. It has been loved by Kyoto's kagai and ryotei restaurants since ancient times.

Honjozoshu	720ml
Alc.	16%

| Water source | Subsoil water from Mt. Kongodoji |
| Water quality | Soft |

Takeno Shuzo

3622-1 Mizotani, Yasaka-cho, Kyotango-shi, Kyoto 627-0111   TEL.0772-65-2021
E-mail: info@yasakaturu.co.jp   https://yasakaturu.co.jp

Passing on the flavor of a fantastic sake rice to the present

Brew sake loved by the local people. A "local brewery" where you can see the faces of the producers.

To be able to see the faces of the brewers, to have people connect to each other, and to gain trust. This is the starting point of Takeno Shuzo. Since its early years, in the late Edo period to the early Meiji period, the brewery has cherished its ties with customers in the region as a local sake brewery, and these ties and mixtures have extended far beyond Tango to the rest of the world. The brand name "Kurabu" is given to the sake produced by going back to the origin of sake such as "Kamenoo". The name "Kurabu" is a combination of the word "kura" meaning "storage" and the word "bu" meaning "passing on the ancient, intangible spirit.Subsoil water from Mt. Kongodoji is used for the brewing, which has been flowing in the brewery since the days of the Yukimachi Shuzo, the original brewery where Takeno Shuzo was founded.

純米吟醸酒

Yasakatsuru Kamenoo Kurabu

"Served well in a wine glass"
Characterized by its sweet, smooth taste and moderate acidity. It is an unfiltered, unpasteurized sake with a smooth, sweet taste and moderate acidity.

Junmai Ginjoshu	720ml
Alc.	14%
Rice type	Kamenoo
Sake meter value	private
Acidity	private

純米酒

Yasakatsuru Iwaikurabu

"Mouth-filling and rich unfiltered sake"
The aroma of lychee and pear, and a hint of acidity in the taste.It has a fresh taste, and is good both cold and warmed.

Junmaishu	720ml
Alc.	14%
Rice type	Iwaimai
Sake meter value	private
Acidity	private

Saito Shuzo

Water source	Shiragiku water system
Water quality	Medium-hard

105 Yokooji Misusu Yamashiro Yashiki-cho, Fushimi-ku, Kyoto-shi, Kyoto 612-8207　TEL.075-611-2124
E-mail: sake@eikun.com　https://www.eikun.com/

Top-level sake rice
Our commitment to top-grade sake rice

Sake rice greatly influences the taste of sake. A light that was once extinguished has been revived in the Heisei era.

In 1895, Sotaro Saito, the ninth-generation head of the family, changed his business from a draper to a sake brewer, and began its history as a sake brewery. The brewery focuses on the rice called "Iwai", which greatly affects the taste of sake. The rice is suitable for Ginjoshu, with its large grains and clear heart white, and the water in Kyoto is ideal for producing a fine, soft, and full flavor. The rice was once discontinued at the end of the 1960s, but revived in 1992, and as a result of the constant refinement of the brewing techniques of Kyoto's sake brewers, it is now regarded as one of the most suitable for sake brewing. Saito Shuzo is dedicated to using this variety, and use 40% of the total crop.

Eikun Ichigin
Junmai Daiginjo

"Excellence made with passion and best skills"

This Junmai Daiginjoshu is made from 100% Yamadanishiki produced in Kyoto Prefecture and polished to 35%. It won a gold medal at the National New Sake Competition.

Junmai Daiginjoshu	720ml
Alc.	15%
Rice type	Yamadanishiki
Sake meter value	+4.0
Acidity	1.1

ShouhinmeiaEikun Junmai
Daiginjo Kotosennen

"Highly acclaimed Daiginjo overseas"

This Daiginjo is brewed in the famous water of Fushimi, Kyoto, to bring out the characteristics of the Kyoto-produced rice "Iwai", and also to produce a deep and mellow ginjo aroma.

Junmai Daiginjoshu	720ml
Alc.	15%
Rice type	Iwai
Sake meter value	+4.0
Acidity	1.1

Water source	Shiragiku
Water quality	Medium-hard

Fujioka Shuzo

672-1 Ima-machi, Fushimi-ku, Kyoto-shi, Kyoto 612-8051 TEL.075-611-4666
E-mail: kyoto@sookuu.net https://www.sookuu.net

Fresh as the blue sky, with a gentle taste

The history of a brewery that once closed, revived by the fifth-generation of the brewery's own master brewer.

Fujioka Shuzo was founded in 1902 by Eitaro Fujioka, the first-generation head of the brewery, in Higashiyama-ku, Kyoto. In 1910, the brewery was established in Fushimi, and in 1918, it moved to its current location. At that time, the brand name was "Mancho" and it was well known for many years. However, in 1994, the history of sake brewing came to an end with the sudden death of Yoshifumi Fujioka, the third-generation owner. Masaaki Fujioka, the fifth generation of the brewery, studied at various breweries, and in the winter of 2002, he became the chief brewer himself and started brewing sake. The new sake is all pure rice sake and named "Sookuu". The refreshing and gentle taste reminiscent of the blue sky is about to create a new history for Fujioka Shuzo.

Sookuu Junmaishu Miyamanishiki

"A sake that is easy to use"

Junmaishu Miyamanishiki is aimed to create a gentle flavor that does not interfere with the soup stock, always bearing in mind its compatibility with Kyoto cuisine.

Junmaishu	500ml
Alc.	15%
Rice type	Miyamanishiki
Sake meter value	-1.0
Acidity	private

Sookuu Junmai Ginjo Yamadanishiki

"High quality ginjo aroma, and soft mouthfeel"

Junmai Ginjoshu Yamadanishiki aims to be a sake you can easily reach for after you have enjoyed Kyoto cuisine.

Junmai Ginjoshu	500ml
Alc.	15%
Rice type	Yamadanishiki
Sake meter value	−2.0
Acidity	private

Masuda Tokubee Shoten

Water source	Fushimi's subsoil water
Water quality	Medium-hard

135 Shimotobaosada-cho, Fushimi-ku, Kyoto-shi, Kyoto 612-8471   TEL.075-611-5151
https://www.tsukinokatsura.co.jp/

Seasonality is the life of sake

The oldest sake brewery in Fushimi, Kyoto never forgets to take on a new challenge.

Founded in 1675. Masuda Tokubee is one of the oldest sake breweries in Fushimi, located on Toba Kaido, a street lined with old houses along the Kamo and Katsura rivers. "Tsukinokatsura", the representative brand name of the brewery, has been handed down from generation to generation through the painstaking handiwork of the master brewers, with a sense of seasonality and individuality, while constantly taking on new challenges. For example, in 1964, the company developed Japan's first sparkling nigori sake. As a pioneer of old sake, the brewery has been storing Junmai Daiginjoshu in jars since 1964. In 1992, the company began planting rice with the Kyoto-grown sake rice "Iwai" and has cultivated about 20 hectares without using pesticides. Masuda Tokubee also started developing low alcohol sake about 30 years ago. The company is also making efforts to collaborate with other sake breweries and is increasing its efforts to reach out to the world.

Tsukinokatsura Junmai Daigokujo Nakagumi Nigori-shu

"Main brand of the original Nigori-shu"

This is Japan's first sparkling nigori-shu, made in 1964 at the recommendation of Dr. Kinichiro Sakaguchi, a professor of sake. It has a refreshing acidity and a pleasant mouthfeel.

Junmaishu	720ml/300ml
Alc.	17%
Rice type	Gohyakumangoku
Sake meter value	+3.0
Acidity	1.7

Tsukinokatsura Kohakuko Jyu-nen Koshu

"Utopia loved by the Japanese"

This is an old sake reproduced from the book "Honcho Shokkan" (1640). It is a rich and mellow sake with an amber color, aged slowly in a porcelain jar.

Junmai Daiginjoshu	1,800ml/720ml
Alc.	17%
Rice type	Yamadanishiki
Sake meter value	+5.0
Acidity	1.7

| Water source | Subsoil water from Mt. Taiko |
| Water quality | Medium-hard |

Mukai Shuzo

67 Hirata, Ine-cho, Yosa-gun, Kyoto, 626-0423 TEL.0772-32-0003
E-mail: info@kuramo-mukai.jp

Sake brewery in a boathouse in the fisherman's town of Ine

A brewery rooted in the local food culture that adds color to springtime drinking in a fishing town.

This sake brewery was founded in 1754 in Ine Town, Kyoto Prefecture, the home of the boatyards famous for Ine Buri. As a fishing town, Ine is lined with boathouse homes where boats can be stored on two sides of the house, and people have been living in harmony with the sea since ancient times. The building of the Mukai Shuzo is also a boathouse, with a pier extending from the brewery to the sea, an exceptional environment for enjoying sake. Since fishermen work very hard and their food is often rich in flavor, Mukai Shuzo has always tried to make sake with a sharp taste that goes well with rich food. The phrase "whale sashimi and sake is Mukai's Kyo no haru" is also in a folk song, and Mukai's basic policy is to continue brewing sake that is rooted in the local food culture. The brewer's concept is to "make the kind of sake I want to make".

純米酒
Kodai Maishu Ine Mankai

"Bright colors and unique taste"
This sake is made from purple ancient rice and has a bright color and unique flavor. It tastes like a rosé wine, with a harmonious combination of sweetness and acidity.

Junmaishu	1,800ml/720ml
Alc.	14%
Rice type	Murasakikomachi
Sake meter value	−46.0
Acidity	6.3

特別純米酒
Tokubetsu Junmaishu Kyo no Haru Kimoto

"Sharp taste with mild flavor that goes well with meals"
With the help of trainees from Tokyo University of Agriculture, this sake is made using the Tajima method of pressing by foot. It is popular as an evening sake that goes well with fish dishes.

Special Junmaishu	1,800ml/720ml
Alc.	15%
Rice type	Iwai
Sake meter value	+5.0
Acidity	1.8

Ibaraki Shuzo

Water source	Groundwater from Mt. Rokko
Water quality	Medium-hard

1377 Nishioka, Uozumi-cho, Akashi-shi, Hyogo 674-0084   TEL.078-946-0061
E-mail: info@rairaku.jp   https://rairaku.jp

Brewed by the 9th-generation owner and master brewer
Sake brewed with modern sensibilities

Sake is for experts who love sake. Sake Abelia is for women and beginners.

Ibaraki Shuzo was founded in 1848 at the end of the Edo period. The owner and Toji (master brewer) Mikihito Ibaraki, the ninth-generation chief, brews sake by hand with a modern sensibility, and the name "Rairaku" is derived from the Analects of Confucius. The sake Abelia, made with flower yeast, is recommended for women and beginners who want to drink it like a glass of wine or a cocktail. The Ibaraki Shuzo hopes to make sake brewing more accessible to the public, so they offer a variety of events such as the "Rairaku Brewing Party", where you can experience sake brewing, the "Shuzo Yose" held every spring and fall, "Rice planting", and the "Nara Zuke Party". If you're looking for a local sake in Akashi City, Hyogo Prefecture, you'll want to remember Ibaraki Shuzo's Rairaku.

Rairaku
Junmai Nama Genshu

"The presence of acidity is also strong"

An unfiltered, unpasteurized sake made with Yamadanishiki for the koji and Gohyakumangoku for the kake rice, both from Hyogo Prefecture.

Junmai Nama Genshu	720ml
Alc.	17%
Rice type	Yamadanishiki/Gohyakumangoku
Sake meter value	+1.0
Acidity	1.8

Rairaku Hananokura Abelia
Junmai Nama Genshu

"Sweet like a white peach"

Yamadanishiki is used for the koji and Gohyakumangoku is used for the kake rice, both from Hyogo Prefecture, and flower yeast isolated at Tokyo University of Agriculture is used.

Junmai Nama Genshu	720ml
Alc.	17%
Rice type	Yamadanishiki/Gohyakumangoku
Sake meter value	+2.0
Acidity	1.7

Water source	Deep well
Water quality	Slightly-soft

Okada Honke

1021 Yoshino, Noguchi-cho, Kakogawa-shi, Hyogo 675-0017 TEL.079-426-7288
E-mail: okadahonke.ad@gmail.com http://www.okadahonke.jp/

The only sake brewery in Kakogawa Brewed by father and son for two generations

Okada Honke thought about closing down the business, but made a fresh start by making local sake.

Okada Honke is the only sake brewery in Kakogawa and was established in 1874. For a long time, the brewery was a subcontractor to major manufacturers for mass production, but with the popularity of beer and wine, the demand for sake itself decreased. The owners thought about closing the brewery down, but in 2010, father and son decided to make a fresh start, brewing local sake in Kakogawa. They want people to acknowledge that "there is good sake in Kakogawa" and are working hard to enhance the brand power. The brand name "Seiten" is the name of the sake that has been produced for generations. It is described as "a light, sweet, spicy, and moderate sake made with subsoil water from the Kakogawa River.

大吟醸酒
Seiten Daiginjo

"A sake with a delicious aroma and good sharpness"
This Daiginjoshu is made from 35% polished Yamadanishiki produced in Hyogo Prefecture, and slowly brewed at low temperatures for more than 40 days.

Daiginjoshu	1,800ml/720ml
Alc.	16%
Rice type	Yamadanishiki
Sake meter value	+4.0-+5.0
Acidity	private

特別純米酒
Seiten Tokubetsu Junmai

"A dry and refreshing sake that can be enjoyed with meals"
It is dry and refreshing, with a hint of rice flavor at the end. The rice used is Gohyakumangoku, a sake rice grown in the company's own fields.

Special Junmaishu	1,800ml/720ml
Alc.	15%
Rice type	Gohyakumangoku
Sake meter value	+4.0
Acidity	private

Okamura Shuzojo

Water source	Mukogawa River
Water quality	Soft

340 Kozuki, Sanda-shi, Hyogo 669-1412 TEL.079-569-0004
E-mail: kozuki@ares.eonet.ne.jp   http://www.eonet.ne.jp/~okamura-sake/

There is good sake in Sanda, Hyogo

Sake made by combining the abundance of local rice, water, people, and nature.

In the first year of the Bunkyu Era (1861), Kohei Okamura, the first generation of the family, started brewing soy sauce in Kouzuki, Takahira Village, Arima County (now Sanda City) in Hyogo Prefecture, and Eikichi Okamura, the second-generation head, established a sake brewery in 1889. Today, Sanda City is a bedroom community in the Hanshin area, which has recorded the highest population growth rate in Japan for 10 consecutive years. The Okamura Sake Brewery is a small sake brewery blessed with a rich climate and a relaxing rural landscape that has not changed since the old days. It is said that the name "Sanda" comes from "Sanpukuden" ("Onden", "Hiden" and "Kyoden") which were found in the womb of the seated statue of "Mirokubosatsu Zazo", a national cultural asset. The rice grown in the rich natural environment of Sanda is used for brewing, with the care of the head brewer himself. These elements create a deep and rich flavor, which is the pride of this brewery.

純米大吟醸酒
Junmai Daiginjo Sanpukuden

"Authentic taste of Yamadanishiki"

The sake is light, sweet, and elegant, with the flavor of the rice coming through. It goes well with lightly flavored dishes such as sashimi and soba noodles.

Junmai Daiginjoshu ...	1,800ml/720ml
Alc.	15.5%
Rice type	Yamadanishiki
Sake meter value	-2.0
Acidity	1.7

純米酒
Junmaishu Sanda Ichi

"The rich and deep taste of rice"
Characterized by a rich and delicious taste and aroma. It is especially good with dark-flavored dishes made with game meat.

Junmaishu	900ml
Alc.	6.5%
Rice type	Gohyakumangoku
Sake meter value	±0
Acidity	1.6

| Water source | Subsoil water from the Chikusa River |
| Water quality | Soft |

Okuto Shoji

1419-1 Sakoshi, Ako-shi, Hyogo 678-0172 TEL. 0791-48-8005
E-mail: okuto@image.ocn.ne.jp

The land of Chushingura with a deep history
400 years in the port town of Ako

Handmade sake passed down from generation to generation using traditional methods.

From the end of the Edo period to the Meiji period, Sakoshi flourished as a port for shipping salt from Ako. The Okuto family, who also served as the headmen of the town and headmen of ships, started brewing sake in 1601. In the past, they also served as the official liquor store for the Asano family, the lords of the Ako domain. The sake brewery is more than 300 years old, and you can feel the atmosphere of the past as well as the streets of Sakoshi. They use the best ingredients for sake, such as water from the clear Chikusa River and sake rice from Harima, and make sake in a careful, traditional way. Ako is known for Chushingura, and the name of Okuto's main brand is also "Chushingura". The goal is not to make a sake that is easy to drink. They focus on a rich flavor that is worth drinking. It is the kind of sake that makes you want to say, "I'll have another cup".

大吟醸酒
Chushingura Daiginjo

"A high-spec sake"

This Daiginjo is brewed from Yamadanishiki in Hyogo Prefecture, polished to 40%, with manual processing of ingredients and careful temperature control.

Daiginjoshu	1,800ml/720ml
Alc.	7.5%
Rice type	Yamadanishiki
Sake meter value	+2.0
Acidity	1.5

純米酒
Chushingura Yamahai Junmai

"Rich, mellow flavor"

This brewing process is called Yamahai, and it waits for a natural lactic acid to form. This traditional method is time-consuming and also requires great care.

Junmaishu	1,800ml/720ml
Alc.	16.5%
Rice type	Hyogoyumenishiki
Sake meter value	+2.0
Acidity	2.2

Kikumasamune Shuzo

Water source	Subsoil water of Mt. Rokko range
Water quality	Hard

1-7-15 Mikagehonmachi, Higashinada-ku, Kobe-shi, Hyogo 658-0046 TEL.078-854-1043
https://www.kikumasamune.co.jp/

Tradition and Innovation From Nada to the World

The first packaged sake to become "Great Value Champion Sake".

Kikumasamune Shuzo was founded in 1659 and became popular as a "kudari zake" (sake sent from Nada to Edo) in line with the development of Edo (present-day Tokyo), and developed during the Bunka-Bunsei period in the latter half of the Edo era. Under the vision of "tradition and innovation," Nada continues to carry on the traditional Nada sake brewing technique of "Kimoto zukuri", which has been used since the Edo period. In 2019, it became the first packaged sake to win a prize in the SAKE category of the International Wine Challenge (IWC), the world's largest wine competition. "Hyakumoku", the brewer's first new brand in 130 years, has grown into a popular sake.

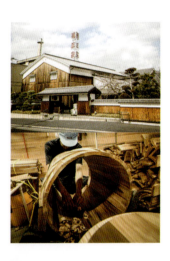

本醸造酒
Kikumasamune Jyosen

"A dry sake recognized around the world"

This is a dry sake brewed using the traditional "Ikatashouzukuri" method, which slowly cultivates strong yeast with the help of living lactic acid bacteria.

Honjozoshu	1,800ml
Alc.	15%
Rice type	Domestic rice
Sake meter value	+5.0
Acidity	1.4

純米酒
Kikumasamune Junmai Taruzake

"The refreshing scent of Yoshino-cedar"

This dry sake brewed using the traditional "kimoto zukuri" method is packed in sake barrels made of Yoshino-cedar and bottled when it is at its best.

Junmaishu	720ml
Alc.	15%
Rice type	Domestic rice
Sake meter value	+5.0
Acidity	1.4

| Water source | Subsoil water from Rokko Nagao mountain range |
| Water quality | Soft |

Konishi Shuzo

2-13 Higashi-Arioka, Itami-shi, Hyogo 664-0845 TEL.072-775-0524
http://www.konishi.co.jp/

The birthplace of Sei-shu
A long-established sake brewery in Itami

Passing on 470 years of history and tradition to the next generation.

Established in 1550 in Itami City, which is known as the "birthplace of sei-shu". Based on the idea of "creating taste and happiness, single-mindedly," the company has existed for a long time with the climate of Itami. In the days of the founder, Shinemon Konishi, the company mainly made nigori-zake, and it was around 1612 that they started making sei-shu. The sake barrels were transported to Edo on horses, and so-called "kudari zake" was served. In 1635, while transporting sake to Edo, Sotaku, the second generation of the family, was impressed by the nobility of Fuji, which was covered with snow, and named the sake "Shirayuki" which continues to be produced now in the Reiwa era. Konishi Shuzo has always valued the customer's point of view, and had pride and hope as part of the sake brewing industry. Now the brewery is aiming for "innovative management that is always in fashion", with an eye on the next generation.

Tokusen Shirayuki
1,8L Binzume

"Deeply prepared honjo-shu"

This is a traditional brand of sake made in Itami, the birthplace of sei-shu. The taste is rich and mellow.

Honjozoshu	1,800ml
Alc.	15-16%
Rice type	Yamadanishiki
Sake meter value	+2.0
Acidity	1.4

Cho-tokusen Shirayuki Edogenroku no sake
(Revival) Genshu 720ml Binzume Keshobakoiri

"A secret sake that unravels a long history"

This sake is based on a recipe from 1702 and Konishi family's secret book, "Sake Eidai Oboecho". The water used in the brewing process is less than now, and the color is amber.

Junmaishu Genshu	720ml
Alc.	17-18%
Rice type	Domestic rice
Sake meter value	−35.0
Acidity	3.3

Konotomo Shuzo

Water source	Chugoku mountain range
Water quality	Slightly-soft

508 Yanase-machi, Santo-cho, Asago-shi, Hyogo 669-5103 TEL.079-676-3035
http://konotomo.jp/

Careful sake brewing is possible Because we are a small brewery

The natural water of Mt. Awaga in the Chugoku Mountains and the skills of the Tajima Toji (master brewer) are poured into the brewing process.

Founded in 1690 in Tajima, the home of Tajima Toji, one of the four major Toji in northern Hyogo Prefecture. The Tajima region, which occupies about a quarter of Hyogo Prefecture, is bordered by the Sea of Japan to the north, Harima and Tanba to the south, Kyoto to the east, and Tottori to the west. The coastal area is designated as Sanin Kaigan National Park. Using sake rice grown under contract with local farmers (Yamadanishiki from Hyogo Prefecture and Gohyakumangoku from Tajima), natural water from Mt. Awaga in the Chugoku Mountains, and the skills of Tajima Toji, the brewery makes sake rooted in its hometown. Says the brewery: We want to preserve and pass on the traditional taste of sake by carefully brewing sake, which we can do because we are a small brewery. With this in mind, the Toji and brewers put all their knowledge and skills into brewing sake during the cold season. The brewery's daiginjo has won the gold medal at the National New Sake Competition for seven consecutive years.

大吟醸酒
Daiginjo Tajima Gokujo

"Splendid ginjo flavor and smooth taste"

This Daiginjoshu is made using the Tajima method of long-term brewing at low temperatures. The entire amount of Yamadanishiki from Hyogo Prefecture (38% milling ratio) is used.

Daiginjoshu	1,800ml
Alc.	16%
Rice type	Yamadanishiki
Sake meter value	−3.0
Acidity	1.2

純米吟醸酒
Junmai Ginjo Kakoya "Ya"

"Fragrant and fruity"

A fruity, fresh, highly aromatic Junmai Ginjoshu. It is easy to drink and can be recommended to women.

Junmai Ginjoshu	1,800ml
Alc.	15%
Rice type	Yamadanishiki
Sake meter value	+2.5
Acidity	1.3

| Water source | Subsoil water from Mt. Rokko |
| Water quality | Medium-hard |

Sakuramasamune

5-10-1 Uozaki-minamimachi, Higashinada-ku, Kobe-shi, Hyogo 658-0025 TEL.078-411-2101
E-mail: sakura@sakuramasamune.co.jp https://www.sakuramasamune.co.jp/

Aiming to make sake that the family can enjoy together

A sake that has been enjoyed in Nada, Kobe for a long time and passed down from generation to generation.

Sakuramasamune located in Nada, Kobe, was founded in 1625 and has a history of about 400 years. It is the originator of the sei-shu "Masamune". The company has contributed to the development of Nada through the discovery of "Miyamizu" and the brewing of sake using highly polished rice from a water mill. After the Great Hanshin-Awaji Earthquake struck, they renewed their belief that sake brewing is only possible with the support of the local community, and opened the Sakuramasamune Memorial Museum, "Sakuraen", with the aim of helping to revitalize the area. The company holds a "kurabiraki" event every November to show their appreciation to the local community. With the support from the local community association, the event has become an annual event that is enjoyed by the local people. The company's corporate philosophy is to put quality first and contribute to the local community, and it aims to continue brewing sake as a brewer loved by the community.

大吟醸酒
Sakuramasamune Kinmare Daiginjo Genshu

"The whole ginjo brewing process is handcrafted"

This is a sake with a splendid aroma and a clear taste. If you drink it chilled, it will go well with carpaccio of Akashi sea bream or boiled hamo.

Daiginjoshu	720ml
Alc.	17%
Rice type	Yamadanishiki
Sake meter value	±0
Acidity	1.2

純米大吟醸酒
Sakuramasamune Kinmare Junmai Daiginjo 40

"The gentle taste of Yamadanishiki"

This is a Junmai Daiginjoshu with a fruity aroma and harmonious rice plumpness. Good served with grilled oysters from Ako.

Junmai Daiginjoshu	720ml
Alc.	15%
Rice type	Yamadanishiki
Sake meter value	-2.0
Acidity	1.4

Sawanotsuru

Water source	Miya-mizu
Water quality	Hard

5-1-2 Shinzaike Minami-machi, Nada-ku, Kobe-shi, Hyogo 657-0864 TEL.078-881-1234
https://www.sawanotsuru.co.jp/site/

The symbol ※ on the label indicates our commitment to rice

A long-established brewery that has maintained traditional sake brewing for 300 years.

Sawanotsuru was founded in 1717, when the first generation of the company, who ran a rice store, started brewing sake as a side business. For this reason, the brewery has been committed to producing only Junmaishu, and the symbol ※ derived from the Japanese character for rice is engraved on the product labels. Nada in Kobe where the brewery is located, has a favorable climate for kanzukuri. The cold wind from Mt. Rokko (called Rokko Oroshi) and the influence of the inland sea, has greatly helped sake brewing in the area since ancient times. Nada was also blessed with a favorable environment; it was located in the coastal area, which was advantageous for transporting sake to Edo (now Tokyo) by sea, and nearby Tanba provided labor. Also the ingredients, "Miya-mizu"(a hard water) and "Yamadanishiki" (the king of sake rice), both suitable for sake brewing, were close by. Since its founding, Sawa No Tsuru's motto has been "Make the most of the rice, examine the rice, and focus on the rice".

Sawa No Tsuru Daiginjo Shunshu

"Craftsmanship full of elegance and dignity"

It has an elegant and graceful aroma like a fruit, and a smooth and sharp taste. Goes well with Akashi sea bream and hamo-suki sashimi.

Daiginjoshu	1,800ml
Alc.	16.5%
Rice type	Yamadanishiki
Sake meter value	+5.0
Acidity	1.4

Sawa No Tsuru Tokubetsu Junmaishu Jitsuraku Yamadanishiki

"The Yamadanishiki of the Yamadanishiki"

This product is brewed using Nada's traditional "Kimoto-zukuri" method, using only Yamadanishiki rice produced in Jitsuraku, certified as "toku-A" area. You can enjoy a sharp aftertaste.

Special Junmaishu	720ml
Alc.	14.5%
Rice type	Yamadanishiki
Sake meter value	+2.5
Acidity	1.8

| Water source | Subsoil water from Mt. Rokko range |
| Water quality | Medium-hard |

Hakutsuru Shuzo

4-5-5 Sumiyoshi Minami-machi, Higashinada-ku, Kobe-shi, Hyogo 658-0041   TEL.078-822-8921
https://www.hakutsuru.co.jp/

The heir to Japanese sake culture

The origins of the original sake rice "Hakutsuru Nishiki". Continuous evolution of the Hakutsuru brand.

"We continue to produce tasty sake". Since its establishment in 1743, Hakutsuru has continued to carry on this belief and commitment. Nada in Hyogo, known as Japan's number one sake brewing region, is home to a cold wind called "Rokko Oroshi", the mineral-rich subsoil water of the Mt. Rokko range, and the best sake rice, Yamadanishikiki. A Tanba Toji, one of the three major schools of Toji in Japan, is the master brewer. In 1995, as a new challenge, Hakutsuru Shuzo started to develop a new type of sake rice by crossing Yamadanishikiki's mother "Yamadabo" with "Wataribune No. 2". Eight years later, the new sake rice "Hakutsuru Nishiki" was born, and it has become the rice of choice for many breweries. Hakutsuru Shuzo is determined to improve quality of their products to develop a reliable brand.

普通酒
Josen Hakutsuru

"Highly polished sake"
The more you drink, the more familiar you become with Hakutsuru's traditional "Umakuchi-sake". Brewing techniques have been passed down from generation to generation since Hakutsuru was founded in 1743.

Futsushu	1,800ml
Alc.	15-16%
Rice type	Domestic rice
Sake meter value	+1.0
Acidity	1.4

純米酒
Hakutsuru Josen Junmai Nigori-sake Sayuri

"Very popular overseas"
"Junmai Nigori-sake" is carefully brewed with only rice, rice malt and water. It has a refreshing sweet taste and a clean aftertaste.

Junmaishu	300ml
Alc.	12-13%
Rice type	Domestic rice
Sake meter value	-11.0
Acidity	1.6

Tajime

Water source	Well water in the brewery
Water quality	Weakly-soft

545 Yanase-machi, Santou-cho, Asago-shi, Hyogo 669-5103 TEL.079-676-2033
E-mail: info@chikusen-1702.com   http://www.chikusen-1702.com

We want to live long with everyone's smiles and the bountiful nature of our hometown

"A single grain of rice has infinite power". Careful sake brewing with a strong sense of tradition.

Tajime was founded in 1702, the 15th year of the Genroku era. The brewery is located in Santo-cho in the Tajima region where the ruins of Takeda Castle (famous as "the castle in the sky") are located. 300 years of history and tradition have been preserved, and the brewery aims to coexist and prosper with the local Tajima region by using local rice and water. They also aim for an integrated brewing process with the local community so that sake production can actually make people happy, as well as protect nature in their hometown. The main brand name of Tajime is "Chikusen". The first character comes from the Takenokaga, a river upstream of the Maruyama River, and the second character is taken from "Izumi no kuni", the birthplace of their ancestors. The name "Chikusen" expresses gratitude and respect for ancestors and the precious water.

Chikusen Junmai Ginjo Konotori

"Strong rice taste and sharpness"

Made from rice cultivated by "Konotori Hagukumi Noho (Stork Nurturing Farming Method)" which does not use agricultural chemicals or chemical fertilizers. It has a delicious acidity and a soft mouthfeel.

Junmai Ginjoshu	1,800ml/720ml
Alc.	16%
Rice type	Gohyakumangoku
Sake meter value	+5.0
Acidity	2.0

Chikusen Junmai Daiginjo Kou No Tori

"Gentle, reassuring taste"

This sake was once served in the first class of ANA international flights. Made from Yamadanishikii cultivated using the "Konotori Hagukumi Noho (Stork Nurturing Farming Method)".

Special Daiginjoshu	1,800ml/720ml
Alc.	15%
Rice type	Yamadanishiki
Sake meter value	+7.0
Acidity	1.2

Water source	Miya-mizu
Water quality	Hard

Tatsuuma Honke Shuzo

2-10 Tateishi-cho, Nishinomiya-shi, Hyogo 662-8510 TEL. 0798-32-2727
E-mail: customer@hakushika.co.jp https://www.hakushika.co.jp/

Over 350 years of tradition Sake with a rich culture

"We believe that sake is not something to be made, but something to be nurtured."

Founded in 1662 in Nishinomiya in Hyogo. It is said that the first brewer, Kichizaemon Tatsuuma dug a well at his residence and found sweetness in the water, so he began to brew sake. Miya-mizu, good quality rice, Sekkai (Osaka bay) and Rokko Oroshi. Hakushika has been deeply connected to the city of Nishinomiya with the blessings of the richness of its natural environment for more than 350 years. The name "Hakushika" comes from a Chinese legend about a long-lived deer. Based on their beliefs, "sake is not something to be made, but nurtured". The brands "Kuromatsu Hakushika" and "Hakushika" were named to express their wishes; they hope the sake can be enjoyed with a sense of generosity, and they pray for long life, which will ultimately bring smiles to the faces of those who enjoy drinking. With these thoughts in mind, they continue to pursue the path of sake brewing.

純米大吟醸酒
Kuromatsu Hakushika Junmai Daiginjo

"Combination of deep taste and sharp aftertaste"

A good body and a clean finish with a refreshing acidity. The aroma and taste are reminiscent of youthful melons.

Junmai Daiginjoshu	720ml
Alc.	16-17%
Rice type	Yamadanishiki
Sake meter value	+4.0
Acidity	1.6

純米吟醸酒
Kuromatsu Hakushika Junmai Ginjo

"Splendid flavor and soft sweetness"

It has a splendid flavor and a soft sweetness. A gentle, splendid and sweet flavor spreads in the mouth. Best served cold.

Junmai Ginjoshu	720ml
Alc.	14-15%
Rice type	Hitomebore
Sake meter value	+1.0
Acidity	1.4

Nadagiku Shuzo

Water source	Subsoil water from the Ichikawa River
Water quality	Somewhat soft

1-121 Tegara, Himeji-shi, Hyogo 670-0972　TEL.079-285-3111

The Harmony of Sake and Food Culture

We focus on locally grown sake rice. and walk together with the local food culture.

Kawaishi Shuzo was founded in 1910 by Mikisaku Kawaishi, the first-generation head of the brewery. The company was later reorganized into the current Nadagiku Shuzo in 1958. The name of the company is a combination of "Nada" of Harimanada which stretches in front of Himeji, and chrysanthemum. The company carefully selects rice from Hyogo, such as "Yamadanishikiki" and "Hyogo Yume Nishiki". In particular, "Hyogo Yume Nishiki", a specialty of Nishi-Harima, is grown under contract with a farmer in Ichikawa-cho, Kanzaki-gun. The company opened a liquor store in the underground mall of Himeji Station in 1964, and a kushiage club called "Kura" in 1984, and established its motto, "Harmony of Sake and Food Culture". Currently, there are three restaurants in Himeji where you can enjoy Himeji Oden, the local soul food, and Nadagiku sake.

大吟醸酒
Daiginjo "Mikinosuke"

"The highest quality sake is named after the founder's younger brother"

This is the highest quality of sake from Nadagiku Shuzo, with a graceful aroma and a deep, mellow flavor. Best served cold.

Daiginjoshu	1,800ml/720ml
Alc.	15%
Rice type	Yamadanishiki
Sake meter value	+4.0

特別純米酒
Tokubetsu Junmai "MISA"

"Unfiltered namazake named after the Toji"

A new type of sweet sake that combines sweetness and a refreshing acidity. It can be served cold or on the rocks, or with a squeeze of lemon.

Special Junmaishu	1,800ml/720ml
Alc.	16%
Rice type	Hyogo Yumenishiki
Sake meter value	-12.0

| Water source | Subsoil water from Mt. Rokko range |
| Water quality | Hard |

Nihonsakari

4-57 Yogai-cho, Nishinomiya-shi, Hyogo 662-8521　TEL.0798-32-2501
https://www.nihonsakari.co.jp/

More delicious, more beautiful

A perfect combination of sake rice, excellent water, and the cold wind of Rokko Oroshi.

Nihonsakari located in Nishinomiya, Hyogo was founded in 1989 by a group of volunteers who wanted to create a business that would contribute to the development of their hometown, and started a sake brewing business. Nada Gogo has excellent conditions for sake brewing, including Yamadanishiki, the best rice for sake brewing, Miya-mizu, a famous water suitable for sake brewing, and Rokko Oroshi, which brings cold air necessary for brewing. In the sake brewing industry, 132 years in business is still a fledgling so the company is constantly taking on challenges including running a cosmetics business, producing bottle cans for Namagen-shu and making health-conscious sake.

Chotokusen Junmai Ginjo Sohana

"The tradition of fine sake is here to stay"

This famous sake is brewed with rice suitable for sake brewing (Yamadanishiki and Gohyakumangoku), essential water (Miyamizu), and the traditional techniques of the Tanba Toji.

Junmai Ginjoshu	720ml
Alc.	15%
Rice type	Yamadanishiki/Gohyakumangoku
Sake meter value	-4.0
Acidity	1.6

Chotokusen Junmai Daiginjo Sohana

"A balance of flavor and aroma"

The famous sake "Sohana" has been made into a Junmai Daiginjo, the highest quality of sake. Made from 100% Yamadanishiki and polished to 38%.

Junmai Daiginjoshu	720ml
Alc.	15%
Rice type	Yamadanishiki
Sake meter value	-6.0
Acidity	1.6

Hakutaka

Water source	Miya-mizu
Water quality	Hard

1-1 Hama-cho, Nishinomiya-shi, Hyogo 662-0942　TEL.0798-33-0001
E-mail: soumu@hakutaka.jp　https://hakutaka.jp/

Fine, mellow beauty Japanese sake

Dignified and elegant sake brewing in the name of the sacred bird "Hakutaka".

The company was founded in 1862 by Etsuzo Tatsuuma, the first-generation head of the Tatsuma family (Hakushika Shuzo). The brewery is committed to "quality first" and received a "Kamon prize" at a first domestic exhibition in Japan in 1877. They have been offering Hakutaka to the Ise Grand Shrine continuously since they were selected as the official sake by the shrine in 1924. The brewery is located in the Miyamizu area, which is considered to be the best place for sake brewing. Using Miya-mizu water from eight wells owned by the brewery, as well as Yamadanishiki produced in Yoshikawa-cho certified as "toku-A" area, the company pursues high-quality sake brewing with a traditional kimoto-zukuri method. The founder's words, "a man of virtue will naturally attract admirers; good wine speaks for itself", is still the company's motto, and they are committed to serving their best products to customers.

Daiginjo Junmai Gokujo Hakutaka

"Exquisite sharpness with enriched taste"

This sake has an elegant finish with minimal Ginjo aroma. Special taste can only be made with the kimoto method. The sharpness makes it ideal for sipping.

Junmai Daiginjoshu	1,800ml
Alc.	16-16.9%
Rice type	Yamadanishiki
Sake meter value	+2.0
Acidity	1.6

Etsukura Tokubetsu Junmai Hitotsubi

"Specialty named after the founder"

This is a sake that makes the most of the characteristics of Hakutaka. You can enjoy the strength of Junmaishu with its robustness.

Special Junmaishu	1,800ml
Alc.	16-16.9%
Rice type	Yamadanishiki
Sake meter value	+3.0
Acidity	1.9

| Water source | Subsoil water from the Ibo River |
| Water quality | Medium-hard |

Honda Shoten

361-1 Takata, Aboshi-ku, Himeji-shi, Hyogo 671-1226 TEL.079-273-0151
E-mail: info@taturiki.com https://www.taturiki.com/

Using the highest quality Yamadanishiki from "toku-A" area in Hyogo

High quality sake that has won a total of 20 gold medals. at the National New Sake Competition.

Honda Shoten was founded in 1921 in Hyogo, the largest producer of Yamadanishiki, a rice suitable for sake brewing. A Honda ancestor was the general director of the Banshu Toji, and the brewery is based on the methods used in the Banshu region, where sake brewing has been active since ancient times. The main brand "Tatsuriki", is of the highest quality and uses "Yamadanishiki" from toku-A district. Their mission is to convey the true charm of rice, the difference in taste depending on harvest area, and history. Since the 1970s, the brewery has been making daiginjo, using rice suitable for sake brewing and milling all rice in-house. They are also working to revive "Shinriki", a fantastic sake rice originating from the local area. The brewery has won a total of twenty gold medals at the National New Sake Competition.

Tatsuriki Daiginjo Kome No Sasayakii

"Using Yamadanishiki produced in toku-A district in Hyogo"

This sake pursues the pinnacle of daiginjo, which exceeds the expectations of each individual. It has an excellent sharpness, and cold sake goes well with fish dishes.

Daiginjoshu	1,800ml
Alc.	17%
Rice type	Yamadanishiki
Sake meter value	+4.0
Acidity	1.4

Tatsuriki Tokubetsu Junmai Kimoto Jikomi

"The flavor of the rice and the acidity of kimoto"

This sake won the gold medal in "Kanzake (heated sake) Contest" which determines the best in Japan. This sake uses kimoto-shubo method which naturally produces lactic acid.

Special Junmaishu	1,800ml
Alc.	16%
Rice type	Yamadanishiki
Sake meter value	+1.0
Acidity	1.8

Yaegaki Shuzo

Water source	Subsoil water from the Hayashida River in the Ibo River system
Water quality	Soft

681 Mukudani Hayashida-cho, Himeji-shi, Hyogo 679-4211　TEL.079-268-8080
https://www.yaegaki.jp/

Simple and strong as a result of 350 years of history

A sake brewery in Banshu Hayashida, Hyogo. Surrounded by beautiful nature and mythology.

This long-established sake brewery located in Banshu Hayashida on the northern outskirts of Himej, Hyogo, was founded in 1666. In 1881, the brewery named its sake "Yaegaki" after Japan's oldest waka poem. The company also brews sake in the U.S., and group companies are engaged in the health-food ingredient and food additives business, so people tend to think of it as a modern company, yet all the koji made in the brewery uses traditional manual techniques. Yamadanishiki, the king of sake rice grown in the toku-A district, subsoil water from the Hayashida River in the famous "Shikagatsubo", and the skill and pride of the brewers who master the art of "Kanjikomi" are the three pillars that give Yaegaki its mellow flavor. The Toji belong to the Tajima Toji Union, but there are also many other breweries in the neighborhood that are alumni of agricultural universities, and they work hard to produce sake while inspiring each other.

純米大吟醸酒
Junmai Daiginjo Aonomu

"The flavor of fresh green apples"

Fresh green apple flavor spreads naturally, and the aftertaste is long and well-balanced. Great with grilled white fish or ayu.

Junmai Daiginjoshu	720ml
Alc.	15%
Rice type	Yamadanishiki/Gohyakumangoku
Sake meter value	+1.0
Acidity	1.2

特別純米酒
Yaegaki Tokubetsu Junmai Yamadanishiki

"A harmony of gentle flavor and refreshing acidity"

This Junmaishu has a sweet and gentle flavor of koji, and a delicious taste of rice and acidity. Versatile sake that goes well with all kinds of food.

Special Junmaishu	720ml
Alc.	15%
Rice typ	Yamadanishiki
Sake meter value	+2.0
Acidity	1.5

Umenoyado Shuzo

27 Higashimuro, Katsuragi-shi, Nara 639-2102 TEL. 0745-69-2121
E-mail: info@umenoyado.com https://www.umenoyado.com

In the land of hidden myths and legends
Brewed with the blessings of nature and human hands

Since the 2000s, a series of sake-based liqueurs have been developed.

The company was founded in 1893, in the southwest of the Yamato-bonchi, at the foot of the Katsuragi peaks, the site of many myths and legends since the time of Kojiki Manyo. Since obtaining a license to produce liqueurs in 2001, the company has developed a series of sake-based liqueurs, including plum wine, which are popular among people of all ages. Sake is made with Yamadanishiki, a rice suitable for sake brewing, and water from the local Katsuragi area, characterized by a fine, soft taste. While the sake brewing process is becoming more automated, the Umenoyado Shuzo believes in high quality sake brewed with the blessings of nature and the hands of people, while maintaining a process that spares no effort and does not compromise in the sake-brewing process.

純米大吟醸酒
Umenoyado Katsuragi Junmai Daiginjo

"A splendid scent swells up"
Made with 100% "Yamadanishiki" produced in toku-A district in Hyogo. This sake is the specialty of Umenoyado, into which the skill and soul of the brewer are poured. Best served cold or at room temperature.

Junmai Daiginjoshu	720ml
Alc.	16%
Rice type	Yamadanishiki
Sake meter value	+3.0
Acidity	1.1

本醸造酒
Umenoyado Honjozo

"Change the temperature and it tastes different"
A brewer's evening drink tastes deeper by changing the temperature. Characterized by a slight aroma and moderate taste.

Honjozoshu	720ml
Alc.	15%
Rice type	Domestic rice
Sake meter value	+3.0
Acidity	1.5

Kita Shuzo

Water source	Subsoil water from the Yoshino River
Water quality	Soft

8 Gobo-cho, Kashihara-shi, Nara 634-0062 TEL. 0744-22-2419
E-mail: miyokiku@miyokiku.com   http://www.miyokiku.com

Japanese Sake Hidden in the capital of Nara

Our commitment to sake brewing dates back 300 years. Sake that conveys tradition and joy.

Kita Shuzo was founded in 1718 by Rihee as a brewing business in Yamatogobo village in Kashihara, Nara, surrounded by the Yamato Sanzan mountains. It is said that Rihee was a stickler by nature and selected the best water and rice, and did not start the brewery until he was satisfied with the quality. In 1940, the brewery was moved to its current location as part of the expansion of the Kashihara-jingu, a project to commemorate the 2600 year of history according to emperors' reigns. Inheriting the intention of the founder, the brewery continues to focus on "delicious sake", even after 300 years. The brewer's idea of good sake is one that conveys the brewer's thoughts and feelings to the drinker. They have pride in "Miyokiku", loved as a beautiful sake of Yamato that conveys tradition and joy.

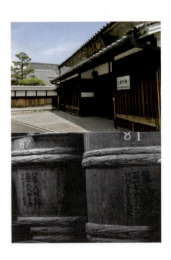

純米酒

Junmai Miyokiku Mizumoto Jikomi

"Old but new sake, made in Nara"

This Junmaishu is a reproduction of the secret "mizumoto" method of Shoryakuji on Mt. Bodai, which was used about 500 years ago.

Junmaishu	1,800ml
Alc.	15.5%
Rice type	Hinohikari by Nara
Sake meter value	-6.0
Acidity	1.6

純米吟醸酒

Junmai Ginjo Rihee

"Junmai Ginjo Genshu with a high alcohol content (20%)"

Gives the impression of distilled sake on the palate. It has an unassertive sweetness that lingers and ends with a sharp taste.

Junmai Ginjoshu	720ml
Alc.	20%
Rice type	Gohyakumangoku by Fukui
Sake meter value	+6.5
Acidity	2.9

| Water source | Mountain water from the mountain behind the house |
| Water quality | Soft |

Kuramoto Shuzo

2501 Tsugehayama-cho, Nara-shi, Nara 632-0231 TEL.0743-82-0008
E-mail: kuramoto-shuzou@comet.ocn.ne.jp https://kuramoto-sake.com

Inheriting and preserving the tradition of sake brewing

Japan's oldest "Yamato no Sake" revived in the land of Mahoroba.

Founded in 1871, Kuramoto Shuzo is a small company that carefully brews sake by hand, a common method in Nara. Their main strength is the bodaimoto brewing method, which is unique to Nara. This method was first developed in the Muromachi period, and uses "Soyashi-mizu" which is made by soaking raw rice in water and fermenting it with lactic acid, to make high-quality sake even during the warm season. This is the basis of Nara sake, and Kuramoto Shuzo is reviving and passing on this technology with the help of volunteer breweries in the prefecture in order to spread the technology. In addition, the brewery hopes to create a new sake culture by applying and creating new technologies and ideas from the past using modern techniques in everything from rice cultivation to raw material processing and fermentation.

純米酒
Junmaishu Kuramoto

"Slightly fruity and refreshing"
This sake is brewed slowly at low temperatures in the severe winter cold using mountain water and rice grown in the company's own fields. Best served cold with Japanese dishes simmered in soup stock.

Junmaishu	720ml
Alc.	15%
Rice type	Yumesansui
Sake meter value	-2.0
Acidity	1.8

純米酒
Bodaimoto Tsugenohimuro Junmaishu

"Mild flavor and acidity of lactic acid bacteria"
A rich sake with a good balance of sweetness and acidity and a mild flavor. Excellent synergy with dishes using dairy products.

Junmaishu	720ml
Alc.	15%
Rice type	Hinohikari
Sake meter value	-2.0
Acidity	2.0

Ozaki Shuzo

Water source	Subsoil water from the Kumano River
Water quality	Soft

3-2-3 Funa-machi, Shingu-shi, Wakayama 647-0002 TEL.0735-22-2105
E-mail: ozakisyuzou@joy.ocn.ne.jp http://ozakisyuzou.jp/

Sake from the land of the Gods, "Kumano"

A sake with its origins in "Okukumano". A hidden region of faith, history and romance.

In the 1869, Ossaki began brewing sake in Sanbon-sugi, Shingu city. Ozaki Shuzo is the southernmost brewery in Honshu, located in the heart of "Kumano Sanzan" mountains on the "Kumano Kodo" World Heritage Site known as a sanctuary of the resurrection of the soul. Just north of the brewery is the Kumano River, which originates in Okukumano. The high-quality subsoil water and the northerly wind that crosses the river surface continue to preserve traditional handmade brewing. The brewery's sake was the favorite of prewar naval commanders and local writers such as Haruo Sato and Kenji Nakagami. The brewery continues to brew sake that is loved by the local community, and is often chosen as the sake of choice for local shrines and wedding ceremonies.

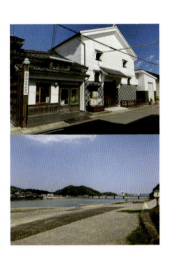

純米酒
Taiheiyo Junmaishu

"Popular local Junmaishu in Kumano"
Brewed with great care using traditional handcrafting techniques. Rich, gentle taste that will keep you coming back for more.

Junmaishu	1,800ml/720ml/300ml
Alc.	15%
Rice type	Locally produced rice
Sake meter value	+1.0~+3.0

本醸造酒
Taiheiyo Honjozo

"Honjozo served hot is also delicious"
Slightly dry Honjozoshu that has a deep and sharp taste. Can be served chilled or warmed, and easily enjoyed in a wine glass.

Honjozoshu	1,800ml/720ml
Alc.	15%
Rice type	Domestic rice
Sake meter value	±0~+2.0

Water source	Hayatsuki gorge
Water quality	Slightly-hard

Takagaki Shuzo

1465 Ogawa, Aridagawa-cho, Arida-gun, Wakayama 643-0142   TEL.0737-34-2109
E-mail: takagakishuzo@hotmail.com   http://takagakishuzo.com

Handcrafted, one by one with love and care

Brewing water from the sacred water of Elixir of Life. A long-established brewery at the foot of Mt. Koya.

The long-established brewery, founded in 1840 at the foot of Mt. Koya, is located in the mountains of Hayatsuki gorge along the road to the sacred Mt. Koya range, which has been selected as a World Heritage Site, and is completely surrounded by tangerine fields. In the Hayatsuki gorge, one of the headwaters of the Arita river, there is a spring of fresh water that people call "Kukai-sui", which is good for longevity. Takagaki Shuzo uses this sacred water for its brewing with the motto of "harmony and good sake". They enjoy brewing sake with manual labor although it tends to be hard work. Sake brewed with the local Wakayama yeast has a good aroma and a strong impact on the taste of the rice. There are many delicious fruits in Wakayama, so people want to make liqueurs using sake to promote the area as a fruit-rich prefecture in the future.

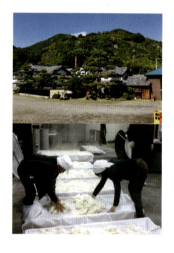

大吟醸酒
Daiginjo Kiseitsuru

"This is the main brand of the brewery"

This is a cold-brewed Daiginjoshu that is matured at low temperatures for a long time using traditional techniques in a climate suitable for sake brewing.

Daiginjoshu	720ml
Alc.	16%
Rice type	Domestic rice
Sake meter value	+4.0
Acidity	1.1

純米吟醸酒
Junmai Ginjo Genshu Kinosake

"Rice, water, and technique"

This Junmai Ginjoshu is brewed by making the most of the rice power, and almost all the processes are carried out by hand.

Junmai Ginjoshu	720ml
Alc.	19%
Rice type	Domestic rice
Sake meter value	−4.5
Acidity	1.2

Yoshimura Hideo Shoten

Water source	Subsoil water from the Kino River
Water quality	Slightly-soft

72 Hatake, Iwade-shi, Wakayama 649-6244 TEL.0736-62-2121
https://nihonsyu-nihonjyou.co.jp/

Famous sake nurtured by the Kino River

Brewing sake that enriches food by inheriting traditional techniques.

Many sake breweries have flourished on the Wakayama Kaido (Yamato Kaido) along the Kino River since the middle of the Edo period. Hideo Yoshimura Shoten was established in 1915 as a sake brewery by the youngest son of a family that had been involved in the raw silk industry. The source of the Kino River is Odaigahara, which is part of Yoshino-Kumano National Park. It is called Yoshino River in Nara and Kino River in Wakayama. The rich water nurtures history and culture, and is also a treasure for sake brewing. The sake brewed by Yoshimura Hideo Shoten is also brewed with subsoil water from the Kino River. The name of the main brand "Kurumazaka" comes from the "Legend of Oguri Hangan", which has been handed down along the Kumano-kodo. The belief of the brewery is to achieve both flavor and sharpness, and a deep flavor that can be enjoyed with meals. In recent years, the sake brewed using the traditional Yamahai method has won awards in European competitions and is highly regarded in the wine culture sphere.

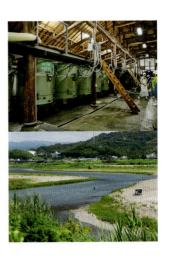

純米酒
Kurumazaka
Yamahai Junmaishu

"Firm structure and a rich taste"

A grainy, spicy aroma and a sharp flavor. It has a rich, smooth mouthfeel from aging. The thick acidity makes it great for heating up.

Junmaishu	1,800ml/ 720ml
Alc.	16.6%
Rice type	Gohyakumangoku
Sake meter value	+4.0
Acidity	2.3

純米大吟醸酒
Kurumazaka
Yamahai Junmai Daiginjoshu

"From appetizer to the last, all you need is one bottle"

It has a beautiful flavor with a pleasant acidity, a rich and deep taste, and an herbal aroma.

Junmai Daiginjoshu	1,800ml/ 720ml
Alc.	16.6%
Rice type	Yamadanishiki
Sake meter value	+2.0
Acidity	1.7

Guide to Sake Breweries and Famous Sake

Chugoku Region

Tottori, Shimane, Okayama, Hiroshima, Yamaguchi

Food Culture of Chugoku Region

Tottori: A large fishing port bordering the Sea of Japan and sandy soil provide an abundance of foodstuffs, with large harvests of snow crabs and natural oysters. In the sand dunes and sandy soil along the coast, crops suited to the sandy soil such as rakkyo, leeks, and 20th Century pears are grown.

Shimane: Lake Shinji is where fresh and sea water mix, and many kinds of fish and shellfish, such as clams and sea bass, can be harvested. On the Sea of Japan side, it is also a source of nodoguro, horse mackerel, flounder, and iwanori. The area is also famous for loach farming, as shown by the "loach scooping dance" in the folk song "Yasugibushi".

Okayama: Vegetables, peaches, grapes, and other fruits are grown in the mild climate, and "barazushi", "mamakari" and "saba-zushi" are dishes made with the abundant seafood landed in the Seto Inland Sea. Oysters and nori are also cultivated.

Hiroshima: Oyster cultivation in Hiroshima Bay thrives, and "tai somen" a dish with a single sea bream, has been handed down as a celebratory dish. Also, since fish such as sharks and fluffy fish are called "Wani" in the northern part of the prefecture away from the sea, various dishes such as sashimi and soup were were once part of "Wani" cuisine.

Yamaguchi: The prefecture's specialty is natsu-mikan, which have been cultivated since the Edo period, and there are a variety of fish dishes landed on the Sea of Japan and Seto Inland Sea sides. In Shimonoseki, Hagi and Nagato, puffer fish, horse mackerel, sea urchin, and squid are famous, and there is also a dish in which salted whale tail fin meat is soaked in boiling water and eaten with vinegared miso.

Inata Honten

Water source	Spring water from Mt. Daisen
Water quality	Soft

325-16 Yomi-cho, Yonago-shi, Tottori 683-0851 TEL. 0859-29-1108
E-mail: info@inata.co.jp https://www.inata.co.jp/

Sake brewing with heart

Not just with inherited traditions. But with the spirit of trying things out.

In the first year of Enho (1673), Kaemon Inata, the third-generation head of the Inata family, established a sake and soy sauce brewing business in Ogishi, Kishimoto-cho (present-day Hoki-cho). At that time, the company's name was "Inabaya". They later moved to various places including Konya-cho and Kume-cho, then finally located in Yomi-cho in 1987. When in Konya-cho (in the middle of the Genroku era), the name was changed to "Inataya". Surrounded by the abundant nature at the foot of Mt. Daisen, the Inata Honten focuses on good water, good rice, and a passionate brewer. In particular, the brewery cooperates with local farmers to grow rice suitable for sake brewing, relying as little as possible on chemical fertilizers. While boasting a history of 350 years, the brewery has not been afraid to take on new challenges ahead of the rest of the country. In recent years, they have opened an izakaya called "Inataya" where you can enjoy Inata Honten's sake and dishes that go well with the sake.

純米大吟醸酒
IKU'S SHIRO

"Junmai Daiginjo Sojun Ichidan Jikomi"
A new balance of sugar and acidity to create a sake that anyone can enjoy, created under the concept of "brewing, new and interesting."

Junmai Daiginjoshu	500ml
Alc.	9%
Rice type	Yamadanishiki
Sake meter value	-52
Acidity	4.3

純米吟醸酒
Junmai Ginjo Inatahime Goriki

"Light and dry with good aroma and good taste"
Made from a fantastic rice called "Gorikimai", and made in the same way as Daiginjo. It has an elegant fruity aroma, a delicious rice flavor, and a good sharpness.

Junmai Ginjoshu	1,800ml/ 720ml
Alc.	15%
Rice type	Goriki
Sake meter value	+2.5
Acidity	1.4

| Water source | Subsoil water from Mt. Jubo range |
| Water quality | Water quality: Super-soft |

Yamane Shuzojo

249 Otsubo, Aoya-cho, Tottori-shi, Tottori 689-0518　TEL.0857-85-0730
https://hiokizakura.jp/

Our goal is to make.
"a sake that does not interfere with food"

Strive for sake that is not only good by itself, but also good with food.

The company was founded in 1887. Only rice suitable for brewing is used, and it is contracted directly from farmers. Rice of each farmer is brewed separately in its own barrel, and the name of the rice producer is indicated on the sake label. This is an expression of the spirit of the Yamane Shuzojo, "brewing is farming". The fifth-generation brewer, Mr. Masanori Yamane, aims for the ideal of "sake that does not interfere with food". This was inspired by the words of his grandfather, Masanori Yamane, the third-generation brewer, "Don't make sake that interferes with food". The sweetness of the unrefined sake, made from the best rice, is thoroughly consumed by the yeast, and the astringent sake is slowly aged to produce a multi-layered flavor. This is what Yamane Shuzojo is aiming for, "a sake that is not so sweet that it interferes with food".

純米酒
Hioki Zakura Junmaishu

"Suppressing sweetness and bringing out the flavor"

Focus on bringing out the flavor of the rice. The yeast is allowed to fully ferment the rice, making it into a moderately sweet sake.

Junmaishu	1,800ml/ 720ml
Alc.	15%
Rice type	Tamasakae
Sake meter value	+13.0
Acidity	2.1

純米吟醸酒
Hioki Zakura Junmai Ginjo "Densho Goriki"

"Unique taste derived from "goriki-mai""

This ginjo distinguishes itself from trendy flavors and competes with its strength and suppleness. It has a deep richness and a dense, concentrated flavor.

Junmai Ginjoshu	1,800ml/ 720ml
Alc.	15%
Rice type	Goriki
Sake meter value	+14.0
Acidity	1.8

Akana Shuzo

Water source: Kobe River system

23 Akana, Iinan-cho, Iishi-gun, Shimane 690-3511   TEL.0854-76-2016
E-mail: akanasakebrewing@vega.ocn.ne.jp   https://kinunomine.localinfo.jp/

Walking together with the hometown, Iinan

A brewery that made a fresh start despite repeated crises.

After Akana Shuzo General Partnership Company went bankrupt, Akana Shuzo was established by a new management team including the current president who purchased the company while retaining the license. The brewery uses locally grown rice as much as possible and produces only Junmaishu. The new brewery aggressively expanded its sales channels both domestically and overseas, and within a year-and-a-half of its establishment, began exporting to Thailand. The mayor of Iinan came to the new product launch in Bangkok to promote the sake made with rice produced in his town. The brewer has also been asked to offer sake at the Emperor's Birthday Reception hosted by the Embassy of Japan in Thailand three times in a row.

Kinu No Mine Junmai Daiginjo 35

"Carefully brewed deep taste"
This sake is characterized by its mellow aroma, soft sweetness, and sharpness. It is made from Iinan-cho rice, a high-quality production area in Shimane.

Junmai Daiginjoshu	1,800ml/ 720ml
Alc.	16%
Rice type	Gohyakumangoku
Sake meter value	+2.0
Acidity	1.6

Kinu No Mine Junmai Ginjo 55 Muroka Genshu

"Genshu stored at -10°C"
Soft and sweet scent with a hint of fruit. Dry and sharp when tested. When opened and exposed to air, the taste and scent become softer.

Junmai Ginjoshu	1,800ml/ 720ml
Alc.	16%
Rice type	Shimane rice
Sake meter value	+6.0
Acidity	1.9

| Water source | Well water on the premises |
| Water quality | Soft |

Oki Shuzo

174 Harada, Okinoshima-cho, Oki-gun, Shimane 685-0027 TEL.08512-2-1111
E-mail: info@okishuzou.com https://okishuzou.com

No limit for improvement of sake quality

Passing on the tradition of sake brewing to the future. in an island of history and tourism, "Oki".

Oki is known as a source of high-quality obsidian, used as stone tools since ancient times, and its history dates back 30,000 years. The place is described as "Ame no oshikorowake" or " Oki no mitsugo no shima " in "A Record of Ancient Matters" and "Chronicles of Japan", and has a historical background. Both Emperor Gotoba and Godaigo were exiled to the island in the Middle Ages. The island has unique rocks and reefs, and is a tourist island designated as a national park. It is registered as a UNESCO Global Geopark. The area has dense forests nurtured by the rain and fog of the Tsushima warm current, and high-quality water gushes out all over the island. Some of the water has been selected as "100 best waters" by the Ministry of the Environment, and the limpid water is essential for brewing sake. In this rich natural environment, Oki Shuzo continues to brew a wide variety of sake.

Okihomare Junmai Daiginjo Tobin Kakoi

"Collected in a bottle, drop by drop"

This sake is made by collecting the drops of sake that naturally drip from the unrefined sake suspended in a bag. It has a splendid flavor, a soft and elegant sweetness, and a clean aftertaste.

Junmai Daiginjoshu	1,800ml/ 720ml
Alc.	16%
Rice type	Yamadanishiki
Sake meter value	-2.0
Acidity	1.3

Okihomare Muromachi No Junmaishu 90

"Sake with a taste reminiscent of noble late harvest wine"

This amber-colored sake is based on a document from the Muromachi period, and has a very rich, sweet and sharp taste. Brewed using the traditional "kimoto" method.

Junmaishu	1,800ml/ 720ml
Alc.	17%
Rice type	Kannomai
Sake meter value	-60.0
Acidity	3.0

Okuizumo Shuzo

Water source	Subsoil water from Mt. Tamamine range
Water quality	Soft

1380-1 Kamedake, Okuizumo-cho, Nita-gun, Shimane 699-1701   TEL. 0854-57-0888
E-mail: f-terado@okuizumosyuzou.com   https://okuizumosyuzou.com/

The smile of the drinker makes the brewer smile

Sake brewed with locally grown rice in Okuizumo, where rice cultivation has flourished since ancient times.

The Izumo region of Shimane has been known for its sake brewing since ancient times, and the history of sake brewing can be found in the Chronicles of Japan. Okuizumo-cho, where the Okuizumo Shuzo is located, is also known as the site of the legendary Yamata-no-orochi, and is famous for the episode where Yamata-no-orochi drank Yashiori-no-sake. Okuizumo Shuzo was established in 2004 as a third-sector industry funded by Okuizumo-cho. Rice cultivation in the area originated from ancient tatara iron manufacturing (which is certified as Japanese Nationally Important Agricultural Heritage System), and various varieties of excellent rice are produced. While the brewery uses excellent rice for brewing sake, they have also launched a brand that uses Koshihikari (which is not primarily used for brewer's rice in general) and sell the product as part of the region's publicity activities.

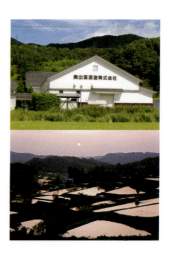

Okuizumo No Itteki
One drop

"Junmaishu for enjoying the difference in the taste of rice"

Currently test marketing. From the spring of 2021, three types will be released.

Junmaishu	1,800ml/ 720ml
Alc.	15%
Rice type	Prefectural rice
Sake meter value	±0
Acidity	1.7

Nitamai Koshihikari
Junmai Daiginjoshu

"For refreshing appetizers and vinegared dishes"

This sake has a soft ginjo aroma, a delicious rice flavor, and the sharpness of the rice. Best served cold.

Junmai Daiginjoshu	1,800ml/ 720ml
Alc.	15%
Rice type	Nitamai Koshihikari
Sake meter value	+1.0
Acidity	1.4

| Water source | Misumi River system |
| Water quality | Medium-soft |

Nihonkai Shuzo

80 Minatoura, Misumi-cho, Hamada-shi, Shimane 699-3224 TEL.0855-32-1221
E-mail: info@kan-nihonkai.com https://www.kan-nihonkai.com/

"The birthplace of Japanese sake" Sake loved by the locals

Contributing to the local community and sending out information through sake brewing. The motto is "sake that connects the hearts of people to people".

Nihonkai Shuzo was established in 1888. Shimane is said to be the "birthplace of sake" and the "land of the gods," and has been closely connected to local traditions and food culture since ancient times. The brewery is located in the Iwami area of western Shimane, on the coast of the Sea of Japan (Nihonkai), where the company name comes from. Misumi-cho, where the subsoil water of the Mt. Yaune flows into the Misumi River, has long been known as a land of clear water (Mizusumi No Sato)" and a place of scenic beauty. Their motto is to make "sake that connects the hearts of people" aiming to contribute to the community and send out information through sake brewing. While using mainly locally grown Gohyakumangoku rice as the raw material, the brewery produces sake with a wide range of shades of flavor and sharpness, which is loved by the local community.

Kannihonkai Junmai Daiginjo Mizusumi No Sato

"Daiginjo brewed in Misumi-icho, the land of clean water"
Made from Yamadanishiki polished to 40%, and characterized by its mellow aroma and soft texture.

Junmai Daiginjoshu	1,800ml/720ml
Alc.	16%
Rice type	Yamadanishiki
Sake meter value	+3.0
Acidity	1.2

Junmai Ginjo Iizu Yamadanishiki

"A vortex of aroma, a vortex of flavor"
Junmai Ginjoshu with a mellow aroma that spreads in the mouth, a robust flavor, and a deep rich taste.

Junmaiginjoshu	1,800ml/720ml
Alc.	16%
Rice type	Yamadanishiki
Sake meter value	+2.0
Acidity	2.0

Fuji Shuzo

Water source	Subsoil water from the Hii River
Water quality	Soft

1403 Imaichi-cho, Izumo-shi, Shimane 693-0001 TEL. 0853-21-1510
E-mail: sake@izumofuji.com http://www.izumofuji.com/

Like the beloved Mt. Fuji, the most beloved sake in Japan

Respect the traditional techniques of the Izumo Toji. and produce local sake that connects people to people and people to food.

Fuji Shuzo was established in 1939 in Izumo, where many of Japan's ancient secrets lie. The area is also known as the birthplace of sake that is made from rice in Japan. It is surrounded by the Japan Sea to the north, the Chugoku Mountains to the south, Lake Shinji to the east, Lake Zinzai to the west, and the Hii River, the setting of the legendary Yamata-no-orochi, flowing through the center. The name "Izumofuji" comes from a passionate desire to produce the sake in the land of Izumo that is loved like Mt. Fuji. Now they also have the idea of "brewing Izumo and aspiring to become Fuji". Respecting the traditional techniques of the Izumo Toji, they aim to produce honest and pure sake as Izumo's local brewer to connect people with people and people with food.

Izumofuji Junmai White Label

"Best served cold or lukewarm"

This Junmaishu is white and clean with the soft, high-quality flavor of Shimane's Yamadanishiki, and gently connects people and food.

Junmaishu	1,800ml/720ml
Alc.	15%
Rice type	Yamadanishiki
Sake meter value	+7.0
Acidity	1.8

Izumofuji Tokubetsu Junmai Black Label

"Special Junmaishu that connects people and food"

This is a Special Junmaishu made from Sakanishiki rice, grown in Izumo under special contract, to maximize the richness and sharpness of the taste.

Special Junmaishu	1,800ml/720ml
Alc.	15%
Rice type	Sakanishiki
Sake meter value	+6.0
Acidity	1.8

| Water source | Well water |
| Water quality | Medium-hard |

Rihaku Shuzo

335 Ishibashi-cho, Matsue-shi, Shimane 690-0881 TEL.0852-26-5555
E-mail: rihaku@rihaku.co.jp http://rihaku.co.jp/

Tradition and new technology From Shimane to the World

Protecting the beloved sake culture and continuing to strive for overseas sales.

Rihaku Shuzo was established in 1882. With the business philosophy of "spreading the culture of sake and passing it on to future generations in a correct manner", the company continues to take on challenges not only in Japan, but also in the world. Overseas sales, which account for 30% of the company's current sales, include Hong Kong, the United States, South Korea, and Singapore. Yuichiro Tanaka, the president of the company, says "Even tradition needs progress". There is no compromise on any grade of sake, from daiginjo to regular sake. The brand name "Rihaku" was named for the Chinese poet Li Bai who wrote many poems in praise of sake. Lihaku Shuzo is now focusing on brewing sake by combining new age technology with the traditional techniques of Izumo Toji, such as using flower yeast isolated by the Department of Brewing Science of Tokyo University of Agriculture. The challenge of staying ahead of the times will continue to advance the tradition of this brewery.

純米吟醸酒
Rihaku Junmai Ginjo
WANDERING POET

"Sake enjoyed worldwide"

Made from 55% polished Yamadanishiki and fermented at low temperature. It has a full, mellow flavor, and is refreshing and sharp.

Junmai Ginjoshu	1,800ml/720ml
Alc.	15-16%
Rice type	Yamadanishiki
Sake meter value	+3.0
Acidity	1.6

純米大吟醸酒
Rihaku Junmai Daiginjo

"Junmai Daiginjoshu that goes well with food"

Made from 45% polished Yamadanishiki, and fermented at low temperatures for a long time. It has a mild aroma and is a perfect match for food

Junmai Daiginjoshu	1,800ml/720ml/300ml
Alc.	16-17%
Rice type	Yamadanishiki
Sake meter value	+4.0
Acidity	1.5

Isochidori Shuzo

Water source	Underground water
Water quality	Soft

306 Shinjo, Satosho-cho, Asakuchi-gun, Okayama 719-0302 TEL.0865-64-3456
E-mail: info@isochidori.co.jp https://www.isochidori.co.jp

Kibi sake as mentioned in Manyoshu

In the birthplace of Japanese sake. Preserving 250 years of traditional taste.

The company was founded in 1751, the first year of the Horeki era, in the middle of the Edo period. Various cultures have flourished in Okayama, known as Kibi-no-kuni (land of kibi) since ancient times. In terms of sake, there is a description of the Manyoshu, an ancient book of poetry, mentions "Kibi no sake". Kibi is said to be the birthplace of sake that differs from that made in Nara. In fact, it is said that sake from this area was carried to Kyoto in Bizen ware pots during the Heian period. The sandy soil of Okayama, which contains a lot of granite, is said to be suitable for producing high-quality rice, and the sake made from this rice is characterized by its pleasant taste and easy drinking. The name of the brewery, Isochidori Shuzo, is said to have been named after the scenic beauty of the surrounding sea, where plovers (chidori) used to fly. The brewery has maintained its traditional taste for over 250 years.

Nama Chozoshu Shizen No Megumi

"Dry sake stored raw"

Sake is stored at low temperatures in its raw state and heated only once when it is bottled. The taste is light and dry. Best served cold.

Honjozoshu	720ml
Alc.	15-16%
Rice type	Akebono
Sake meter value	+3.0
Acidity	1.2

Daiginjo Seto No Sazanami

"Aromatic and mellow. An excellent sake"

Daiginjoshu made with Asahi rice polished to less than 50%. It is a light and dry sake with a mellow aroma similar to wine.

Daiginjoshu	1,800ml
Alc.	15-16%
Rice type	Asahi
Sake meter value	+2.0
Acidity	1.3

Water source	Subsoil water from the Takahashi River
Water quality	Medium-soft

Kamikokoro Shuzo

7500-2 Yorishima-cho, Asakuchi-shi, Okayama 714-0101 TEL.0865-54-3101
E-mail: info@kamikokoro.co.jp   https://www.kamikokoro.co.jp/

The pursuit of "Kome Umakuchi" by a brewery that sells quality

Better koji and stable fermentation enable sake brewing that retains the flavor of the rice.

Kamikokoro Shuzo was founded in 1913 by Chojuro Fujii, the first-generation head, in Yorishima, Okayama. Their brewing is different from the sweet sake made with the "Sanbai zojo (triple brewing) method" which was rapidly popularized after the war. Kamikokoro focuses on "Kome umakuchi", and has inherited the attitude of "a brewery that sells quality". There are two main characteristics of Kamikokoro Shuzo. One is the careful brewing of sake that brings out the flavor of rice. The rice produced in Okayama is soaked in units of 60 kg for a few seconds, and the right amount of water is absorbed to produce a better koji. The second is to keep the air cool and clean at all times to promote stable fermentation, which makes it possible to brew sake that retains the flavor of the rice. The brewery is committed to pursuing new possibilities for the traditional taste "Kome Umakuchi".

Kamikokoro Junmai Ginjo Muroka Namazake Fuyu No Tsuki

"Pay close attention to the characteristics of brewing rice"

Uses freshly harvested rice and their own yeast, "Okayama Hakuto Yeast", which has an excellent sharpness. This sake is like silk dripping down from the moon.

Junmai Ginjoshu	1,800ml/720ml
Alc.	16%
Rice type	Akihikari by Okayama
Sake meter value	-3.0
Acidity	1.6

Kamikokoro Junmai Ginjo Binkakoi

"Bring out the original taste of rice"

Based on Kamikokoro Shuzo's quintessential "umakuchi" style, which brings out the true flavor of the rice along with exquisite acidity. You can enjoy the contrast of umami and sharpness.

Junmai Ginjoshu	1,800ml/720ml
Alc.	15%
Rice type	Akebono by Okayama
Sake meter value	-2.0
Acidity	1.4

Sanko Masamune

Water source	Subsoil water from the Kojiro River
Water quality	Medium-hard

951 Kamikojiro, Tessei-cho, Niimi-shi, Okayama 719-3702 TEL.0867-94-3131
E-mail: info@sake-sanko.co.jp http://www.sake-sanko.co.jp

Bitchu's local sake with a focus on rice

Focusing on technique, water and rice. The three "classy elements" of sake brewing.

In 1902, Shigegoro Miyata, the founder of the company, took a leap of faith and went to the United States by himself to start a strawberry farm. Although he had many hardships, he made a lot of money and returned to Japan in 1911. In 1913, he used his wealth to establish Miyata Shuzo (now Sanko Masamune) in Kamikojiro, Tessei-cho, northern Okayama. The brewery was born of the "American dream", which was rare at the time. Sanko Masamune's sake is made with local flavors and ingredients, with the rice used for brewing produced in Okayama. In particular, the sake rice "Yamadaniishiki" and "Omachi" are grown in the Yada district of Tessei-cho using a special "Nagatani farming method".

大吟醸酒
Sanko Masamune Daiginjo

"Splendid Ginjo flavor, sharp and dry"
Brewed with a perfect balance so that the flavor is not too flamboyant even when eating. The acidity is striking.

Daiginjoshu	1,800ml/720ml
Alc.	16%
Rice type	Yamadanishiki

本醸造酒
Hiyashite Nomu Kan-irino Nama Genshu

"Nama Genshu, drink from a chilled can"
A sake that is rich and dry, and popular for its fresh aroma. It has the full-bodied flavor of unpasteurized sake, and is delicious when frozen.

Honjozoshu	200ml
Alc.	20%
Rice type	Domestic rice

Water source	Asahi River
Water quality	Soft

Miyashita Shuzo

184 Nishigawara, Naka-ku, Okayama-shi, Okayama 703-8258 TEL.086-272-5594
E-mail: info@msb.co.jp https://www.msb.co.jp/

Famous Sake of Okayama Kiwami Hijiri

Contributing to the local community as a comprehensive sake manufacturer.

The founders, Miyashita brothers (Kamezo and Motosaburo), lost their parents at an early age, and were left in the care of relatives who ran a sake brewery. The brothers gained experience there. In 1967, they established a company in Okayama. In 1967, the brewery moved to the banks of the Asahi River, one of the three major rivers in Okayama, and began brewing sake according to traditional Bitchu Toji technicues, using rice grown in Okayama and subsoil water from the Asahi River. Omachi, one of the best rice varieties for sake brewing, is grown near the brewery and was once called "phantom rice" because of the difficulty in growing it. Miyashita Shuzo's main brand is "Kiwami Hijiri", a Junmai Daiginjoshu made from "Takashima Omachi" harvested in the Takashima district, the birthplace of Omachi. Today, Miyashita Shuzo is aiming to become a comprehensive manufacturer of all types of alcoholic beverages.

Daiginjo Kiwami Hijiri Mukashi Shibori Tobindori

"A refreshing, slightly dry sake"
This sake is made by a young Toji under the guidance of Mr. Akio Nakahama, a modern master brewer. The brewery pursues a balance of Ginjo aroma and taste, freshness and mellowness.

Daiginjoshu	1,800ml/720ml
Alc.	17-18%
Rice type	Yamadanishiki
Sake meter value	+5.0
Acidity	1.4

Kiwami Hijiri Junmai Daiginjo "Takashima Omachi"

"Slightly dry sake with umami"
This Junmai Daiginjoshu is made with the fantastic sake rice, Takashima Omachi. It features a splendid, elegant ginjo aroma and a full, mellow flavor.

Junmai Daiginjoshu	1,800ml/720ml
Alc.	16-17%
Rice type	Omachi
Sake meter value	+1.0
Acidity	1.3

Ochi Shuzojo

Water source	Subsoil water from the Bitchu River, a tributary of Asahi River
Water quality	Medium-hard

664-4 Shimoazae, Maniwa-shi, Okayama 716-1433　TEL.0866-52-2311
E-mail: tisntr26@ka2.so-net.ne.jp

Delivering deliciousness
that can be experienced with the five senses

Sake brewing with a focus on traditional techniques and flavors, with quality the top priority.

Ochi Shuzojo was founded in 1893 by Shintaro Ochi, the first-generation head, in Maniwa, Okayama, in search of high-quality and delicious water. The focus here is on rice and water produced in Okayama. The brewery mainly uses "Asahi" which is the origin of many common rice varieties, and the mineral-rich subsoil water of the Bitchu River to make a strong and sharp Junmaishu. The secret of its even better taste lies in its tank storage and thorough temperature control within the range of minus 5°C to plus 5°C, according to the quality and condition of the sake. The fourth-generation brewer, Yasuaki, says "I would like to continue to convey the goodness and excitement of sake to as many people as possible by using the data I accumulate every year as well as my five senses."

画像提供「写真の新田」

特別純米酒

Taisho No Tsuru RISING 60 Tokubetsu Junmai

"Aiming for sharpness and the umami of rice"

This special Junmaishu is brewed by Noboru Ochi by making the best use of the flavor of "Asahi", the most important rice in Okayama.

Special Junmaishu	1,800ml/720ml
Alc.	16%
Rice type	Asahi
Sake meter value	+2.0
Acidity	1.5

特別純米酒

Tokubetsu Junmai (Genshu) Taisho No Tsuru

"A bottleful of the brewer's ideas"

This unfiltered, unpasteurized sake has a gentle, fruity aroma and flavor drawn from Asahi and medium-hard water.

Special Junmaishu	1,800ml
Alc.	18%
Rice type	Asahi
Sake meter value	+4.0
Acidity	1.8

Water source	Underground water of the Yoshii River system
Water quality	Soft

Toshimori Shuzo

762-1 Nishikarube, Akaiwa-shi, Okayama 701-2215 TEL.086-957-3117
E-mail: hitosuji@sakehitosuji.co.jp https://www.sakehitosuji.co.jp/

To produce an authentic sake Reviving a phantom rice

The spirit of reviving phantom rice. A true local sake with a focus on the local area.

The Karube area of Akaiwa in Okayama used to be the place where the highest quality of "Omachi", a rice suitable for sake brewing, could be harvested, but in the latter half of the 1960s, the area under cultivation was reduced as agriculture was modernized. Mr. Tadayoshi Toshimori, the current president of the company, saved Omachi from becoming an extinct rice. He promoted the cultivation of Omachi by offering income guarantees to farmers, and eventually worked with agricultural cooperatives and the government to revive Omachi, which is now known throughout Japan. Moreover, he attempted to brew sake using a jar, the way it was made 500 years ago. The sake brewed in a 500-liter jar has a rich flavor and taste that comes from the time and effort put into it. There is no hesitation in the spirit of Toshimori Shuzo who believes "the true local sake can only be made using the local rice, water and climate."

Akaiwa Omachi Junmai Daiginjo

"With a phantom rice, with all my soul"

This Junmai Daiginjoshu is brewed with Omachi rice with the soul of the brewery. Enjoy the unique flavor of Omachi that spreads in the mouth.

Junmai Daiginjoshu	720ml
Alc.	15.5%
Rice type	Omachi
Sake meter value	+4.0
Acidity	1.3

Sake Hitosuji Junmai Ginjo (Kinrei)

"Brewing with techniques from the Edo period"

This is a Junmai Ginjoshu with a clean taste and a strong rice flavor.

Junmai Ginjoshu	720ml
Alc.	15.5%
Rice type	Omachi
Sake meter value	+3.0
Acidity	1.6

Watanabe Shuzo Honten

Water source	Subsoil water from the Takahashi River
Water quality	Soft

170 Kamejima Shinden, Tsurajima-cho, Kurashiki-shi, Okayama 712-8002　TEL.086-444-8045
E-mail: minenohomare@mx8.kct.ne.jp　https://watanabeshuzou.shop/

Old-fashioned sake brewery located in Tsurajima, Kurashiki

The name "Kurashiki Junmaishu Shinjidai" (New Era Kurashiki Junmaishu) is used to promote the confidence of the company's sake, both in Japan and abroad.

Watanabe Shuzo was established in 1909 in Kamejima Shinden, Tsurajima, Kurashiki, which was opened due to land reclamation during the Edo period. In 2021, after 112 years of establishment, they still retain the appearance of a traditional sake brewery, with its namako walls, brick chimneys, and cedar balls hanging from the storefront. The name "Minenohomare", which became the main brand of the brewery in the early Showa period, was chosen to represent the majestic appearance of the sun rising at the foot of Mt. Fuji. and also as a wish for the prosperity of the business. Mr. Hideki Watanabe, the fourth generation of the brewery, has been practicing good sake brewing at low cost by using a variety of products in small quantities. In the last year of the Heisei era, he started a new attempt at sake brewing called "Kurashiki Junmaishu Shinjidai" (New Era Kurashiki Junmaishu), and is confidently developing sake for domestic and overseas markets.

純米大吟醸酒
Junmai Daiginjo Yume Kurashiki 39

"The best in the history of Watanabe Shuzo"

This Junmai Daiginjoshu is made with rice, koji (malted rice) and subsoil water from the Takahashi River, a clear stream flowing into Kurashiki. "Omachi" rice polished to 39%.

Junmai Daiginjoshu	1,800ml/720ml/300ml
Alc.	16-17%
Rice type	Omachi
Sake meter value	-1.0
Acidity	1.8

純米酒
Kimoto Junmaishu Yume Kurashiki 60

"The taste of nature lives"

This sake is made using the traditional "kimoto" method of the Edo period. It has a full-bodied flavor that brings out the five tastes; pungency, acidity, sweetness, bitterness, and astringency.

Junmaishu	1,800ml/720ml/300ml
Alc.	15-16%
Rice type	Asahi
Sake meter value	+3.0
Acidity	2.2

Water source	Well in the brewery
Water quality	Soft

Imada Shuzo Honten

3734 Mitsu, Akitsu-cho, Higashi Hiroshima-shi, Hiroshima 739-2402　TEL.0846-45-0003
E-mail: info@imada-shuzo.co.jp　https://fukucho.jp

Aspirations for "Hyakushi Senakai" (One Hundred Trials and One Thousand Revisions)

Continuous efforts to make delicious sake. in the hometown of Ginjoshu, Akitsu.

Imada Shuzo was founded in 1868 in Mitsu, Akitsu, the home of the Hiroshima Toji (master brewer), overlooking the Seto Inland Sea. Although the brewery is small in size, it brews only high-quality sake with a thorough focus on Ginjo. The main ingredients are sake rice produced in Hiroshima, and the brewery is making a special effort to revive "Hattanso" the oldest indigenous sake rice variety in the region. The town of Akitsu, where the brewery is located, is the birthplace of Senzaburo Miura, who is called "the father of Ginjoshu" because he was the first person in Japan to brew sake with soft water. Senzaburo Miura devoted himself to the training of Toji on the island of Hiroshima, and Imada Shuzo's flagship brand "Fukucho" was named by him. Senzaburo Miura's motto was "One hundred trials, one thousand revisions". Inheriting his passion and the brand, Imada Shuzo is introducing the appeal of "Hiroshima Ginjo" to the world.

Fukucho Junmai Ginjo Hattanso

"Sake that brings out the taste of phantom rice"

This sake is made from the oldest indigenous variety of rice, "Hattanso". The balance of clean acidity and umami is exquisite. It has a unique, clean aftertaste.

Junmai Ginjoshu	1,800ml/720ml
Alc.	16%
Rice typ	Hattanso
Sake meter value	private
Acidity	private

Kanemitsu Shuzo

Water source	Unknown
Water quality	Medium-hard

1364-2 Nominoo, Kurose-cho, Higash Hiroshima-shi, Hiroshima 739-2622 TEL.0823-82-2006
E-mail: info@kamokin.com https://www.kamokin.com

In search of deliciousness that remains in your heart

Brew sake with all our heart and soul, to give our drinkers a sense of excitement and comfort.

The brewery was founded in 1880 (Meiji 13). At the time of its founding, the brewery had a number of brands, including "Kamo No Tsuyu" "Kisui" and "Sakura Fubuki" but as time went on, it became difficult to hire seasonal brewers, which resulted in an introduction of automated plants to drastically reduce the number of brewers. In the early Heisei period, Hideki Kanemitsu, the son of the brewer, couldn't find a sake he liked, including those from his own brewery. However, one day, when he drank sake from a brewery in Tohoku, he couldn't help but say, "It's delicious". Then he had the desire to produce inspiring sake at his own brewery and switched to a hand-crafted method. Now, in order to produce high quality sake, mainly Junmaishu, the brewery uses more than 80% locally grown Omachi and Hiroshima rice, aiming to become a sake brewery that the region can be proud of.

特別純米酒
Kamokinshu Tokubetsu Junmai

"Fresh and firm taste"

The aroma of the rice is moderate, and the koji (malted rice) is rich and sweet. The refreshing yet swelling umami complements food.

Special Junmaishu	1,800ml/720ml
Alc.	16%
Rice type	Omachi/Hattannishiki
Sake meter value	+3.0
Acidity	1.4

純米吟醸酒
Kamokinshu Junmai Ginjo Omachi

"Junmai Ginjoshu with 100% charm of Omachi"

This Junmai Ginjoshu has a beautiful Ginjo aroma and the deep flavor of the "Akaiwa Omachi" rice, making it a perfect match for a variety of foods. It has a refreshing acidity and a pleasant aftertaste.

Junmai Ginjoshu	1,800ml/720ml
Alc.	16%
Rice type	Omachi
Sake meter value	+3.0
Acidity	1.5

| Water source | Subsoil water from the Chugoku Mountains |
| Water quality | Soft |

Koizumi Honten

3-3-10 Kusatsu Higashi, Nishi-ku, Hiroshima-shi, Hiroshima 733-0861 TEL.082-271-4004
E-mail: hon-ten@muf.biglobe.ne.jp http://www2u.biglobe.ne.jp/~mi-yuki/

Rooted in the climate of Hiroshima More than 180 years of history

Trusted by locals for brewing Omiki (sacred sake) for Itsukushima Shrine.

Although it is unclear when Koizumi Honten was founded, records show that it was already in existence during the 1830s, and has a history of over 180 years. They have been brewing sake rooted in the climate of Hiroshima, such as making sacred sake for Itsukushima Shrine on Miyajima, one of the three most scenic spots in Japan. Since its founding, the brewery has had several names for its sake, including "Enjugiku" (longevity chrysanthemum), but the name "Miyuki" was chosen to commemorate the visit of the Meiji Emperor. In 2007, after 20 years of planning, the Miyuki Gallery (a guest house and research center for information on Japanese sake culture) was established, much to the delight of sake lovers. Hiroshima's sake is characterized by its fine texture, smooth taste, richness, and deliciousness, which you never get tired of. The people of Hiroshima have been training their taste buds, and are determined to keep making delicious sake.

大吟醸酒
Miyuki Koizumi Daiginjo

"Daiginjo brewed with locally grown rice"

Made from Senbon Nishiki, a rice grown in Hiroshima. When Prime Minister Koizumi was appointed, this sake was presented to him.

Daiginjoshu	720ml
Alc.	16-17%
Rice type	Senbonnishiki
Sake meter value	+3.0
Acidity	1.5

本醸造酒
Miyuki Hiroshima Tokusen

"Mellow taste"

Brewed with abundant fresh water and carefully selected sake rice. This gem is the result of a blessed climate and the skills of toji (master brewer). Best served warm.

Honjozoshu	720ml
Alc.	15-16%
Sake meter value	+5.5
Acidity	1.5

Saijotsuru Jozo

Water source	Subsoil water from Mt. Ryuo
Water quality	Medium-hard

9-17 Saijo Honmachi, Higashi Hiroshima-shi, Hiroshima 739-0011 TEL.082-423-2345
E-mail: saijoutsuru4232345@saijoutsuru.co.jp https://saijotsuru.co.jp/

We want to bring you happiness and good taste and good fortune

The sake brewery and its main building, which have been used since its founding, are designated as Tangible Cultural Properties of Japan.

Saijotsuru Jozo was founded in 1904 by Ichimatsu Inomoto. The brewery was named "Saijotsuru" by combining "Saijo" the name of the place, and "Tsuru" a word of congratulations. The brewery and main building, which have been in use since its establishment, were designated as Tangible Cultural Properties of Japan in 2016. In the fall of 2006, following the retirement of the previous Toji (master brewer), the job of seasonal master brewer was discontinued and the employees took on the role of brewers. Inheriting the traditional skills of Hiroshima Toji, Mitsuyoshi Miyaji, became Toji at the age of 38. He strives to make the best use of his skills under the theme, "Local sake is compassionate sake, bringing happiness and good taste and good fortune to the customers who drink Saijotsuru". Miyaji says, "I don't like to be categorized, so I want to express that in my sake. I hope to make sake with a taste that can't be found anywhere else, and that will become the taste of our Saijotsuru." The brewery may be small, but their passion for making sake is on the rise.

Saijotsuru Junmai Daiginjo Genshu Shinzui

"The best sake rice, Senbon Nishiki"

With well-balanced aroma, flavor, and acidity, you can taste the true pleasure of sake in just one glass. This gem represents Hiroshima.

Junmai Daiginjoshu	1,800ml/720ml
Alc.	16%
Rice type	Yamadanishiki/Senbonnishiki
Sake meter value	±0
Acidity	2.0

Saijotsuru Junmai Daiginjo Hibi Shinjo Sakewokamosu

"A sake that can be enjoyed at a variety of temperatures"

The sake rice "Nakatesenbon-nishiki" is polished to 50%. This premium sake is brewed with the traditionally used Kyokai No. 6 yeast".

Junmai Daiginjoshu	1,800ml/720ml
Alc.	16%
Rice type	Nakateshinsenbon
Sake meter value	-2.0
Acidity	2.1

Water source	Subsoil water from the Kamo River
Water quality	Medium-hard

Nakao Jozo

5-9-14 Chuo, Takehara-shi, Hiroshima 725-0026 TEL.0846-22-2035
E-mail: sake@maboroshi.co.jp   http://www.maboroshi.co.jp

The pursuit of beautiful sake symbolizes the brewer's sincerity

The best sake produced through research in a historical sake brewing town.

The area of Takehara prospered as the estate of Shimogamo Shrine in Kyoto beginning in the Heian period, and has long hours of sunlight. Due to the abundance of good quality rice and subsoil water from the Kamo River, sake brewing has flourished in the area since early times. Nakao Jozo was established in 1871 under the name of Hiroshimaya. The famous sake "Seikyo" was created. The name means a mirror that reflects the brewer's sincerity. A typical example of the brewer's sincerity is the proprietary apple yeast. In 1940, Kiyomaro Nakao, the fourth-generation chief, discovered that yeast extracted from the skin of apples produced a high level of aroma during fermentation. Seven years later, he developed a high temperature saccharification method for making sake malt and succeeded in utilizing 100% of its performance. With its acidity, fruity aroma and strong fermentation, Nakao Jozo sake is highly popular both in Japan and overseas.

Seikyo Junmai Ginjo Omaohi

"The true value of the deep Omachi"

This sake is brewed with Omachi, the oldest sake brewing rice in Japan. It takes fifty-two hours to make the koji that most affects the taste, in order to bring out the depth.

Junmai Ginjoshu	720ml
Alc.	15.4%
Rice type	Omachi (pesticide-free cultivation by a local contract farmer)
Sake meter value	+3.0
Acidity	1.4

Seikyo Maboroshi Kurobako Junmai Daiginjo

"Splendid scent of apple yeast"

This is the pinnacle of the brewing process, from rice washing to storage. It has a mellow aroma from the apple yeast and a delicious taste from the Yamadanishiki.

Junmai Daiginjoshu	720ml
Alc.	16.5%
Rice type	Yamadanishiki
Sake meter value	±0
Acidity	1.6

Nakano Mitsujiro Honten

Water source	Well water in the brewery
Water quality	Super-soft

2-7-10 Yoshiura Nakamachi, Kure-shi, Hiroshima 737-0853   TEL.0823-31-7001
E-mail: suiryu@rapid.ocn.ne.jp   http://jizake-suiryu.jp/

A genuine local sake loved by the locals

Inheriting the name of Mitsujiro Nakano. Carrying the responsibility and history for the succession to their predecessor's name.

This brewery was founded in 1871. One night, Mitsujiro Nakano, the first generation of the brewery, who was living in the village of Yoshiura at that time had a dream that a dragon jumped into the land of Yoshiura and sake gushed out of it. Based on this dream, Mitsujiro dug a well, established a sake brewery and named the brand "Suiryu". The fourth-generation head of the brewery, Mitsujiro Nakano, who was involved in the family business from a young age, joined the brewery after graduating from the Department of Brewing at Tokyo University of Agriculture and working at the National Brewery Research Institute in Saijo. The brewery's sake is said to be "shipped almost completely face-to-face" within the local community. Since around 80% of sake is consumed in Kure alone, it is considered to be a genuine jizake (local sake). With a strong desire to bring enjoyment to people's lives through local sake, Suiryu Nakano Mitsujiro Honten has been brewing sake rooted in the local community.

吟醸造り本醸造酒原酒
Suiryu Genshu Hiya (Summer)

"Rich flavor that is absorbed into the body"
Rich and delicious flavor. It is good to enjoy on the rocks before the ice melts away.

Honjozoshu Genshu	1,800ml/720ml
Alc.	18-18.9%
Rice type	Domestic rice
Sake meter value	-1.0
Acidity	1.5

吟醸造り本醸造酒
Suiryu Kuromatsu

"Standard sake brewed with well water in the brewery"
A standard sake of Suiryu, it has the same "local sake" taste it has had since its founding, with a sake content of ±0.

Honjozoshu	1,800ml/720ml
Alc.	15-15.9%
Rice type	Domestic rice
Sake meter value	±0
Acidity	1.2

| Water source | Subsoil water from the upper reaches of the Kamo River |
| Water quality | Soft |

Fujii Shuzo

3-4-14 Honmachi, Takehara-shi, Hiroshima 725-0022 TEL.0846-22-2029
E-mail: info@fujiishuzou.com   http://www.fujiishuzou.com

Brewed in dialogue with nature Traditional Junmaishu

Sake brewed in the same brewery building since its founding is a perfect match for seafood from the Seto Inland Sea.

Fujii Shuzo was established in 1863 in Takehara, Hiroshima, a town facing the Seto Inland Sea and where the streets of the Edo period remain. The founding brand "Ryusei" was named after the excellent sake produced by brewing with well water from the foot of "Ryuzusan", the mountain behind the brewery. In 1907, "Ryusei" was awarded the honor of being the best sake in Japan at the first national sake competition, which made the excellence of Hiroshima sake known. This achievement was largely due to the work of Senzaburo Miura, who established a unique method of brewing with soft water at the time. The Fujii Shuzo currently brews only Junmaishu, and the brewing method "Dento Kimoto" (traditional yeastless) has immeasurable appeal. It is best served cold or heated, and is delicious with seafood from the Seto Inland Sea.

Ryusei Bekkakuhin Kimoto Junmai Daiginjo

"Kimoto Jikomi, a limited quantity"

The noble, refreshing, and transparent fruit aroma is pleasant, and the deep, dignified flavor spreads in the mouth, making it a masterpiece.

Junmai Daiginjoshu ... 1,800ml/720ml	
Alc.	17%
Rice type	Yamadanishiki
Sake meter value	+11.0
Acidity	2.2

Yoru No Teio Forever Tokubetsu Junmai

"Pushes the limits of Junmaishu"

A high-alcohol content with a sweet, rich taste that pushes the limits of Junmaishu. It is delicious on the rocks or with water.

Special Junmaishu ... 1,800ml/720ml	
Alc.	20.5%
Rice type	Yamadanishiki
Sake meter value	-8.0
Acidity	2.4

Houken Shuzo

Water source	Spring water in the brewery
Water quality	Soft

1-11-2 Nigata Honmachi, Kure-shi, Hiroshima 737-0152   TEL.0823-79-5080

Houken and Tetsuya Doi are in Hiroshima

Brew sake that the Toji (master brewer) judges to be delicious using Hiroshima-grown "Hattan Nishiki" as the main ingredient.

Founded in 1871, the Houken Shuzo is located in Nigata, Kure, Hiroshima, surrounded by mountains and the sea. In the brewery, the filtered subsoil water "Houken Meisui" springs from Mt. Noro. Tetsuya Doi, the seventh generation of the Doi family, is currently the head of the brewery. Although he was a novice when he took over the brewery at the young age of 21, he was inspired by the reputation of Houken at a tasting event for sake brewers. Today, he is known as the one and only Hiroshima Toji. "Hattan Nishiki", a rice grown in Hiroshima is the main ingredient in Houken Shuzo. This is part of the project conducted by the local city of Kure to deal with fallow rice fields, with which Houken Shuzo has cooperated. The quality of the harvested Hattan Nishiki was superb, and Tetsuya named it "Kure Miki Mai". Houken Shuzo brewed the rice and sold it under the brand name "Kure Miki Mai Hattan Nishiki Junmaishu" to contribute to the local community.

純米酒

Junmaishu Houken

"This is a sake that has a wide range of flavors and goes well with meals"

The goal is to make a sake that can be enjoyed with meals. It is mild on the palate and has a good balance of flavors overall.

Junmaishu	1,800ml/720ml
Alc.	15%
Rice type	Hattannishiki
Sake meter value	+5.0–+7.0
Acidity	1.5

純米吟醸酒

Junmai Ginjo Houken Hattan Nishiki

"A calm, mature taste"

A delicate and clean rice flavor with a subdued aroma. The finish is sharp with a slight spiciness that complements the food.

Special Ginjoshu	1,800ml/720ml
Alc.	15%
Rice type	Hattannishiki(Kuremikimai)
Sake meter value	+4.0–+6.0
Acidity	1.5

| Water source | The mountain behind the brewery |
| Water quality | Soft |

Sumikawa Shuzojo

611 Nakaogawa, Hagi-shi, Yamaguchi 759-3203 TEL.08387-4-0001
https://toyobijin.jp

Charming sake
The brewery of "Toyo Bijin"

Miraculously overcoming the torrential rain disaster and aiming to make sake rooted in the local community.

The brewery, founded in 1921, will celebrate its 100th anniversary this year in 2021. The brewery was swallowed up by nearly two meters of sediment and suffered devastating damage in the torrential rains that hit eastern Hagi on July 28, 2013. After receiving a lot of support from inside and outside the prefecture, and their own hard work, the brewery has miraculously resumed brewing. The current fourth-generation brewer, Takashi Sumikawa, who also serves as Toji (master brewer), is one of the best young Toji in Japan, having trained at Takagi Shuzo, the brewer of the famous "Juyondai". He brews "Toyo Bijin" mainly using local rice from Hagi and Yamaguchi, in the hope that it is "water that has passed through rice". Recently, they have been involved in sake related activities rooted in the local community, such as releasing products jointly planned with six local sake breweries in Hagi and participating in research projects conducted by local high school students.

Toyo Bijin Ichiban Matoi Junmai Daiginjo

"Famous sake that Russia's President Putin also enjoyed"
Fruity, clear and soft on the tongue. The aftertaste is clean and elegant. It is named after the first brewer's love for his late wife.

Junmai Daiginjoshu	720ml
Alc.	15.8%
Rice type	Yamadanishiki
Sake meter value	±0
Acidity	1.4

Toyo Bijin Junmai Daiginjo Princess Michiko

"Elegant scent and elegant sweetness"
The toyo bijin Junmai Daiginjoshu is characterized by the long-lasting floral aroma unique to "Princess Michiko", flower yeast of Tokyo University of Agriculture.

Junmai Daiginjoshu	720ml
Alc.	16%
Rice type	Yamadanishiki/Saitonoshizuku
Sake meter value	-5.0
Acidity	1.6

Nakashimaya Shuzojo

| Water quality | Medium-soft |

2-1-3 Doi, Shunan-shi, Yamaguchi 746-0011 TEL.0834-62-2006
E-mail: tnk1774@ccsnet.ne.jp https://nakashimaya1823.jp/

"Kanenaka" made by the kimoto-zukuri method Awarded in the IWC 2020 Junmaishu category

Brewed in Shunan, Yamaguchi for 200 years with single-minded focus.

Founded in 1823, the brewery has been brewing sake for 200 years; from Kunigoro, the fourth-generation head, to the eleventh generation today. Shunan is located in the center of Yamaguchi on the Seto Inland Sea side. The brewery is located in Tomita Doi, at the confluence of the Kojiro and Tomita Rivers, which run through the foothills of Eigenzan. The name of the sake "Kotobuki" used to be "Kotohogi" and it was used in ancient times for celebrations and ceremonies to wish for long life and longevity. The local people call this sake, "Tsuru, Kame, Kotobuki" (crane, turtle, Kotobuki) because of its label design, and it has long been a favorite of the people. In recent years, sake such as "Nakashimaya" (named after the brewery, from which you can feel their commitment) and "Kanenaka", which uses the kimotozukuri method from the Edo period (which is rare nowadays), have been attracting attention. The company's motto is "Shu-chu-zaishin" meaning "there is heart in sake". They will continue to brew sake with a single-minded focus.

Nakashimaya Junmai Daiginjo

"Splendid and elegant aroma"
100% Yamadanishiki produced in Yamaguchi is used. It has a splendid and elegant aroma and is soft on the palate. A hint of sweetness spreads.

Junmai Daiginjoshu	720ml
Alc.	16%
Rice type	Yamadanishiki
Sake meter value	±0
Acidity	1.3

Kanenaka Kimoto Junmai Yamadanishiki

"The aroma and taste are rich in character"
This sake has a unique aroma and taste due to its traditional sake brewing process "kimoto", and is characterized by its umami and rich acidity. It won a trophy in the IWC2020 Junmaishu category.

Junmaishu	720ml
Alc.	15-16%
Rice type	Yamadanishiki
Sake meter value	+6.0
Acidity	1.7

| Water source | Asa River |
| Water quality | Hard |

Nagayama Shuzo

367-1 Oaza-Asa, Sanyo Onoda-shi, Yamaguchi 757-0001　TEL.0836-73-1234

The brewery that led the Yamaguchi sake boom

Sake brewing rooted in the local climate, reviving phantom sake rice with farmers.

Established in 1887. The brewing process in the Asa area of Yamaguchi where the brewery is located, is characterized by brewing water from the Asa River flowing from Akiyoshidai. It is a medium-hard water that produces sake with a firm taste. In addition, the brewery is closely connected with local farmers and grows "Kokuryomiyako", a fantastic sake rice grown by contract farmers in Yamaguchi. This was the catalyst for the birth of "Yamazaru". When the veteran toji retired, Gentaro Nagayama, the sixth-generation brewer, took over the role. As the youngest toji in the prefecture, he has developed a sake brewing style that takes advantage of Yamaguchi's climate and food culture, and has won awards at the Yamaguchi Prefecture New Sake Competition, the National New Sake Competition, and the U.S. National Sake Appraisal. The aim is to create a sake with a flavor and sharpness that matches the sweet soy sauce used in Yamaguchi. Since it goes well with sweet soy sauce, it can also be used to accompany Western dishes, such as those using demi-glace or tomato sauce.

Tokubetsu Junmaishu Yamazaru

"Rich and mellow taste"

This Junmaishu is made from "Kokuryomiyako", a fantastic sake rice produced in Yamaguchi. It has a delicious sharp taste that complements a wide range of meals.

Special Junmaishu	1800ml/720ml
Alc.	15-16%
Rice type	Kokuryomiyako
Sake meter value	+3.0
Acidity	1.7

Junmai Shochu Netaro 43 degrees

"So-called Japanese brandy"

A pure Shochu Genshu (rice liquor) that is distilled under reduced pressure from mash made with sake yeast. It is characterized by a clear mouthfeel and the rich aroma of yeast.

Junmai Syochu	1,800ml/720ml
Alc.	43%
Rice type	Domestic rice

Horie Sakaba

6781 Hirose, Nishiki-machi, Iwakuni-shi, Yamaguchi 740-0724 TEL.0827-72-2527

Water source	Mt. Mizunoo
Water quality	Hard

Sake can be more delicious

Weaving the latest technology into the family's traditional techniques to spread its wings out into the world.

Founded in 1764, the middle of the Edo period. Horie Sakaba is the oldest brewery in Yamaguchi and located in Nishiki-machi, Iwakuni (western part of the Chugoku mountain range). The area is surrounded by 1,000-meter-high mountains including Mt. Jakuchi, which is the highest mountain in Yamaguchi. The Nishiki River (the largest river in Yamaguchi) runs through the center of the town, making it a place of "Sanshi-suimei" which means beautiful scenery surrounded by mountains and water. The abundance of good quality water and the climate with a large difference in temperature between day and night are ideal for sake brewing as well as for growing rice. The brewery's main brand, "Kinsuzume" is made using only the highest quality sake rice that has passed the brewery's own strict standards, with minimal use of agricultural chemicals. The name "Kinsuzume" is derived from the fact that the sparrow is revered as a divine messenger symbolizing a good harvest. In recent years, the company has started to produce a super-premium brand, "Premium Kinsuzume" which is highly regarded around the world.

Premium Kinsuzume
Junmai Daiginjo

"Excellent sake to be proud of worldwide"

This is the world's best sake for two consecutive years at the IWC, the world's most prestigious sake competition. It was praised for its fruitiness and balanced acidity and sharpness.

Junmai Daiginjoshu	750ml
Alc.	16-17%
Rice type	Yamadanishiki

Guide to Sake Breweries and Famous Sake
Shikoku Region
Tokushima, Kagawa, Ehime, Kochi

Food Culture of Shikoku Region

Tokushima: The warm plains of Tokushima are a major producer of sudachi and yuzu. The coastal areas of Tokushima produce wakame seaweed and laver throughout the year, and the Seto Inland Sea is rich in marine products such as sea bream, crabs, and shrimps. Ayu (sweetfish) dishes from the Yoshino River are also famous, and "Naruto Kintoki", a very sweet yam, is used in wasanbon (refined sugar) as well as wagashi (Japanese-style confections).

Kagawa: The climate is mild with little rainfall, and olive cultivation flourishes on Shodoshima. Sanuki udon, for which the prefecture is famous, is made from wheat grown in the plains of the Sanuki region. Kagawa is the first prefecture in Japan to successfully farm yellow tail fish, and it is served on celebratory occasions as "Shusseuo".

Ehime: Taking advantage of the characteristic climate of mountains and plains, mandarin oranges are grown in terraced fields facing the Seto Inland Sea, and traditional vegetables and kiwis are also produced. The fishing industry is thriving too, and mackerel, sea beam, and "jakoten" (deep-fried fish paste made from whole small fish) are famous. Ippon-zuri (handline fishing with a single baited hook) is well-known as the way to catch hana-aji (horse mackerel) and hana-saba (mackerel) at Cape Sada.

Kochi: Crops such as eggplant, ginger, yuzu, and citrus maxima grow well here, thanks to the t mountainous terrain, plains along Tosa Bay and the warm climate of the Kuroshio Current. Katsuo (bonito) fishing thrives, so there are many bonito dishes including "Shuto" a local delicacy. It is made by salting the offal left over from making dried bonito flakes and seasoning it with sake and mirin, then leaving it for about six months. There are many clear streams in the area, such as the Shimanto River, and grilled ayu (sweetfish) is a popular dish.

Tsukasagiku Shuzo

Water source	Subsoil water from Mt.Ryuo, Shikoku mountain range
Water quality	Soft

93 Aza-Myoken, Mima-cho, Mima-shi, Tokushima 771-2106 TEL.0883-63-6061
E-mail: tsukasagiku@novil.co.jp http://www.tsukasagiku.co.jp/

Brewed in the clearest stream in Shikoku
Tokushima's Beautiful Sake

Even if you brew under the same conditions using the same method, you will never get the same taste twice, which makes it interesting.

The brewery is located in Mima, western Tokushima, an area full of natural beauty and resources. Mt. Tsurugi is one of Japan's 100 most famous mountains, the Yoshino River is the largest river in Shikoku, and the Anabuki River is one of the clearest rivers in Japan. The brewery was established in 1896 in a place called "Kirai" which dates back to the Kamakura period. The Awa Toji (master brewer) who inherit the traditional techniques have been using the same methods the whole time, sparing no effort. The fourth-generation head brewer says "even if you make sake under the same conditions using the same method every year, it will never taste the same. That's why sake brewing is so interesting and difficult". He assesses the climate and the condition of the ingredients, and brews sake in dialogue with it, all in the same brewery where he and his predecessors have worked for more than 100 years. Only locally grown rice is used to make sake, and as one of the few breweries in Japan that uses only rice and rice malt, the brewery pursues the art of making Junmaishu.

純米大吟醸酒
Junmai Daiginjo Kirai Gin(Watashi-e, Gin)

"Well-earned drink for myself"

The sake is clean, fragrant, and well-balanced with the flavor of the rice. A gentle sake that will make you smile when you sip it.

Junmai Daiginjoshu	720ml
Alc.	16%
Rice type	Yamadanishiki
Sake meter value	+2.0
Acidity	1.4

特別純米酒
Tokubetsu Junmaishu. Anabukigawa

"Sake from the Anabuki River for you"

This sake is brewed by the Toji himself using water from the headwaters of the Anabuki River. The aroma is clear, the taste is fresh, and the flavor is clean.

Special Junmaishu	1,800ml/720ml
Alc.	15%
Rice type	Ginnosato
Sake meter value	±0
Acidity	1.5

| Water source | Subsoil water from the Asan Mountain range |
| Water quality | Somewhat soft |

Honke Matsuura Shuzojo

19 Aza Yanaginomoto, Ikenotani, Ooasa-cho, Naruto-shi, Tokushima 779-0303 TEL.088-689-1110
E-mail: shop@shumurie.co.jp   https://narutotai.jp/

More than 200 years since establishment
We just keep on going

Delivering sake and sake culture to the world while contributing to the development of Tokushima.

The company was founded in 1804 by Naozo Matsuura, the second-generation head of the family, and has continued to brew in the same area. The brewery's main brand "Naruto Tai" was named in 1886 by Akira Sakai, the prefectural governor at the time, and Kuhei Matsuura, the fifth-generation brewer, with the hope that it would be as beautiful and elegant as the king of the fish, "Tai" (sea bream). The current sake brewing system uses rice from Tokushima, local water, and craftsmen from Tokushima, making it a local sake in both name and reality. The brewery also holds a monthly event called "KuraKura Tachikyu" which is a place for locals to gather and talk. In addition, the brewery works with local shrines, temples and companies with an eye on the future, aiming to create a tourist attraction with the brewery at its core for customers who love sake.

純米大吟醸酒
Naruto Tai Junmai Daiginjo

"An exquisite sake that complements delicate Japanese food"

This sake has an elegant aroma reminiscent of fruit, which is unique to Junmai Daiginjo. You can taste the rich rice flavor and fine acidity.

Junmai Daiginjoshu …	1,800ml/720ml
Alc.	16-17%
Rice type	Yamadanishiki
Sake meter value	2.5

純米大吟醸酒
Naruto Tai Onto the table

"Best served chilled and enjoyed in a wine glass"

This sake has a mellow aroma reminiscent of ripe bananas. If you cool it a little, sharpness is added and the taste will increase.

Junmai Daiginjoshu …	720ml/180ml
Alc.	15%
Rice type	Tokushima rice
Sake meter value	±0
Acidity	-1.0

Miyoshikiku Shuzo

Water source	Subsoil water from the Yoshino River
Water quality	Soft

1661 Sarada, Ikeda-cho, Miyoshi-shi, Tokushima 778-0003 TEL.0883-72-0053
https://miyoshikiku.shop/

Local water, Local rice Local yeast

The upper reaches of the Yoshino, one of Japan's three major rivers, is a cold region ideal for sake brewing.

Miyoshikiku Shuzo is located in Awaikeda, upstream of the Yoshino River which is one of Japan's three major rivers. This location has an ideal cold climate for sake brewing; it is surrounded by the Asan mountain range in the north as well as peaks of Mt. Tsurugi and the Shikoku mountain range in the south. Sake is brewed with spring water from the subsoil of the Yoshino River, and with the cooperation of local farmers, the sake rice is made from Yamadanishiki, Omachi, Gohyakumangoku, and other varieties grown in Tokushima. The goal is to create a sake with the flavor of water and rice. The local Ikeda High School offers a fermentation course, and Miyoshikiku Shuzo has been inviting students to participate in its annual sake brewing program for the past twelve years. I wonder what Miyoshikiku Shuzo's sake will taste like when they grow up. The ideal local sake can be found here. There is a variety of labels at the Miyoshikiku online shop "that don't look like sake" and which are fun to look at.

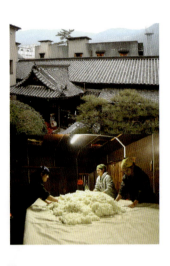

純米大吟醸酒

Miyoshikiku Junmai Daiginjo Ayane

"Sharp taste as a result of polishing Yamadanishiki to 50%"

This sake has a fruity, acidic taste and aroma thanks to the use of Tokushima yeast, which was jointly developed with Tokushima Prefecture.

Junmai Daiginjoshu	720ml
Alc.	16%
Rice type	Yamadanishiki
Sake meter value	+5.0
Acidity	1.6

純米吟醸酒

Miyoshikiku Junmai Ginjo Orie

"Unfiltered, unpasteurized"

It has a unique taste with sourness, spiciness and bitterness in the mouth and a soft aftertaste.

Junmai Ginjoshu	720ml
Alc.	15%
Rice type	Yamadanishiki
Sake meter value	private
Acidity	private

Water source	Subsoil water from the Saita River
Water quality	Medium-hard

Kawatsuru Shuzo

836 Motodai-cho, Kanonji-shi, Kagawa 768-0022 TEL.0875-25-0001
E-mail: kura@kawatsuru.com https://kawatsuru.com/

Just like the flow of a river, with an honest heart

A sake that draws out the unique flavor of the rice ingredients and vitality for tomorrow.

Kawatsuru Shuzo is located in Kagawa on Shikoku Island. Since its establishment in 1891, they have been carrying on the spirit of "impressing the drinker with an honest heart, just like the flow of a river". The rice is harvested in the paddy fields of the Sanuki Plain, and the water used to brew "Kawatsuru" is the subsoil water of the Saita River, where fireflies fly around. It is mellow and full of flavor, powerful and refreshing, with a deep and pleasant aftertaste. The rice used is "Ooseto" and "Sanuki Yoimai" produced in Sanuki, and a contract-grown "Yamadanishiki", each of which has its own unique flavor. Not only does Kawatsuru brew sake, but they also do their best to reward the efforts of all the people involved in the brewing process, including the producers of the ingredients that go with the sake and the chefs who prepare it.

Setouchi KAWATSURU
Junmai Ginjo Sanuki Olive Kobo Jikomi

"Excellent chemistry with ingredients produced in Setouchi"

This sake is made with "Sanuki olive yeast" harvested from olives grown in Shodoshima, and Oseto rice grown in Kagawa.

Junmai Ginjoshu	1,800ml/720ml
Alc.	14%
Rice type	Ooseto
Sake meter value	±0
Acidity	2.0

Kawatsuru Junmai Daiginjo Sanuki Yoimai

"Sake brewing rice from Sanuki"

"Sanuki Yoimai" brewing rice is used to make this Junmai Daiginjo, which can be enjoyed at any time of the day.

Junmaishu	1,800ml/720ml
Alc.	15%
Rice type	Sanukiyoimai
Sake meter value	±0
Acidity	1.9

Ishizuchi Shuzo

Water source	Subsoil water from the Ishizuchi mountain range
Water quality	Soft

402-3 Himi-hei, Saijo-shi, Ehime 793-0073　TEL.0897-57-8000
E-mail: sake@ishizuchi.co.jp　https://www.ishizuchi.co.jp

Sake brewing that brings out the taste of food

A passion for sake that can only be conveyed by handcrafting, which is not possible with large-scale brewing.

Ishizuchi Shuzo was founded in 1920. The brewery abolished the Toji (master brewer) system and shifted to family-oriented sake brewing in 1999. The goal of Ishizuchi Shuzo is to "brew sake that makes the most of food". The brewery's slogan is to "brew for customers who love Ishizuchi". They aim to "make sake that begins to taste even better from the third glass" with a focus on Junmaishu and Junmai Ginjoshu. The city of Saijo in Ehime, where the brewery operates, is located at the foot of Mt. Ishizuchi, the highest mountain in western Japan, and a place with reputation for its water. Of course, the brewery uses the water from the Ishizuchi mountains for brewing. In addition, the town is located near the grain-growing areas of the Saijo and Shuso plains, and has a climate suitable for sake brewing. The sake is made by hand, which is not possible with large scale brewing, and the brewer's attitude and passion are expressed in the sake.

純米吟醸酒
Ishizuchi Junmai Ginjo Green Label

"Great sake awarded a Silver at IWC 2018"

This is a standard sake recommended by the brewery. It has a light, dry taste with a gentle yet dignified elegance. It can be served cold or at room temperature.

Junmai Ginjoshu	1,800ml/720ml
Alc.	16%
Rice type	Yamadanishiki
Sake meter value	+4.0
Acidity	1.6

純米吟醸酒
Ishizuchi Junmai Ginjo Yamadanishiki 50

"Served in ANA First Class"

A popular brand of "Ishizuchi" series. You can enjoy the elegant, broad, deep taste unique to Yamadanishiki.

Junmai Ginjoshu	1,800ml/720ml
Alc.	16%
Rice type	Yamadanishiki
Sake meter value	+4.0
Acidity	1.6

| Water source | Groundwater |
| Water quality | Soft |

Kondo Shuzo

1-11-46 Shinsuka-cho, Niihama-shi, Ehime 792-0802 TEL.0897-33-1177
E-mail: info@kondousyuzou.com https://www.kondousyuzou.com/

The only sake brewer in Niihama, Ehime Prefecture

Chikaramizu for the local "Taiko Matsuri". "Hanahime Sakura" has won two gold medals.

Kondo Shuzo is a small brewery founded in 1878, the only one remaining in Niihama, Ehime. The brewery produces Niihama's only locally brewed sake using soft water drawn from a deep well 110 meters below ground, and local rice "Matsuyamamii", produced by contract farmers in the town. At New Year's, the brewery holds a sake-offering festival and serves it to the local community. Every year, the brewery holds study sessions for elementary school students who visit the brewery, and summer school seminars for junior high and high school students. Yoshiro Kondo, the president of the company, has been the chairman of the Niihama City Tourism Association since 2018, and is actively involved in PR activities with the local media. The brand name "Hanahime Sakura" is indispensable for participants at the annual Taiko Festival. It has won two gold medals and three awards at the National New Sake Competition.

Hanahime Sakura Junmaishu

"Can be enjoyed as Hiyazake (cold) and also as Okan (moderately warm)"

This sake is made from 100% Matsuyamii, a rice variety grown in Niihama, and has a clean, crisp taste.

Junmaishu	1,800ml
Alc.	15–16%
Rice type	Matsuyamamii
Sake meter value	+2.0
Acidity	1.3

Hanahime Sakura Daiginjoshu

"This sake has a refreshing throat feel"

This ginjo-shu is made by polishing Yamadanishiki to 40%, processing with a limited water supply, making koji for ginjo-shu, followed by long-term low-temperature fermentation.

Daiginjoshu	1,800ml
Alc.	15-16%
Rice type	Yamadanishiki
Sake meter value	+3.0
Acidity	1.2

Sakurauzumaki Shuzo

Water source	Subsoil water from the Takanawa mountain range
Water quality	Soft

71 Hattanji-Kou, Matsuyama-shi, Ehime 799-2424 TEL.089-992-1011
E-mail: uzumaki@mocha.ocn.ne.jp https://www.sakurauzumaki.com

Sake brewing is
blessings from heaven, the earth, and people

The company was named "Sakura Uzumaki" after a famous cherry blossom spot and a newspaper novel that the brewer loved.

The Shinohara family, whose origins go back to the priests of Kunitsuhiko-no-mikoto Shrine, started brewing sake in 1871. In 1951, the company name was changed to the present-day Sakurauzumaki Shuzo. The name "Sakura Uzumaki" comes from the fact that the mountain forest they owned, Mt. Hachiku, was famous for its cherry blossoms, and also the third- generation head loved reading a newspaper novel called "Uzumaki". The brewery is located in the northern part of Matsuyama, right in the center of Ehime, where the subsoil water of the Tateiwa River and the groundwater of Mt. Takanawa meet, an enormous good fortune for sake brewing. Sakura Uzumaki believes that "sake is made with the blessings of heaven, the earth, and people". They work hard and never forget to be thankful for the environment they were given.

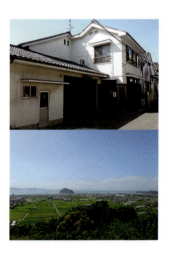

大吟醸酒
Daiginjo Sakano Ueno Kumo

"The highest quality sake"
This Daiginjo is made from 35% polished Yamadanishiki, carefully cultivated by the local farmers in the Tateiwa district in Ehime.

Daiginjoshu	720ml
Alc.	17%
Rice type	Yamadanishiki
Sake meter value	+5.0
Acidity	1.2

純米大吟醸酒
Sakura Junmai Daiginjo

"Bottled unfiltered and unheated"
This Junmai Daiginjoshu is made from 48% polished Yamadanishiki, and bottled unfiltered and unheated for a full-flavored sake.

Junmai Daiginjoshu	1,800ml/720ml
Alc.	16%
Rice type	Yamadanishiki
Sake meter value	+4.0
Acidity	1.4

| Water source | Subsoil water from the Ishizuchi mountain range |
| Water quality | Soft |

Suto Shuzo

312-2 Ooto-Kou, Komatsu-cho, Saijo-shi, Ehime 799-1106 TEL.0898-72-2720
E-mail: suto@sukigokoro.co.jp http://sukigokoro.co.jp/

Great chemistry of brothers brewing together

A brewery of brothers that does not allow compromise, even in the friendly atmosphere.

Founded in 1901, Suto Shuzo is a brewery where three brothers work together to make sake. It is located in Iyo Saijo, at the foot of Mt. Ishizuchi, the highest mountain in western Japan. The main brand "Sukigokoro" is carefully brewed using the famous water from the Ishizuchi mountain range, which is drawn from the brewery's well. The brewing process at Suto Shuzo is Sandan-shikomi (three-stage process); "soe", "naka", and "tome". While a normal brewery would efficiently proceed with each stage of the process for multiple barrels, this brewery finishes brewing one type of sake before it starts brewing the next barrel. This is because they want to make delicious sake to be proud of and which the locals love. So they lovingly watching over the brewing process themselves. They continue efforts to make theirs even better sakes to drink with local cuisine.

Sukigokoro Shizukuhime
Junmai Ginjo

"Soft and juicy like fruit"
A fruity and elegant aroma, slightly sweet in spite of its alcohol content. This sake has low acidity and a deep, soft finish.

Junmai Ginjoshu	1,800ml/720ml
Alc.	15%
Rice type	Shizukuhime

Sukigokoro Omachi
Junmai Ginjo

"A perfect balance of sweet and sour"
Soft in the mouth, but the sweet and sour taste of the rice are well combined to create a solid finish. This is Sukigokoro's most popular sake.

Junmai Daiginjoshu	1,800ml/720ml
Alc.	16%
Rice type	Omachi

Seiryo Shuzo

Water source	Subsoil water from the Ishizuchi mountain range
Water quality	Weakly-soft

1301-1 Shu, Saijo-shi, Ehime 799-1371 TEL.0898-68-8566
E-mail: info@seiryosyuzo.com http://www.seiryosyuzo.com

To brew sake is to brew the heart of a person

A sake brewery that sticks to traditional methods through the teamwork of its brewers.

Since its establishment in 1877, Seiryu Shuzo has survived the whirlwind of history with the help of many people under the motto, "Sake is made with dreams and the heart". The times constantly change and civilization evolves. However, in the shadow of this, the culture built up by our ancestors is gradually lost. With the belief that "the past is what makes the present, and the present is what makes the future", the Toji and four others work as a team to make sake. With a brewing style that incorporates a lot of old-fashioned manual labor, they brew sake with dreams and heart, watching the daily temperatures and thinking only of making the best sake every year with the rice harvested that year. In the spring and fall, the brewery opens its doors, for kura-biraki (admission is free), where you can enjoy new sake in the spring and aged sake in the fall.

純米大吟醸酒

Iyo Kagiya Muroka Junmai Daiginjo Green Label

"Can be enjoyed as Hi-ya-zake (cold) and also as Nurukan (lukewarm)"

This is a slightly luxurious sake for those who prefer Junmaishu, with a full-bodied flavor that can be enjoyed both cold and lukewarm.

Junmai Daiginjoshu	1,800ml
Alc.	16-17%
Rice type	Shizukuhime
Sake meter value	+5.0
Acidity	1.6

大吟醸酒

Seishu Miyosakae Daiginjo

"For celebratory occasions"

This Daiginjo is the most popular gift for celebratory occasions. It boasts a balance of high aroma, sharpness in the throat, and rich flavor.

Daiginjoshu	1,800ml
Alc.	15-16%
Rice type	Yamadanishiki
Sake meter value	+5.0
Acidity	13

Water source	Groundwater from the production site
Water quality	Soft

Kameizumi Shuzo

2123-1 Izuma, Tosa-shi, Kochi 781-1142　TEL.088-854-0811
E-mail: contact@kameizumi.co.jp

The sake of Tosa, in the southern part of Japan
supported by the people of Kochi, who are great drinkers

Hearts and technologies inherited from the predecessors. Sake brewed with rice and water produced in Kochi.

Kameizumi Shuzo is located in Tosa, at the mouth of the clear Niyodo River, which is blessed with both sea and mountain produce. The fresh water that gushes from the side of the Sukumo Highway that runs nearby, never runs dry even in times of drought, and is known as "Manten no Izumi" (the perennial spring). Since this water was used for preparation, the brewery was named "Kameizumi". In 1897, eleven comrades founded the company as the "Fumoto Sake Shop". Since then, they have continued to preserve the Kameizumi brand name and brew a wide variety of sake using only rice, yeast and water produced in Kochi. They have cultivated high-level brewing techniques over the years. These require passion to carry out, especially when the brewery is located in a warm region in southern Japan considered to be difficult for sake brewing. It is obvious that the brewery has inherited the founder's spirit of "We brew our own sake to drink ourselves."

特別純米酒

**Kameizumi
Tokubetsu Junmai**

"Tosa dryness that you never tire of"
The balance of sweetness, sourness, bitterness, and umami is well balanced, and the elegant dryness makes it a sake for everyone.

Special Junmaishu	1,800ml
Alc.	15%
Rice type	Tosanishiki by Kouchi
Sake meter value	+5.0-+7.0
Acidity	1.4-1.5

純米吟醸酒

**Kameizumi Junmai Ginjo
Nama Genshu CEL-24**

"Tastes like fragrant white wine"
This fruity, unpasteurized sake has a perfect balance of acidity and sweetness. It is recommended for women, and people who are not big fans of sake.

Junmai Ginjoshu	720ml
Alc.	14%
Rice type	Hattannishiki by Hiroshima
Sake meter value	-5.0--16.0
Acidity	1.5-2.2

Takagi Shuzo

Water source	Monobe River system
Water quality	Medium soft

443 Akaoka-cho, Konan-shi, Kochi 781-5310 TEL.0887-55-1800
E-mail: takagi@toyonoume.com   https://toyonoume.com/

Local Sake in which you can feel Tosa and the charm of Kochi

Using Kochi yeast and Kochi-grown sake rice. We aim to make the best 100% Kochi sake.

Akaoka is located about 20 km east of Kochi and once prospered as a merchant town. It is a small but unique town where nationwide festivals such as "Dorome Matsuri" and "Ekin Matsuri" are held and used to promoted tourism and town revitalization. Takagi Shuzo, founded in this town in 1884, is a sake brewery that works together with the local community. For example, the light and dry "Toyo No Ume" that is poured into a large cup at the Dorome Festival in April, is a traditional sake from Takagi Shuzo. Kochi is famous for its sake, and the brewery has been making efforts to breed yeast unique to the prefecture and rice suitable for sake brewing. In terms of sake rice, the breeding of "Ginnoyume" and "Tosaurara" have been successful. Takagi Shuzo continues to brew sake with these Kochi products, aiming to brew the best and most sophisticated sake with 100% Kochi ingredients.

純米大吟醸酒

Toyo No Ume
Junmai Daiginjo Ryuso

"Enjoy the aroma even more by drinking it from a wine glass"

The aroma of green apples and the taste of rice can be enjoyed in a wine glass. The bittersweet taste of the rice is accentuated ". The name "Ryuso" comes from a tornado that once hit the brewery.

Junmai Daiginjoshu … 1,800ml/720ml	
Alc.	16%
Rice type	Ginnoyume
Sake meter value	-1.0
Acidity	1.8

特別純米酒

Tosa Kinzo
Tokubetsu Junmaishu

"Fine taste of 'Tosaurara' produced in Kochi"

A light banana-like ginjo aroma, a clear taste and a sharp acidity. It can be enjoyed cold or lukewarm.

Special Junmaishu … 1,800ml/720ml	
Alc.	14%
Rice type	Tosaurara
Sake meter value	+3.0
Acidity	2.0

Water source	Niyodo River system
Water quality	Soft

Tsukasabotan Shuzo

1299 Sakawa-cho-Kou, Takaoka-gun, Kochi 789-1201 TEL.0889-22-1211
E-mail: ainet@tsukasabotan.co.jp   https://www.tsukasabotan.co.jp/

Tsukasa Botan is the best of all botan (peonies), king of flowers

For more than 400 years, the brewery has been rooted in Tosa and has devoted itself to the art of being the king of sake.

Sake brewing in Tosa's Sakawa district began in 1603, when Katsutoyo Yamauchi, who had been given 240,000 koku of Tosa land, and his vassals established merchant houses, including liquor stores. In 1918, when the company was established, Count Mitsuaki Tanaka, a reformer from Sakawa, named the company "Tsukasabotan", saying "Peonies are the king of all flowers, and Tsukasa Botan should be the best of all peonies". Today, Tsukasa Botan Shuzo uses Yamadanishiki and other varieties of rice grown using the "Nagata Farming Method", which does not place a burden on the environment, and spring water from the Niyodo River system. The Niyodo River, which originates from the Shikoku mountain range, boasts water clarity that surpasses that of the Shimanto River, famous as "Japan's last clear stream" and is said to be "the cleanest river in Japan". This water has been a factor in Sakawa's prosperity as a sake brewing town since ancient times.

Tsukasa Botan
Junmai Senchu Hassaku

"The power to complement any dish"

An elegant, natural aroma, a smooth, full-bodied taste, and a clean, crisp finish. A high-quality sake enjoyed during meals, one that boasts outstanding sharpness.

Special Junmaishu	1,800ml/720ml
Alc.	15.4%
Rice type	Yamadanishiki
Sake meter value	+8.0
Acidity	1.4

Tsukasa Botan
Daiginjo Kuroganeya

"Sake that brings supreme harmony to the palate"

This is the ultimate daiginjo from which only the best parts of the highest-ranked daiginjo are extracted. It has a gorgeous ginjo aroma and an incomparable mellowness.

Daiginjoshu	720ml
Alc.	17.8%
Rice type	Yamadanishiki
Sake meter value	+2.0
Acidity	1.2

Guide to Sake Breweries and Famous Sake

Kyushu Region

Fukuoka, Saga, Nagasaki, Kumamoto, Oita, Miyazaki, Kagoshima

Food Culture of Kyushu Region

Fukuoka: During its history of commerce, the cuisine of Fukuoka has been influenced by the Korean Peninsula and other countries . Mizutaki with chicken bone soup and ramen noodles, as well as "Okyuto", a dish made with ogonori, which has been popular since the Edo period. The area also has good catches of fish, such as sea bream (tai) and puffer fish (fugu).

Saga: Famous for the Mutsugoro dishes (a type of fish found in the tidal flats of the Ariake Sea). The Genkai Sea is rich in mackerel, shrimp, and abalone. In addition, zaru-tofu (tofu in a bamboo basket) has been made and handed down since the Edo period.

Nagasaki: Famous for "Champon" and "Castella", which came out of the culture introduced through foreign trade at Dejima Island. The mild climate produces loquat. In the sea, longtooth grouper (kue), puffer fish (fugu) and swimming crabs (watarigani) are caught, and suppon turtles are farmed as well. In addition, "karasumi", dried salted mullet ovaries, is used for celebrations.

Kumamoto: Takana, tomatoes, and dekopon are grown in the volcanic ash soil, and the local cuisine includes "spicy lotus root" and "horse meat sashimi".

Oita: Pumpkins were introduced from Portugal and are called "Sorin pumpkins". Oita has a thriving fishing industry, with Seki mackerel, Seki horse mackerel and Shiroshita flounder all caught in Seki. Many of the fish dishes eaten by fishermen have been handed down as local cuisine.

Miyazaki: "Obiten" made of fish paste, is one of the local dishes in the Nichinan area. Vegetables, fruit and tropical mangoes are grown in the warm climate, and livestock is a major industry, including Jitokko chicken and Miyazaki beef.

Kagoshima: Sweet potatoes which are resistant to volcanic ash, the Makurazaki coast's skipjack tuna, kibinago, sardines, and flying fish catches thrive. The cuisine is also diverse. Kagoshima is known for ivestock products such as Satsuma shamo and Kurobuta pork.

Water source	Chikugo River system
Water quality	Soft

Asahikiku Shuzo

403 Itchoubaru, Mizuma-machi, Kurume-shi, Fukuoka 830-0115 TEL.0942-64-2003
E-mail: asa2003@ruby.ocn.ne.jp   https://www.asahikiku.jp/

The best sake is junmai, and it's even better heated

Sake brewing that sticks to the flavor of the rice not influenced by trends.

The city of Kurume in Fukuoka has has the water of the Chikugo River, the rice of the Chikugo Plain, and water transportation, making it an area where many sake breweries have been located. In addition, the Jojima area of Kurume is considered one of the top three sake breweries in Japan, along with Nada and Fushimi. The first generation of the Asahikiku Shuzo was established in the Joshima area in 1900, the last year of the 19th century. He named the brewery "Asahikiku" in the hope of producing a sake with the same vigor and sharpness as the morning sun. Since 1994, the brewery has been brewing sake using Yamadanishiki, an agricultural chemical-free rice, with the goal of producing sake that is not influenced by the changing times or trends, but rather by the flavor of the rice. Every year in February, eight breweries in the area hold the Jyojima Shuzo Opening, and in recent years more than 110,000 people have visited the event, making it a major sake event in western Japan.

特別純米酒
Asahikiku Ayaka
Tokubetsu Junmai Binkakoi

"Delicacy that blooms softly on the tongue"

Special Junmai with a mellow aroma and a gentle taste. With bottling and storage, it has the unique flavor and soft richness of Yamadanishiki.

Special Junmaishu	1,800ml/720ml
Alc.	15%
Rice type	Yamadanishiki
Sake meter value	+5.0
Acidity	1.4

特別純米酒
Asahikiku Daichi
Tokubetsu Junmaishu

"The taste of rice and yeast is a great combination"

This sake is brewed using only Yamadanishiki, a chemical-free rice grown under contract in the Itoshima area of Fukuoka. The natural flavor and acidity are harmonized to create a deep taste.

Special Junmaishu	1,800ml/720ml
Alc.	15%
Rice type	Yamadanishiki
Sake meter value	+5.0
Acidity	1.6

Ishikura Shuzo

Water source	Matsubara water in Chiyo
Water quality	Soft

1-30-1 Katakasu, Hakata-ku, Fukuoka-shi, Fukuoka 812-0043 TEL.092-651-1986
E-mail: info@ishikura-shuzou.co.jp   https://www.ishikura-shuzou.co.jp

The only sake brewery in Hakata, nicknamed "Hakata 100-year Brewery"

Brewed year-round, which is rare. You can enjoy fresh, unpasteurized sake at any time.

Ishikura Shuzo, which has passed down the atmosphere of a traditional sake brewery for 150 years since the Meiji era, is often called "Hakata 100-year brewery" by Hakata locals. In 2020, it celebrated its 150-year anniversary. The old-fashioned appearance still remains, such as white walls, reddish-brown brick chimney, and cedar balls (sakabayashi) a symbol of a sake brewery. In 2011, it was registered as a Tangible Cultural Property of Japan. The brewery is used not only for sake brewing, but also for a wide range of other purposes such as receptions, parties, concerts and cultural activities. Ishikura Shuzo's image of sake is "fresh". This is because they brew year-round, which is rare for a sake brewery, so you can enjoy freshly squeezed nama-zake at any time. Hakata's only sake brewery will continue to brew delicious sake, loved by the people of Hakata.

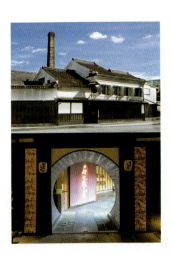

純米大吟醸酒
Junmai Daiginjo Hyakunengura

"Unique aroma of Daiginjoshu"
This is a Junmai Daiginjoshu brewed with luxurious Yamadanishiki produced in Itoshima, Fukuoka, and fermented at low temperatures for a long time. It is great for gifts.

Junmai Daiginjoshu	1,800ml/720ml
Alc.	16%
Rice type	Yamadanishiki
Sake meter value	-1.0
Acidity	private

スパークリング
Sparkling Seishu Awayura

"This sake can be used in toasts at wedding receptions"
Sparkling sake brewed with "Fukuoka Dream Yeast". The alcohol content is 7 degrees, about half that of regular sake.

Sparkling	250ml
Alc.	7%
Rice type	Fukuoka rice
Sake meter value	-85.0
Acidity	private

| Water source | Subsoil water from the Minouchi mountain range |
| Water quality | Slightly-soft |

Isonosawa

1-2 Nishikumanoue, Ukiha-machi, Ukiha-shi, Fukuoka 839-1404 TEL.0943-77-3103
https://isonosawa.com/

There is good sake in Ukiha, the home of famous water

All tap water in the town is natural water. "Where there is great water, there is great sake".

Ukiwa Town, located on the eastern edge of the vast Chikushi Plain, is one of the few "water villages" in Japan, where all the town's tap water is supplied by natural water from the Chikugo River system, which originates from Mt. Aso and the Mina mountain range. Since its establishment in 1893, Isonosawa has been brewing sake with water drawn from the brewery's well. Eighty percent of sake is made up of water. "Where there is great water, there is great sake". This is the reason why their sake is so good. In 2021, the brewery was reorganized with the addition of veteran Toji Watanabe to the management team. In the future, the brewery will strengthen its ties with local and other companies around Japan, and contribute to the local community not only by brewing sake, but also by promoting sake culture and creating opportunities for interaction with tourists.

純米酒
Shun Junmaishu

"Junmaishu polished to 60%"
Junmaishu made by polishing Yamadanishiki to 60%, pursuing a subtle aroma, the flavor of Yamadanishiki and sharpness. Ideal as a sake for sipping.

Junmaishu	720ml
Alc.	16%
Rice type	Yamadanishiki
Sake meter value	+2.0
Acidity	1.4

大吟醸酒
Densho Daiginjo

"Filled with traditional skills"
This Daiginjo is made entirely from carefully selected Itoshima-produced Yamadanishiki, and luxuriously polished to 35%. Traditional techniques from the time of the brewery's founding are still used today to make this sake.

Daiginjoshu	720ml
Alc.	16-17%
Rice type	Yamadanishiki
Sake meter value	+6.0
Acidity	1.2

Takahashi Shoten

| Water source | Yabe River |
| Water quality | Medium-hard |

2-22-1 Motomachi, Yame-shi, Fukuoka 834-0031 TEL.0943-23-5101
E-mail: info@shigemasu.co.jp http://www.shigemasu.co.jp

Rice grown in Yame 300 years of tradition

Training young brewers to preserve and refine inherited skills.

Takahashi Shoten, known by the brand "Shigemasu", was founded in 1717 and has a history of over 300 years. The brewery is widely loved by the local people, especially in the Yame and Chikugo regions, and their tradition is to brew sake with a dry taste. The subsoil water of the Yabe River used for brewing is medium-hard water with moderate amounts of potassium, phosphoric acid, and magnesium, making it suitable for dry sake. For the sake rice, the brewery uses only rice from Fukuoka, such as Yamadanishiki, Omachi, Ginnnosato, and Yumeikkon. In order to maintain the tradition of this famous sake, the brewery is not only seeking efficiency, but training young brewers with well-honed sensibilities while maintaining the traditional wooden koji rooms and other facilities. By nurturing young people, the entire brewery is pushing its skills to a higher level.

大吟醸酒
Daiginjo Hako Iri Musume

"Carefully raised sake, like a beloved daughter"
The fruity ginjo aroma, elegant taste, and sharpness characteristic of daiginjo that go well with Fukuoka's light, refreshing sashimi.

Daiginjoshu	1,800ml/ 720ml
Alc.	16%
Rice type	Yamadanishiki
Sake meter value	+4.0~+5.0
Acidity	1.2-1.3

特別純米酒
Shigemasu Classic Tokubetsu Junmaishu

"Slightly dry with fullness of umami"
With a delicious rice flavor and high acidity, the sweetness becomes stronger when heated to around 45 degrees, creating a perfect balance with the acidity.

Special Junmaishu	1,800ml/ 720ml
Alc.	16%
Rice type	Yumeikkon
Sake meter value	+1.0~+2.0
Acidity	1.4-1.5

Water source	Chikugo River system
Water quality	Soft

Hiyokutsuru Shuzo

466-1 Uchino, Jojima-machi, Kurume-shi, Fukuoka 830-0204 TEL.0942-62-2171
E-mail: info@hiyokutsuru.co.jp https://www.hiyokutsuru.co.jp/

The Sake of Happiness Hiyokutsuru

This sake gently nestles against the heart of the drinker, As close as a happily married couple.

The Hiyokutsuru Shuzo, located in Kurume, Fukuoka, was established in 1895. At that time, the brewery was called "Ninomiya Meishu Jozobu", and "Hiyokutsuru" was the name of the brand. When the company became a joint-stock corporation in 1919, the name was changed to "Hiyokutsuru Shuzo". The name "Hiyokutsuru" comes from the crest of the Kamachi family, the ancestors of the brewery, which has been handed down since the Kamakura period. Hiyokutsuru is another name for a married couple of cranes based on the legend of Yang Guifei, who said, "If you wish to be in heaven, you will be a bird of wings; if you wish to be on earth, you will be a branch of trees". The local sake rice is carefully milled in house, and the subterranean water of the Chikugo River drawn from 200 meters below the ground is used for brewing, aiming to produce a sake that is soft on the palate and does not leave the drinker feeling tired.

本醸造酒
Jousen Hiyokutsuru

"Sake for evening drinks that go well with gameni"

This is the representative brand of Hiyokutsuru, a sake with a gentle taste that you will never get tired of drinking. It is best served cold or heated

Honjozoshu	1,800ml/720ml
Alc.	15%
Rice type	Domestic rice
Sake meter value	+2.0
Acidity	1.5

特別純米酒
Tokubetsu Junmai Yamakanbai

"Special Junmaishu that goes well with the flavor of soy sauce"

This popular sake complements food with its robust flavor and pleasant aftertaste. Its sharpness goes well with fried foods.

Special Junmaishu	1,800ml/720ml
Alc.	15%
Rice type	Domestic rice
Sake meter value	±0
Acidity	1.8

Morinokura

Water source	Subsoil water from the Mt. Kora water system
Water quality	Medium

2773 Tamamitsu, Mizuma-machi, Kurume-shi, Fukuoka 830-0112 TEL.0942-64-3001
E-mail: welcome@morinokura.co.jp

Polish up the culture of sake

From a traditional Sake-kasu Shochu brewery to a Jumaishu brewery that uses 100% prefectural rice.

Kurume in Fukuoka, is a vast granary blessed by the Chikugo River, the largest river in Kyushu, and at the same time, it is a place where sake brewing has been flourishing since ancient times. Morinokura was founded in 1898. In 2005, they stopped adding alcohol to any of the sake they produced and became the first sake brewery in Kyushu to produce Junmaishu. Present-day Mori No Kura only brews Junmaishu, and takes pride in three important elements: 100% Fukuoka-produced rice, the clear and pure water pumped from underground, and the skills of Mizuma Toji that have been handed down from generation to generation. Their unique approach, is making sake in a cyclical manner by "making sake rice→making Junmaishu→making Kasutori Shochu (from sake lees)→composting Shochu lees→making sake rice". This is the way it used to be done in local rural culture.

純米吟醸酒
Mori No Kura
Junmai Ginjo Suisui

"Refined sweetness that goes well with light dishes"
An elegant, slightly sweet aroma and a light aftertaste with a soft, refreshing flavor. Best served chilled.

Junmai Ginjoshu … 1,800ml/720ml/300ml	
Alc.	15%
Rice type	Yumeikkon
Sake meter value	+3.0
Acidity	1.4

純米吟醸酒
Komagura Gen Enjuku
Junmai Ginjo

"Perfect match for Fukuoka's famous 'sesame mackerel'"
Matured in a unique way to give a relaxed and soft flavor as well as umami. Warm it up a little for a deeper taste.

Junmai Ginjoshu 1,800ml/720ml	
Alc.	15%
Rice type	Yamadanishiki
Sake meter value	+6.0
Acidity	1.9

| Water source | Subsoil water from the Sefuri mountain range |
| Water quality | Soft |

Amabuki Shuzo

2894 Higashio, Miyaki-cho, Miyaki-gun, Saga 849-0013 TEL.0942-89-2001
E-mail: info@amabuki.co.jp https://www.amabuki.co.jp/

"This sake is delicious" Just to hear those words

300 years of "single-minded focus on taste" Young Toji and brewers are passionate about what they do.

The Amabuki Shuzo was founded during the Genroku era (1688-1704) and has a history of about 300 years. The current brewer is Sotaro Kinoshita, the 11th generation. His eldest brother is the company president, and his younger brother is the Toji (master brewer). The average age of the brewers is 30s, and the brewery is full of enthusiasm, always looking for something new in their research. Amabuki Kimoto produces and sells sake and shochu under the brand "Amabuki", named after the mountain to the northeast. The brewery is also one of the leading sake breweries in Japan, using flower yeast which collected from flowers by Professor Emeritus Hisayasu Nakata of Tokyo University of Agriculture. The Amabuki sake brewed with the flower yeast matures quietly in the refrigerator or underground storage, gaining natural richness and flavor. The only consistent aspect of this brewery is the "single-minded focus on taste". They continue to devote themselves to making this sake delicious.

Amabuki Kimoto Junmai Daiginjo Omachi

"Tastes like a white wine"

This sake is made by brewing Omachi with rhododendron flower yeast. The acidity catches the flavor of the rice and matches well with food.

Junmai Daiginjoshu	1,800ml/720ml
Alc.	16%
Rice type	Omachi
Sake meter value	+3.0
Acidity	1.8

Amabuki Junmai Ginjo Ichigo Kobo Nama

"Sake brewed with strawberry yeast"

As the name suggests, this sake is brewed with yeast isolated and cultivated from strawberry flowers. It goes well with dishes made with fats and oils.

Junmai Ginjoshu	1,800ml/720ml
Alc.	16-17%
Rice type	Yamadanishiki/Omachi
Sake meter value	+2.0
Acidity	1.6

Sachihime Shuzo

Water source	Subsoil water from the Taradake mountain range
Water quality	Soft

Kabuto599 Furueda, Kashima-shi, Saga 849-1321 TEL.0954-63-3708
E-mail: sachi-2@po.asunet.ne.jp http://www.sachihime.co.jp

Carefully brewing sake like raising a beloved daughter

A brewery plays a role in local tourism in a rice producing region, Kashima, Saga.

Since its establishment in 1934, Sachihime Shuzo has been located in Kashima, Saga in Kyushu. Saga has long been a rice-producing region, and many sake breweries have been established in the area. Kashima is no exception to this trend, with six breweries producing their own sake in a small city of only 30,000 people. "Sachihime", the main brand of Sachihime Shuzo, was named by the founder of the brewery with the hope that his only daughter would grow up happily. In recent years, the company has been active in "Kashima Shuzo Tourism", the birthplace of sake brewery tourism, offering tours and tastings for tourists. Due to Covid-19, the company also offers online tours.

純米大吟醸酒
Junmai Daiginjo Sachihime

"Fruity and sweet with sharpness"

This is a Junmai Daiginjoshu made with 100% Yamadanishiki, which has a fruity aroma and a firm sweetness. Best served cold.

Junmai Daiginjoshu	720ml
Alc.	16%
Rice type	Yamadanishiki
Sake meter value	-1.0
Acidity	1.3

純米吟醸酒
Junmai Ginjo Sachihime
DEAR MY PRINCESS

"The aroma is reminiscent of apples and grapes"

A medium-bodied, fruity sweet sake like European liquor that goes well chilled with Saga beef steak.

Junmai Ginjoshu	1,800ml/720ml
Alc.	16%
Rice type	Yamadanishiki
Sake meter value	-1.0
Acidity	1.4

| Water source | Subsoil water from the Tenzan mountain range |
| Water quality | Soft |

Koyanagi Shuzo

903 Ogima-chi, Ogi-shi, Saga 845-0001 TEL.0952-73-2003
E-mail: taka3n5@agate.plala.or.jp http://www.ogi-cci.or.jp/kigyou/koyanagi/

Little Kyoto in Kyushu Sake brewery in Ogi

A sake brewery that retains the atmosphere of the Edo period aiming to create a sake that will delight sake lovers.

The Koyanagi Shuzo, located in Ogi, Kyushu's Little Kyoto with the atmosphere of a castle town, was founded in the Bunka era (1804-1818) of the Edo period. In this brewery, a series of buildings from the brewing process from the Meiji era to the Showa era still exist, and the entire building of Koyanagi Shuzo is registered as a national cultural asset and a Saga Prefecture heritage site. Many people visit this merchant-style building with its dazzling white plaster walls. "Takasago" is the brand name of the brewery's famous sake, which is made with subsoil water from Mt. Tenzan and locally grown rice, all of which are representative of Ogi as a sake brewing region. Inheriting the founder's desire to brew sake that would make sake lovers swoon, the brewery strives to make dry sake to attract young people and women, while keeping in mind that they must not be influenced by trends.

原酒

Seishu Kuradashi Genshu Takasago

"Never miss the best time to drink"

Sake is carefully brewed in the harsh winter, aged slowly, and bottled when it is at its best.

Genshu	1,800ml/900ml
Alc.	19%
Rice type	Reihou
Sake meter value	-2.0-+0.0
Acidity	1.7

大吟醸酒

Daiginjo Takasago Kinsho

"The pinnacle of the 'Takasago' series"

Characterized by a light, dry, elegant, fruity ginjo aroma and a soft, clean taste with a light sharpness.

Daiginjoshu	1,800ml/720ml
Alc.	17%
Rice type	Yamadanishiki
Sake meter value	+3.5-+4.5
Acidity	1.7

Tenzan Shuzo

Water source	Subsoil water from the Tenzan mountain range
Water quality	Medium-hard

1520 Oaza-Iwakura, Ogimachi, Ogishi, Saga 845-0003 TEL.0952-73-3141
E-mail: info@tenzan.co.jp   https://tenzan.co.jp

Sake brewing of "continuity and change"

Brew with the best quality sake rice from the Saga Plain and water known as "famous water of fireflies".

The roots of Tenzan Shuzo can be traced back to 1861, when the Shichida family started a flour milling business using a water wheel in Ogi, Saga. They were sometimes asked by local sake breweries to polish rice. However, in 1875, at the urging of a brewery that was going out of business, they bought all the brewing tools and the warehouse, and word spread that the Shichida family was going to start a sake brewery, so they were forced to start the business. This is a true story. Still, from the time of its founding to today, they have passed on quality-oriented sake brewing in order to make the best use of the rich blessings of local nature, and also to protect the tradition. In recent years, the brewery has been holding biannual brewery openings in the spring and fall to interact with local residents and to promote enjoyment of Tenzan's sake.

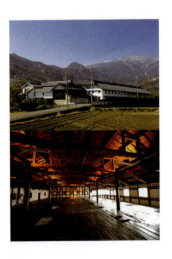

純米吟醸酒
Tenzan Junmai Ginjo

"This sake was served on ANA international flights"

It has a good balance with an aroma of La France pears, the elegant taste and sweetness of Yamadanishiki produced in Saga, and acidity. Best served cold.

Junmai Ginjoshu ... 1,800ml/720ml/300ml
Alc. 16%
Rice type .... Yamadanishiki
Sake meter value .... +1.0
Acidity 1.4

純米酒
Shichida Junmai

"Shichida' main brand, available only at certain shops"

A versatile Junmaishu with a light taste and a delicious rice-derived flavor that goes well with a variety of Japanese, Western and Chinese food.

Junmaishu ... 1,800ml/720ml/300ml
Alc. 17%
Rice type ...... Yamadanishiki/Reihou
Sake meter value .... +1.0
Acidity 1.7

| Water source | Spring water from the Mitsuse Pass |
| Water quality | Soft |

Madonoume Shuzo

1640 Oaza-Shinden, Kubota-cho, Saga-shi, Saga 849-0203   TEL.0952-68-2001
E-mail: koga@madonoume.co.jp   http://www.madonoume.co.jp

Utilize traditions to create unique aspects of Kyushu

More than 300 years of history in Saga Strive for a deep flavor.

Madonoume Shuzo was established in 1688 and has a long history of over 300 years. They are known for brewing sake that has a "deep flavor" which is unique to Kyushu. The brewery also produces unique Mugi Shochu with traditional "Kimotojikomi" method, using a Kabutogama distiller, the only one of its kind in Japan. The current main brand is "Madonoume", but when the company was founded, it was called "Kankiku". The name was changed in 1860 when the lord of the Nabeshima clan praised the quality of the sake, saying, "Year after year, the name has changed, and the world has become fragrant with the aroma of 'Madonoume'". In recent years, "Hananoyoi" has been chosen to be served on international first class flights as well as "Nanatsuboshi", the cruise train of JR Kyushu. Madonoume Shuzo name recognition just keeps going up.

純米大吟醸酒
Junmai Daiginjo Hananoyoi

"Clear, mellow taste"
Made by fermenting Yamadanishiki, the best sake rice, milled to 45% and brewed at low temperaturea for a long time. Won a gold medal in the Fukuoka Prefectural Taxation Bureau's Sake competition.

Junmai Daiginjo	720ml/
Alc.	16%
Rice type	Yamadanishiki (grown in Saga)
Sake meter value	±0
Acidity	1.4

特別純米酒
Madonoume Tokubetsu Junmai

"The taste stands out with Okan (moderately warm sake)"
This sake is served on the JR Kyushu cruise train "Nanatsuboshi" and has a deep, mellow flavor that is typical of Junmaishu.

Special Junmaishu	720ml
Alc.	15%
Rice type	Saganohana (grown in Saga)
Sake meter value	-1.0
Acidity	1.7

Kiyama Shoten

Water source	Subsoil water from the Sefuri mountain range
Water quality	Soft

151 Oaza-Miyaura, Kiyama-cho, Miyaki-gun, Saga 841-0204   TEL.0942-92-2300
E-mail: kihotsuru@gmail.com   https://www.kihotsuru.com

We want local people to drink good sake

Our commitment to sake rice We will continue to grow locally.

Kiyama in Saga, is located at the easternmost tip of the prefecture, on the border with Fukuoka. The Nagasaki Highway runs in front of the brewery. The brewery started the business in 1869, and was incorporated in 1920 as the present-daylimited partnership Kiyama Shoten, now in its 102 year. The Kiyama area is suitable for rice cultivation due to differences in temperature. Sake is brewed from the clear, soft subsoil water flowing from Mt. Ki in the Sefuri mountain range. At the end of the Showa period, the company sought out contract farmers to grow Yamadanishiki. They are willing to increase the number of local contract farmers because of their strong desire for local rice. Eight years ago, Kenichiro Komori took over as the Toji, and he believes in brewing sake by combining scientific knowledge with the five senses.

Kihotsuru Junmai Ginjo Yamadanishiki

"Gentle and smooth flavor"

This award-winning Junmai Ginjoshu is made from 100% Yamadanishiki produced in Saga and Fukuoka, and is recognized worldwide.

Junmai Ginjoshu	1,800ml/720ml
Alc.	16%
Rice type	Yamadanishiki
Sake meter value	-2.0
Acidity	1.2

Tokubetsu Junmaishu Sefuriyusui

"Sharp, crisp taste"

This sake won the Grand Prize in the Junmai Sake category at the U.S. National Sake Appraisal. A signature product that has been brewed for about 30 years.

Junmaishu	1,800ml/720ml
Alc.	15-16%
Rice type	Yamadanishiki
Sake meter value	-2.0
Acidity	1.2

| Water source | Arita Izumiyama Hakuji's spring water |
| Water quality | Soft |

Matsuuraichi Shuzo

312 Kusuku, Yamashiro-cho, Imari-shi, Saga 849-4251 TEL.0955-28-0123
E-mail: info@matsuuraichi.com http://www.matsuuraichi.com

Good sake that lives up to the spirit of handcrafting

A sake brewery with a Kappa mummy that attracts many visitors every year.

For more than 300 years since its establishment in 1716, Matsuuraichi Shuzo has wished to be the best in the Matsuura region. Water comes from Arita Izumiyama Hakuji's springs, the most famous water in Kyushu, and the sake rice is polished to perfection. Brewing techniques are passed down from generation to generation, each time with a new inspiration. Under the watchful eye of the "Mummy of Kahaku", the guardian deity of Matsuura, the brewery devotes itself to making sake with the motto, "Good sake that lives up to the spirit of handcrafting". The brewery is also popular as a tourist spot, attracting many visitors every year. Visitors can tour the brewery and see the tools and agricultural equipment used until the 1950s, but they are also welcomed by the numerous "Kappa" collections. If you visit Imari in Saga, be sure to visit Matsuuraichi Shuzo.

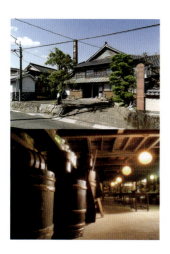

大吟醸酒
Daiginjo Matsuura Ichi

"A clean taste"
This Ginjoshu has a splendid aroma and a refreshing taste. Best served slightly chilled.

Daiginjoshu	1,800ml/720ml
Alc.	18%
Rice type	Yamadanishiki

純米大吟醸酒
Junmai Daiginjo Matsuura Ichi

"Ideal sake for sipping"
This Junmai Daiginjoshu is made with 100% Yamadanishiki rice from Saga. It has a moderate Ginjo aroma and a delicious rice flavor.

Junmai Daiginjoshu	1,800ml/720ml
Alc.	17%
Rice type	Yamadanishiki

Yamato Shuzo

Water source	Subsoil water from the Sefuri mountain range
Water quality	Medium-hard

2620 Niiji, Yamato-cho, Saga-shi, Saga 840-0201 TEL.0952-62-3535
https://sake-yamato.co.jp/

The five genealogies Passing on skills and spirit

The spirit and techniques passed down from generation to generation and the relentless pursuit of innovation.

In 1975, four breweries in Saga City (Mado No Tsuki Shuzo, Akakabe Shuzo, Morita Shuzo, and Kitajima Shuzo) merged to form a capital tieup with "Ozeki", a long-established Nada brewery, and became Yamato Shuzo. The company was established at the former site of the prefectural brewing experiment station in Yamato, inheriting the traditional techniques of the five long-established breweries and introducing modern facilities to produce sake. This was a true fusion of tradition and innovation. In 1984, the brewery started to produce shochu (distilled spirits), and developed an authentic shochu with bamboo charcoal filtration. In 2013, the brewery merged with Tanaka Shuzo, which was founded in 1640, to create the world's only Hishi shochu (Hishi is a lythraceous waterweed. The seeds are edible). In 2014, the company opened a tourist sake brewery, "Gallery Yamato", and holds a one-day brewery opening every April to promote the region with tourists who love Japanese sake.

純米酒
Hizen Toji Junmaishu

"The trinity of Saga's water, rice, and techniques"
This Junmaishu brings out the flavor of rice. It is the best sake for eating, and goes well with not only Japanese food, but also with strong flavors such as Chinese food.

Junmaishu	1,800ml/720ml
Alc.	15%
Rice type	Saga rice
Sake meter value	+5.0
Acidity	1.8

純米大吟醸酒
Nagomi Junmai Daiginjo

"A blend of traditional techniques and new technologies"
With a mellow and elegant aroma and a soft, round taste, it goes well with steamed dishes and boiled fish.

Junmai Daiginjoshu	1,800ml/720ml
Alc.	15%
Rice type	Yamadanishiki
Sake meter value	+5.0
Acidity	1.8

| Water source | Subsoil water from Mt. Unzen |
| Water quality | Soft |

Aimusume Shuzo

1378 Aino-machi Kou, Unzenshi, Nagasaki 854-0301 TEL.0957-36-0025

Like raising a beloved daughter

A sake brewery located at the foot of Mt. Unzen A place of pure water and the skills of a Toji.

Aimusume Shuzo was established in Shimabara, Nagasaki in 1873, and later moved to Aino, at the foot of Mt. Unzen in 1940, where the brewery is currently located. Although the Shimabara Peninsula is rich in water in Nagasaki, Aino is an area with many sake breweries. While the previous generation focused on productivity, the current brewery is trying to increase the variety of high-quality sake, such as Junmai Daiginjo, and trying to produce small quantities of sake carefully by hand in the traditional way. The name "Aimusume" is a reference to the local town of Aino, and the sake was named in the hope that the sake would be widely known and loved, like nurturing a beloved daughter. In fact, 80% of the sake brewed is consumed and loved by the locals. It is brewed with underground water from the subsoil of Mt. Unzen. The rice used in the brewing process is Yamadanishiki grown in the prefecture.

純米大吟醸酒
Junmai Daiginjo Ai

"A fragrant, deep, rich taste"
This is the highest quality sake with a vivid aroma and richness. Yamadanishiki is fermented for a long time, and only the sake that naturally drips from the sake bag is bottled.

Junmai Daiginjoshu	1,800ml/720ml
Alc.	17%
Rice type	Yamadanishiki

特別純米酒
Tokubetsu Junmaishu Unzennokagayaki

"A rich taste you won't get tired of drinking"
A mellow, dry taste with a refreshing aroma. It goes well with grilled fish and meat dishes.

Special Junmaishu	1,800ml/720ml
Alc.	15%
Rice type	Domestic rice
Sake meter value	+1.0

Fukuda Shuzo

Water source	Spring water from a natural broad-leaf virgin forest
Water quality	Medium-hard

1475 Shijiki-cho, Hirado-shi, Nagasaki 859-5533 TEL.0950-27-1111
E-mail: jagatara@vega.ocn.ne.jp https://www.fukuda-shuzo.com

The westernmost sake brewery in mainland Japan

A brewery that inherits the wisdom and efforts of its predecessors with a popular guest house and museum.

Fukuda Shuzo was established in 1688 as a sake brewery for the Lord of Hirado Domain. The brewery, which has been in existence since its establishment, has been designed to keep the temperature below 15 °C without air conditioning even in summer, and the wisdom and efforts of its predecessors have been passed down to this day. In addition to the brewery, there is also a guest house where you can taste sake, and a museum where you can learn about the history of sake brewing, which always attracts many visitors. One of the brands, "Nagasaki Bijin Daiginjo", has won many awards including a gold medal for the best sake at the National New Sake Competition and first place in the Junmai Ginjoshu category at the Toronto International Sake Festival in Canada. In addition to sake, the brewery also produces Shochu (rice liquor), with distinctive brands such as "Jagatara Oharu" and "Kapitan" named after historical words that have been used in Hirado since ancient times.

焼酎
Jagataraoharu

"Recommended as a souvenir of Hirado"

Made from fresh spring potatoes grown in Nagasaki, and has a distinctive potato aroma and a refreshing aftertaste.

Shochu	1,800ml/900ml/720ml
Alc.	25%
Main ingredient	Potato

大吟醸酒
Nagasaki Bijin Daiginjo

"This Daiginjoshu is best served cold"

Made from prefectural-produced Yamadanishiki and fermented slowly at low temperatures. It has a splendid aroma in harmony with the delicious taste of rice.

Daiginjoshu	1,800ml/720ml
Alc.	17-18%
Rice type	Yamadanishiki
Sake meter value	+2.0
Acidity	1.2

| Water source | Spring water from the foot of Saikyoji Temple |
| Water quality | Soft |

Mori Shuzojyo

31 Shin-machi, Hirado-shi, Nagasaki 859-5115 TEL.0950-23-3131
E-mail: hiran@mx7.tiki.ne.jp   https://mori-shuzou.jp/

Exotic atmosphere A Sake Brewery on Hirado Road

Stick to the rice and water in Hirado and strive to create an attractive sake.

Hirado, where Mori Shuzojyo is located, was the first port town in Japan to flourish as a base for overseas trade, and 400 years ago, Portuguese, Dutch, and British trading houses stood side by side there. In such an exotic town, Kokichi Mori, the founder of the company, started the business under the name of "Komatsuya" in 1895. Since then, supported by local rice and the famous water gushing from the foot of Saikyoji Temple, the company has continued to brew sake that attracts people in its hometown while contributing to Hirado tourism by opening up the brewery and supporting local festivals. During the Age of Discovery which, was 400 years ago, Western navigators called Hirado "Firando", and in even older times it was also called "Hiran". In order to let people feel the terroir of Hirado, the main brand name is "Hiran", and in recent years, "Firando" Junmai Genshu, has also been released.

純米酒
Hiran Junmai 40

"Connect and bond through sake brewing"

You can enjoy the taste of 100% Yamadanishiki. This Junmaishu has a splendid aroma and a delicate taste. Best served cold or at room temperature.

Junmaishu	1,800ml/720ml
Alc.	16%
Rice type	Yamadanishiki
Sake meter value	private
Acidity	private

純米原酒
Firando Mumeishu

"Junmai presented to the Pope"
Made from Kasuga Tanada rice, which is registered as a World Heritage. It is as sweet and sour as white wine. Best served well chilled in a glass.

Junmai Genshu	500ml
Alc.	9%
Rice type	Koshihikari
Sake meter value	private
Acidity	private

Yoshidaya

Water source	Well water
Water quality	Hard

785 Yamagawa, Ariecho, Minami Shimabara-shi, Nagasaki 859-2202   TEL.0957-82-2032
E-mail: yoshidaya@bansho.info   http://bansho.info/

Traditional sake brewing at the foot of Mt. Unzen Fugen

The delicate "wooden pressing" method has been handed down from generation to generation.

Arie, located near Mt. Unzen on the Shimabara Peninsula, is blessed with a mild climate and subsoil water from Mt. Unzen Fugen. Since ancient times, the sake, miso, and soy sauce brewing industry has thrived in this area. Yoshidaya was established in 1917 and initially named "Bansho", which means "to be victorious in all things". The subsoil water from Mt. Unzen Fugen is drawn from their own well, which is used for preparation, and the sake is squeezed using an old-fashioned hanegi's tank. There are only a few sake breweries in Japan that use this "hanegi shibori" method. While brewing sake using the flower yeast developed by the Department of Brewing Science at the University of Agriculture, the brewery won gold medals at the Fukuoka Prefectural Taxation Bureau's Sake Competition. In the Junmaishu category, a special Junmaishu brewed with "Himawari" yeast was awarded, and in the Ginjoshu category, a Junmai Daiginjoshu brewed with "Abelia" yeast was awarded. This is a long-established sake brewery that continues to uphold tradition and pass on authentic flavors.

純米酒
Bansho Hanegi Shibori Junmaishu

"A slight aroma and the taste of rice"

A slightly dry Junmaishu brewed with 70% polished rice using rose yeast and pressed slowly with a wooden press.

Junmaishu	1,800ml
Alc.	14%
Rice type	Gohyakumangoku
Sake meter value	+5.0
Acidity	1.4

純米大吟醸酒
Bansho Seisen Sekijou Wo Nagareru Junmai Daiginjo

"This is the brewer's most confident work"

Yamadanishiki rice is polished to 40%. It is carefully brewed with abelia flower yeast and bottled without any alteration after the tank is filled.

Junmai Daiginjoshu	720ml
Alc.	16%
Rice type	Yamadanishiki
Sake meter value	±0
Acidity	1.3

Water source	Natural spring water from Nakao water source
Water quality	Soft

Kameman Shuzo

1192 Tsunagi, Tsunagi-machi, Ashikita-gun, Kumamoto 869-5602　TEL.0966-78-2001
E-mail: info@kameman.co.jp　https://www.kameman.co.jp/

Overcome the disadvantages of southern Japan with skill and spirit

A unique brewing method called "Nantan jikomi" Japanese sake brewing in southern Japan is right here.

Kameman Shuzo has been pursuing high quality sake since 1916 when the founder, Chinju Takeda, decided to produce and consume locally in Kumamoto. However, the warm climate of a southern land is a big disadvantage for sake brewing. This is why the brewery uses an ingenious method called "Nantan jikomi" in which the temperature of the unrefined sake is adjusted while adding large amounts of ice. In Kumamoto, where Shochu culture is more prevalent than sake culture, many breweries rely on seasonal Toji and brewers. After returning from an apprenticeship in Tochigi, the fourth-generation Toji, Ryusuke, was appointed to the position. Kameman Shuzo has a long history and will soon celebrate its 100th anniversary, and the time has come for young talent and sensitivity to flourish. With skills and spirits, they will overturn the disadvantages and brew sake that can only be made in this region.

純米吟醸酒
Manbo

"Delicious sake is made with help from ducks"

This sake is carefully brewed from pesticide-free sake rice grown using the rice-duck farming method. It has an elegant, subtle Ginjo aroma and a light acidity that complements the food.

Junmai Ginjoshu	720ml
Alc.	16%
Rice type	Hananishiki

純米酒
Kameman Nojiro Kinisshiki No.9 Kobo

"The taste of 60% polished Hananishiki is brilliant"

Brewed with Kumamoto's No. 9 yeast and the newly developed "Hananishiki" rice. It has an excellent balance of flavor and aftertaste, making it ideal for sipping.

Junmaishu	1,800ml/720ml
Alc.	16%
Rice type	Hananishiki
Sake meter value	+3.0
Acidity	private

Kawazu Shuzo

Water source	Subsoil water from the headwaters of the Chikugo River in Kyushu
Water quality	Moderately soft

1734-2 Miyahara, Oguni-machi, Aso-gun, Kumamoto 869-2501   TEL.0967-46-2311
E-mail: info@kawazu-syuzou.com   https://kawazu-syuzou.com

Sake brewing is a valuable technology

As a brewery that inherits the culture of sake brewing, we will pass it on to the next 100 to 200 years.

Kawazu Shuzo is a relatively young sake brewery in Kumamoto, started operations in 1932 after taking over a brewery that was founded about 150 years ago. The brewery is located in Oguni, the northern part of Kumamoto, and continues to brew sake using spring water refined by the magnificent nature of Aso and rice grown in the area. Although the production volume is not very large, the brewery takes pride in the delicate sake brewing that can only be done by hand. The most famous sake is "Hanayuki", a Junmai Ginjoshu brewed with Yamadanishiki and Kumamoto yeast, which boasts a sake meter value of -20. It is a popular product that sells out every year, as it has a super sweet taste but is crisp and well suited to the food of Kyushu, which is often sweetly seasoned. Not afraid to take on new challenges, the brewery will continue to brew beautiful sake, handing down the culture of Japanese sake to the next generation.

純米吟醸酒
Junmai Ginjo Shichihoda

"Modest aroma with a splendid fruity taste"

A mild mouthfeel and a full flavor, and is moderately rich and crisp. It has a good reputation among sake beginners.

Junmai Ginjoshu	1,800ml/720ml
Alc.	15%
Rice type	Ipponjime
Sake meter value	+5.0
Acidity	1.5

純米吟醸酒
Junmai Ginjo Hanayuki

"A lingering finish that melts lightly like cotton candy"

An intensely fruity and sweet taste with a sake meter value of -20, but it is sharp and not boring to drink. It also has the delicacy of freshly steamed rice.

Junmai Ginjoshu	1,800ml/720ml
Alc.	15%
Rice type	Yamadanishiki
Sake meter value	-20.0
Acidity	1.8

Water source	Aso headwaters, groundwater on the premise
Water quality	Soft

Zuiyo

4-6-67 Kawashiri, Minami-ku, Kumamoto-shi, Kumamoto 861-4115　TEL.096-357-9671
https://www.zuiyo.co.jp

We brew and enrich your heart

Strive to make high-quality sake that represents the hometown, Kumamoto.

During the Edo period, the Higo Hosokawa clan (present-day Kumamoto) protected the traditional method of making "Akazake" and forbade the production of sake and the influx of sake from other clans. However, with the restoration of imperial rule in 1867, the distribution of people and goods became freer, and the first owner of Zuiyo, Tahachi Yoshimura, decided to make a sake that would represent Kumamoto. However, it was a difficult road. The turning point came in 1903 when a sake expert, Kinichi Nojiro, was assigned to the Kumamoto Taxation Bureau, where he spread awareness of the need for quality improvement in order to develop Kumamoto sake, which led to the establishment of the Sake Brewing Institute. Zuiyo was the first to join the institution, and was responsible for the development of sake in Kumamoto as a major brewery. It continues to pursue high-quality sake rooted in local rice, water, climate and traditional brewing methods.

大吟醸酒
Zuiyo Daiginjo Shizukudori

"Fine, clear taste without any impurities"

This sake is made by hanging the sake bag containing the Daiginjo mash and collecting the drops. Since no pressure is applied, there are very few impurities (undesirable flavors).

Daiginjoshu	1,800ml/720ml
Alc.	17%
Rice type	Yamadanishiki
Sake meter value	+3.0
Acidity	1.1

純米吟醸酒
Junmai Ginjoshu Suugun

"Focus on the local ingredients and quality"

This sake is brewed with sake rice cultivated using natural farming methods. You can enjoy a smooth mouthfeel with just a hint of rice flavor and a splendid Ginjo aroma.

Junmai Ginjoshu	1,800ml/720ml
Alc.	16%
Rice type	Ginnosato
Sake meter value	-3.0
Acidity	1.6

Chiyonosono Shuzo

Water source	Subsoil water from the Aso mountain range
Water quality	Slightly-soft

1782 Yamaga, Yamaga-shi, Kumamoto 861-0501 TEL.0968-43-2161
E-mail: info@chiyonosono.co.jp https://www.chiyonosono.co.jp/

Commitment to the rice and the challenge of sake brewing

Utilize the history of a rice wholesaler to create a new breed, "Kyushu Shinriki".

Yamaga in Kumamoto, is known for its Chibusan burial mounds and other decorative mounds, as well as its lantern festival, High-quality Higo rice is produced, with a view of the Aso volcano to the east. Blessed with water stored by the majestic nature of Aso, Chiyonosono Shuzo was established in 1896. In the Edo period, Yamaga City prospered as a rice trading center and an inn town on the Buzen Highway, and Kikuhachi Honda, a rice wholesaler, started brewing sake here. Since he was a rice wholesaler, he was very particular about the rice he used to brew which lead to the production of new breed called "Kyushu Shinriki". They began producing Junmaishu when there were only a few breweries in Japan, and has continued to take on the challenge of sake brewing by releasing 1844 ml bottles of Nama Genshu and Daiginjo with cork stoppers.

大吟醸酒
Daiginjo Chiyonosono EXCEL

"Best served lightly chilled in a wine glass"
The sake is aged in its own bottle with a cork stopper to give it a rounded, calm taste.

Daiginjoshu	720ml
Alc.	15%
Rice type	Yamadanishiki
Sake meter value	+2.0
Acidity	1.6

純米酒
Junmaishu Shuhai

"Each bottle contains a commitment to Junmaishu"
Junmaishu with a gentle mouthfeel and a refreshing taste. It goes well with meat dishes.

Junmaishu	1,800ml/720ml
Alc.	15%
Rice type	Yamadanishiki
Sake meter value	+2.0
Acidity	1.5

Water source	Shimoyama water source
Water quality	Soft

Tuzyun Shuzo

54 Hamamachi, Yamato-cho, Kamimashiki-gun, Kumamoto 861-3518 TEL.0967-72-1177
E-mail: info@tuzyun.com https://tuzyun.com/

Passing on moisture for 250 years

Almost 100% of the sake rice is grown in Yamato by contract farmers and all the sake is made with "Kumamoto yeast".

Yamato in Kamimashiki, is located on the southern side of the Aso outer rim. In 1770, Seikuro Bizenya, who ran a shipping wholesaler, started sake brewing as an industry to save the village from the heavy annual tax on rice. In 1963, the company name was changed from "Hamacho Shuzo" to "Tuzyun Shuzo", and it has inherited the spirit of placing great value on "the rice, water, and people of this region". The natural environment is cool in summer due to the cold uplands, and very cold in winter due to the snowfalls, making it ideal for sake brewing. Sake rice such as Yamadanishiki and Hananishiki are grown by contract farmers, and "Kumamoto yeast" is used not only for Ginjoshu but also for all other sake, including Junmaishu and ordinary sake. Aiko Yamashita, who is the 13th-generation head of the brewery, grew up in it—the brewery was her playground.. Sake brewing is her livelihood, her hometown, and her family.

純米吟醸酒
Junmai Ginjoshu Semi

"Dry taste with a soft mouthfeel"
The rice used for this sake is "Yamadanishiki" and "Hananishiki". This is the most popular product, slowly fermented and kept in the brewery for one year.

Junmai Ginjoshu	1,800ml/720ml
Alc.	15%
Rice type	Yamadanishiki/Hananishiki
Sake meter value	+6.0
Acidity	1.7

純米酒
Organic Junmaishu Sakuya

"Umami can be tasted in Nurukan (lukewarm)"
Organic Junmaishu "Sakuya" is a single-vintage sake that is engraved with the local Yamato town itself.

Junmaishu	720ml
Alc.	15%
Rice type	Hananishiki
Sake meter value	+4.0
Acidity	1.7

Kuncho Shuzo

Water source	Subsoil water from the Hiko mountain range
Water quality	Soft

6-31 Mameda-machi, Hita-shi, Oita 877-0005 TEL.0973-23-6262
E-mail: info@kuncho.com https://www.kuncho.com/

The artisan's technique nurtured by the water of Hita

Continue to brew sake in the traditional way in Hita, Oita, a place steeped in history.

Kuncho Shuzo is located in the town of Mameda in Hita, Oita. The area flourished as a natural territory under the direct control of the Edo Shogunate, and was once called the "Little Kyoto of Kyushu". It has been selected as an "Important Preservation District for Groups of Traditional Buildings" by the national government, and is a tourist spot visited by 400,000 to 500,000 people a year. The area is also blessed with good quality water, which make sake-brewing businesses thrive. Kuncho Shuzo's five warehouses, including one built in 1702, have been preserved in their original state, which is rare in Japan. The brewery was originally owned by the Chihara family, a wealthy merchant, and was taken over in 1932 by Tomiyasu Shuzo, located in Kurume, Fukuoka at the time. The brewery is still used for sake brewing even as the interior is remodeled, and part of it is open to the public as a museum, contributing to the local economy as a base for tourism.

大吟醸酒
Daiginjo Zuika

"Suitable for a special occasion"
This Daiginjo is brewed with Yamadanishiki under strict temperature control. It features a mild Ginjo aroma and a dry, yet soft and deep flavor.

Daiginjoshu	1,800ml/720ml
Alc.	15%
Rice type	Yamadanishiki
Sake meter value	+3.0
Acidity	1.2

特別純米酒
Tokubetsu Junmaishu Kuncho

"The more you drink, the better it tastes"
With a perfect balance of umami, sweetness, spiciness, and acidity, this sake is perfect for sipping. It can be enjoyed cold or warmed.

Special Junmaishu	1,800ml/720ml
Alc.	15%
Rice type	Hinohikari
Sake meter value	+4.0

Water source	Subsoil water from the Yakkan River
Water quality	Hard

Komatsu Shuzojo

3341 Oaza-Nagasu, Usa-shi, Oita 872-0001 TEL.0978-38-0036
E-mail: koma2sake@gmail.com https://hojun.jimdofree.com/

A desire to reconnect with a history that has died out

Reopening a dormant brewery after twenty years the challenge of a young Toji (master brewer) who is attracting attention in the industry.

The brewery was founded in 1868 by Etsuzo Komatsu, the first generation. The Nagasu district of Usa, where the brewery is located, is blessed with water suitable for sake brewing, rice grown in the Usa Plain, and monsoon winds in winter. Nagasu, a small place, used to have seven sake breweries. Actual brewing, however, was halted in 1988, and since then the breweries have continued to operate while outsourcing production. Junpei Komatsu, the sixth generation of the family, decided to bring back the vitality of the brewery. In the early morning of November 28, 2008, the steam used to cook sake rice was revived for the first time in twenty years. In 2009, he also started to revive "Oitamii", a lost sake rice, with the aim of producing sake that can only be made in Oita. In 2018, the company celebrated the 150th anniversary of its founding and the 10th anniversary of the "Hojun" brand.

Hojun Tokubetsu Junmai
Hojun Karakuchi

"Dry sake with a mellow flavor that is perfect for sipping"

A dry sake with a mellow flavor and a refreshing crispness. It can be served cold or lukewarm, and enjoyed in a wide range of ways.

Special Junmaishu	1,800ml/720ml
Alc.	16%
Rice type	Ginnosato
Sake meter value	+13.0
Acidity	1.7

Hojun Junmai Daiginjo
Oita Miii

"Brewed with a newly revived sake rice"

This sake is brewed with the "Oita Mii", a recently resurrected unique variety that maximizes the flavor of rice. It goes well with Kyushu's sweet soy sauce.

Junmai Daiginjoshu	1,800ml/720ml
Alc.	16%
Rice type	Oita mii
Sake meter value	+3.0
Acidity	1.7

Sanwa Shurui

Water source	Groundwater in Usa-shi, Oita
Water quality	Soft

2231-1 Oaza-Yamamoto, Usa-shi, Oita 879-0495 TEL.0978-32-1431
https://www.sanwa-shurui.co.jp

Carefully and single-mindedly, a little bit of the good stuff

"Waka Botan", the original version of "Iichiko" has been brewed here since its founding.

Sanwa Shurui, headquartered in Usa, Oita, is a brewing company famous throughout Japan for its barley shochu, Iichiko. In 1958, three local sake breweries (one more was added the following year) merged to set up a sake warehouse where they could bottle together and introduce the unified brand name "Waka Botan" to the world. The division of the company currently makin sake, "Kokou no Kura", centers on high-quality, small-lot production and also brews sacred sake for Usa Jingu, the head shrine of Hachiman Shrine. The brewery uses "Hinohikari", a locally loved food rice, as its main ingredient, and works hard to brew sake "painstakingly and single-mindedly" as a thoroughly handcrafting brewery. The concept of sake quality is "to fuse with the local food culture, to value the balance of umami and acidity, and to go well with meals".

純米吟醸酒
Waka Botan Junmai Ginjo Hinohikari 50

"Subtle sweetness and fresh taste"

A delicate balance of sweetness and acidity with a hint of sweetness and freshness.

Junmai Ginjoshu	1,800ml/720ml
Alc.	14-15%
Rice type	Hinohikari
Sake meter value	±0
Acidity	2.0

本醸造酒
Honjozo Waka Botan

"A mellow sake with umami"

Raw materials are carefully processed to minimize impurities (undesirable flavors), and moderate storage is used to express mellow flavors.

Honjozoshu	1,800ml/720ml
Alc.	15-16%
Rice type	Hinohikari
Sake meter value	+2.0
Acidity	1.1

Water source	Groundwater on the premises
Water quality	Medium-hard

Nakano Shuzo

2487-1 Minami-Kitsuki, Kitsuki-shi, Oita 873-0002 TEL.0978-62-2109
E-mail: info@chiebijin.com http://chiebijin.com

All-important brewing water is the sacred springs of Rokugo Manzan

Sake brewing with pride in Oita, Kyushu where shochu is popular.

Nakano Shuzo is located in the city of Kitsuki in the southeastern part of the Kunisaki Peninsula in Oita, the home of Buddhism. In the Edo period, the city prospered as the castle town of the 32,000 koku Matsudaira clan of the Kitsuki domain. The brewery has been operating there since 1874. The main brand since its founding has been "Chiebijin", named after the founder's wife "Chie". The brand name was simplified in 2007, and is currently directly delivered to about fifty sake shops throughout Japan. Kitsuki is blessed with nature, having both sea and mountains, and uses not only the highest quality sake rice, Yamadanishiki, but also locally produced black tea, Nankoume plums, and lemons, the main ingredients for Nakano liqueur.

純米酒
Chiebijin Junmaishu

"Grand prize winner at Kura Master in France"

A clear aroma reminiscent of a beautiful woman, and the gentle sweetness and acidity of the rice are in perfect balance. Best served well chilled.

Junmaishu	1,800ml/720ml
Alc.	16%
Rice type	Yamadanishiki/Domestic rice
Sake meter value	±0
Acidity	1.8

純米吟醸酒
Chiebijin Junmai Ginjo Yamadanishiki

"A sake that the brewery is proud of, and one with a good international reputation"

The fruity, gentle sweetness and clean acidity are appealing. They balance well with the umami of the rice. Best served well chilled in a wine glass.

Junmai Ginjoshu	1,800ml/720ml
Alc.	16%
Rice type	Yamadanishiki
Sake meter value	+1.0
Acidity	1.8

Bungo Meijyo

Water source	Banjo River
Water quality	Hard

789-4 Aza-Kamenoko, Oaza-Yokogawa, Naokawa, Saiki-shi, Oita 879-3105   TEL.0972-58-5855
E-mail: kika-bungo@saiki.tv   http://www.bungomeijyo.co.jp/

Beautiful sake brewed with "water that fireflies drink"

Sake carefully and patiently brewed with water from the Banjo River, one of the clearest in Kyushu.

Bungo Meijyo, located in Oita in Kyushu, was established in 1910. It is located in a mountainous area surrounded by majestic mountains, and has been brewing sake in an environment surrounded by beautiful, lush nature. One of the assets of the brewery is the water from the Banjo River, one of the clearest rivers in Kyushu. In early summer, tens of thousands of fireflies can be seen dancing wildly on its surface, creating a fantastic sight. This "pure water that fireflies drink" is the source of delicious sake. The main brand is "Kakujo". It has a good balance of taste and aroma and is recommended for women. The brewery is known not only for its sake, but also for its shochu. Of course, "water that fireflies drink" goes well with shochu. With patience and care, this brewery will continue to produce delicious sake and Shochu in the future.

純米吟醸酒
Kakujo Junmai Ginjo

"A perfect balance of taste and aroma"

This is a Junmai Ginjoshu with a perfect balance of taste and aroma. The sake is fermented slowly to keep out any unpleasant flavor. Recommended for women.

Junmai Ginjoshu	1,800ml
Alc.	15-16%
Rice type	Domestic rice
Sake meter value	+2.0
Acidity	1.3

麦焼酎
Kogen No Sasayaki

"Barley shochu with fruity aroma"

This shochu is brewed with barley that has been polished to 50%. It is characterized by a fruity citrus aroma reminiscent of the Ginjo aroma of Japanese sake.

Barley Shochu	720ml
Alc.	28%
Rice type	Domestic wheat

| Water source | Subsoil water from the Kuju mountain range |
| Water quality | Medium-soft |

Yatsushika Shuzo

3364 Oaza-Migita, Kokonoe-machi, Kusu-gun, Oita 879-4692 TEL.0973-76-2888
http://www.yatsushika.com/

The ideal sake is one that never fails to please

A brewery in Kokonoe that makes beautiful sake like a beautiful human heart.

Surrounded by the Kuju mountain range, the roof of Kyushu, winters are harsh and the average annual temperature is almost the same as in Niigata. Even though Kyushu is a warm region, it is blessed with a climate suitable for sake brewing. In this location, Yatsushika Shuzo was established in 1864, and has been brewing sake for 150 years. The water from Kuju is indispensable for brewing here. In the winter, the snow that falls on the mountains eventually becomes rich subterranean water that enriches the land. The pure spring water is transformed through the brewers' skills and the brewing process into a unique sake that can only be made here in Kokonoe. In addition to the climate, water, rice, and techniques necessary for sake brewing, Yatsushika Shuzo also strives to make sake with sincerity, based on the spirit of "making beautiful sake means making beautiful hearts.

Yatsushika Junmai Daiginjo [Kin]

"The ultimate Yatsushika"
A delicious taste of rice that spreads slowly and elegantly in the mouth. This Junmai Daiginjoshu masters the original umami of sake rather than the Ginjo aroma.

Junmai Daiginjoshu	1,800ml/720ml
Alc.	17%
Rice type	Yamadanishiki
Sake meter value	+2.0
Acidity	1.3

Yatsushika Sparkling Niji

"Create a splendid atmosphere for a toast on a special occasion"
A new type of sake with delicious rice flavor and the gentle aroma of Junmaishu, together with refreshing carbonation from secondary fermentation in the bottle.

Junmaishu	720ml
Alc.	18%
Rice type	Prefectural rice

Not all companies connected to graduates of Tokyo University of Agriculture are listed here.

Aomori Osanai Shuzoten	**Nagano** Usui shoten
Iwate Sekinoichi Shuzo	Ookuni Shuzo
Ryoban Shuzo	Kametaya Shuzoten
Miyagi Abekan Shuzo	Kiuchi Jozo
Akita Akitaseishu	Koten Shuzo
Kimura Shuzo	Shigaizumi Shuzo
Saiya Shuzo	Takahashi Shosaku Shuzoten
Suzuki Shuzoten	Takasawa Shuzo
Yaesu Meijyo	Tenryohomare Shuzo
Yamagata Haneda Shuzo	Furuya Shuzoten
Fumotoi Shuzo	Maihime Shuzo
Sakata Shuzo	Yoshinoya
Shuho Shuzojo	**Niigata** Onda Shuzo
Terashima Shuzo	Kinshihai Shuzo
Matsuyama Shuzo	Koyama Shuzoten
Rokkasen	Sekihara Shuzo
Fukushima Aizu Shuzo	Taiyo Shuzo
Aizunishiki	Hakuro Shuzo
Shirakawa Jozo	Hiki Shuzo
Shirakawa Meijyo	Fujinoi Shuzo
Taiheizakura Shuzo	Matsunoi Shuzojo
Ibaraki Kowa	Midorikawa Shuzo
Takita Shuzoten	**Ishikawa** Kazuma Shuzo
Morishima Shuzo	Fuse Shuzoten
Tochigi Kojima Shuzoten	**Fukui** Asahi Shuzo
Sampuku Shuzo	Kikukatsura Shuzo
Hokkan Shuzo	Rikisen Shuzo
Gunma Seitoku Meijyo	**Gifu** Ena Jozo
Machida Shuzoten	Otsuka Shuzo
Matsuya Shuzo	Shiraki Kosuke Shoten
Saitama Ishii Shuzoten	Watanabe Shuzojo
Kanbai Shuzo	**Shizuoka** Hananomai Shuzo
Nagasawa Shuzo	**Aichi** Koyumitsuru Shuzo
Yokozeki Shuzoten	Morita
Chiba Azumanada Shuzo	**Mie** Miyazaki honten
Hananotomo	**Shiga** Ikemoto Shuzo
Meshiyoshi honke	Okamura honke
Umeichirin Shuzo	Shiga Shuzo
Tokyo Ishikawa Shuzo	Tsukinosato Shuzo
Ozawa Shuzo	Fujimoto Shuzo

Kyoto	Yosamusume Shuzo
Oosaka	Kotobuki Shuzo
Hyogo	Tsubosaka Shuzo
	Matsuo Shuzo
Nara	Ueda Shuzo
	Naratoyosawa Shuzo
Tottori	Oiwa Shuzo honten
	Nakai Shuzo
Shimane	Migita honten
Okayama	Akagi Shuzo
	Tanaka Shuzojo
	Hiraki Shuzo
Hiroshima	Aihara Shuzo
	Ono Shuzo
	Hakubotan Shuzo
Tokushima	Karakuchi Shuzo
	Naka Shuzo
	Hananoharu Shuzo
Ehime	Kyouwa Shuzo
	Takeda Shuzo
	Chiyonokame Shuzo
	Nakajo honke Shuzo
	Minakuchi Shuzo
Kouchi	Sento Shuzojo
	Tosatsuru Shuzo
Fukuoka	Kobayashi Shuzo honten
	Shiraito Shuzo
	Hananotsuyu
	Wakatakeya Shuzojo
Saga	Azumatsuru Shuzo
	Gochoda Shuzo
	Baba Shuzojo
Nagasaki	Ikinokura Shuzo
	Omoya Shuzo
Ooita	Nikaido Shuzo
	Kamenoi Shuzo
	Sato Shuzo
	Fujii Shuzo

Companies not listed have been omitted at their request.

Sake
National Sake that attracts the world
Tokyo University of Agriculture,
Guide to Sake Breweries and Famous Sake

日本酒　世界を魅了する国酒たち
東京農業大学 蔵元＆銘酒案内

2022年3月14日　第一刷発行

発行人	学校法人東京農業大学 理事長　大澤 貫寿
企画制作	学校法人東京農業大学 経営企画部 〒156-8502 東京都世田谷区桜丘1-1-1 電話: 03-5477-2300 https://www.nodai.ac.jp/hojin/
発行所	世音社 〒173-0037 東京都板橋区小茂根 4-1-8-102 電話/FAX: 03-5966-0649
印刷・製本	株式会社あーす
通販サイト	株式会社農大サポート https://utsusemi-agriport.shop/

© Tokyo University of Agriculture Educational Corporation 2022
ISBN 978-4-921012-49-6　Printed in Japan

●本書の全部、または一部を無断で複写（コピー）することは、著作権法上での例外を除き、禁じられています。
●落丁本・乱丁本はお取り替えいたします。